William L. Graf

Fluvial Processes
in Dryland Rivers

With 143 Figures

THE BLACKBURN PRESS

Reprint of First Edition by Springer-Verlag

Copyright © 1988 by William L. Graf

Fluvial Processes in Dryland Rivers

ISBN-10: 1-930665-51-2
ISBN-13: 978-1-930665-51-4

Library of Congress Control Number: 2002102519

THE BLACKBURN PRESS
P. O. Box 287
Caldwell, New Jersey 07006
U.S.A.
973-228-7077
www.BlackburnPress.com

*For Dorothy and Lister,
who taught true appreciation
of the natural world.*

Preface to the 2002 Reprinted Edition

Almost a decade and a half have passed since the publication of the first edition of *Fluvial Processes in Dryland Rivers*. I wrote the preface in that first edition while on the banks of the Gila River, Arizona, believing that the field is the best place to write the preface to a book. It still seems so. As I write this preface, though, I sit on the bank of the Congaree River, South Carolina. The Congaree is no dryland stream—rather it is a humid-region river teeming with life. Its dark waters slide silently through the dense riparian forest, and rain filters through the multiple canopies to make ever-widening circles on the water surface. I am struck again by the enormous differences between humid-region rivers and those in drylands. True, there are many fundamental similarities; but just as it is diversity that makes human populations interesting, it is the differences among rivers that make their regional study so interesting and important.

After I completed this book, I found that some of the ideas in it were of use to others and to the society that had supported my research. For example, the contamination of rivers and their sediments by heavy metals and radionuclides became a critical scientific and management issue; and the concepts assembled in the book on the subject became useful to me (and I hope to others) in addressing these topics. The portions of the book dealing with erosion and flooding came to the fore in subsequent research as well as in court cases related to riparian properties. The hydrology statements surfaced in cases where, unfortunately, river users lost their lives as a result of flash flooding. And now I find much of my research and advisory time taken up with issues related to dams and associated river management, which I had guessed at the time of writing the book might become important.

Two issues that I did not foresee were the rise in interest in restoration and in the preservation of endangered species. Restoration, mandated by law in many locales, seeks to make rivers more natural through management of processes and forms. This is a fair trick in many dryland settings where decades of technological manipulation of the fluvial system have obscured the "natural" condition, whether it be pre-dam, pre-technology, or pre-human. I also did not foresee the critical connection between the geomorphologic aspects of dryland rivers and the threatened or endangered species that depend on the habitats supported by the geomorphology. I now find myself trying to interpret dryland (and humid) river processes and forms in light of their implications for biodiversity.

Fortunately, this book is now historical and foundational. Since its publication, my colleagues in geography, geology, and hydrology have pursued many new ideas and expanded our horizons in understanding these fascinating river systems. In the United States, for example, we now enjoy major advances in understanding the complexities of change associated with climatic and human induced adjustments. More importantly, we have exciting new work emerging from experiences outside the United States (where most of my perspective developed). Investigations in Australia in particular have brought to light new perspectives on fluvial systems that extend our range of experience and knowledge. I eagerly look forward to the new summary volumes that are now emerging from that work. Also, the analysis of the surfaces of other planets, particularly Mars, is expanding the range of questions we ask about dryland rivers.

I am grateful to Francis Reed and her colleagues at The Blackburn Press who offered to reprint this volume. I have received many requests for copies since 1998 when it went out of print, and I am delighted to be able to supply it to readers again. Rivers are like art, the more we know about them, the more we appreciate them. If this volume aids in that understanding, it will have been successful.

As I write, the Congaree slides by, soft, silent, dark, mysterious, a hydrologic dream for one who spent so much time with dry channels. No matter where they are located, may we always cherish these rivers, mainsprings of nature as well as of societies.

Congaree Swamp National Monument
January 2002

William L. Graf
Educational Foundation University Professor and
Professor of Geography
University of South Carolina
Columbia, South Carolina 29208
Telephone: 803-777-4437
e-mail: graf@sc.edu

Preface

From this high promontory, the landforms of the Sonoran Desert take on awesome proportions. The mesas, buttes, cinder cones, basalt flows, mountains, and valleys, all carved and modified by water in this dry place, must be measured in kilometers and millions of years. Even the colors are dazzling. The volcanic rocks, altered by ancient hydrothermal activity, glow a warm orange and yellow in the October sun. The bright, clear light illuminates each tiny crevice etched by weathering and erosion in the cliffs. Several hundred meters below and about a kilometer away the river rolls, a brown silt-laden ribbon on the floor of the gorge. The water rustles so quietly that I can hear it only when the wind dies. In the distance, sharp mountain peaks rake the bottom of the ocean of air where streamers of clouds stretch over the horizon to the ocean of water.

The field is a fitting place to write the preface for a book on geomorphology because of the importance of the field experience in the development of the science. Dramatic landforms and processes, especially in drylands, have excited the imagination and intellect of artists, writers, and scientists. Each observer has explored a different route to knowing and understanding this thin envelope that is the contact between sky and earth. Descriptions of these striking landscapes have appealed to cataclysmic forces, operation of machine-like processes, and even random occurrences governed by happenstance. The meanings attached to these places have included their definition as the homes of devils and gods, as wastelands or wonderlands, as places of desolation or of beauty. While meaning may reside in the experience of the individual, from a collective scientific perspective it is possible to begin the process of explanation.

If geomorphology is to progress to the stage of a mature and useful science that can successfully explain water-related processes in drylands, it must combine the field experience of perception, classification, description, and measurement with effective theory building. In the analysis of fluvial processes in dryland environments, geomorphology has produced a wealth of basic data in most subject areas (although relatively few geographic areas are included). Enough research is available to make possible the search for generalizations and the development of embryonic theory. That theory can take a variety of forms, but as with all theory it should include a language, a series of law-like statements, and a body of supporting testable data.

The primary purpose of this book is to begin the development of a geographic theory for modern dryland rivers by organizing the generalizations available from previous research into a connected intellectual framework. From this assembly of ideas comes a broad perspective on dryland rivers as manifestations of the interactions of three landscapes: a landscape of energy, a landscape of resistance, and a landscape of geomorphic work. The interactions among these landscapes (or the spatial distributions of relevant physical measures) produce water and sediment processes and resulting landforms. Vegetation and direct human intervention are major modifications to the landscapes that alter the processes and forms. This perspective forms the outline of a general theory and is the outline of this book.

The book is not a literature review. The twelve hundred references following the text do not exhaust the relevant literature even in English. I have attempted to sample research from non-English-speaking parts of the world, but of necessity many works are not included here. I have also included a critical evaluation of some humid-region literature from a dryland perspective because the development of theory for dryland rivers depends in part on the importation of ideas developed for humid regions.

A secondary purpose of this book is to put my own research into some context. After a decade and a half of research on dryland rivers, with the results published in journal articles, book chapters, a monograph, research reports, and legal testimony, it seemed that I had produced a series of sketches that were linked together in my own mind, but perhaps not in anyone else's. This book represents an attempt to assemble and organize those intellectual sketches into a single landscape painting. It was my intention to produce a volume for use by other scientists, engineers, planners, environmental decision-makers, students, and educated laypersons.

None of the ideas expressed in the following pages is strictly my own. Interchange with my colleagues in the geographic, geologic, engineering, agricultural, planning, and legal communities produced this work. A complete list of all those to whom I owe a debt of gratitude would be a lengthy one. I hope that the citations in this work perform three functions for my professional colleagues: first, to recognize this connection of my work to theirs; second, to compliment them; and third, to express my appreciation to them.

Four people early in my career had profound but perhaps not obvious influences on this work. Norbert P. Psuty introduced me to the excitement of geomorphology, and invested his time and effort in me at a very early stage of my career. Donald R. Currey guided my early field research in the American West. George H. Dury educated me in the art and science of scholarly writing. James C. Knox taught me science as practiced by the geomorphologist. I can never adequately thank these people for what they have given me except to say that my life has been richer for having known them, and to offer this book as an expression of appreciation.

Patricia Gober, my wife and colleague on the faculty at Arizona State University, has also contributed to this work through numberless philosophical discussions. Her intellectual and personal support during the production of this volume have been sources of strength and happiness. I am deeply grateful to her for all this and much more.

Superstition Mountains, Arizona
October 1987   WILLIAM L. GRAF

Contents

Part I   Basic Perspectives

1 Introduction

1.1 Fluvial Geomorphology

1.1.1 An Overview

One of the most startling paradoxes of the world's drylands is that although they are lands of little rain, the details of their surfaces are mostly the products of the action of rivers. To understand the natural environments of drylands, deserts, arid, and semiarid regions of the earth is to understand the processes and forms of their rivers. Because over one third of the earth's surface is generally considered arid (Cooke et al. 1982, p. 5), and because at least one fifth of the world's population lives in such areas, the subject of fluvial geomorphology in arid environments has direct significance for human welfare. The purpose of this book is to relate the present understanding of fluvial processes in arid environments in a manner that specifies our present knowledge as well as our uncertainties.

Practical considerations of an economically developing world community are balanced by another major benefit of the exploration of river processes and forms in drylands. The natural beauty of dry ecosystems and the landscapes that support them have stimulated the mind of mankind from earliest times. If it is true that to know and understand a thing is also to admire and appreciate it, then the study of fluvial geomorphology of dry environments is an esthic as well as a scientific pursuit.

Although fluvial geomorphology of dry environments may seem to be a hopelessly limited topic, the explosion of natural scientific knowledge makes the limitation practical and desirable. Studies of the earth-surface processes have been under way since the time of Herodotus almost 25 centuries ago, but the present conception of the field of "geomorphology" developed only after about 1890. According to Roglic (1972) and Tinkler (1985) the word first appeared in German literature in 1858 and later in American works (McGee 1888a, b). Defined as the study of earth surface forms and processes, geomorphology is largely an intellectual child of the twentieth century. For example, a survey of 12 leading earth-science journals that published geomorphologic papers shows that during a 5-year period, there were about 4,540 papers pertaining to geomorphology (Costa and Graf 1984).

With the general subdivisions into fields of study related to dominant processes, more geomorphologists probably study fluvial phenomena (those related to running water) than any other, so that the extant body of literature is now enormous. Fluvial geomorphic literature reflects the input primarily of geographers and geologists, but closely related work also occurs in the fields of engineering, planning, forestry, soils, agriculture, and hydrology. Twenty years ago it was possible to produce a volume largely subsumming the knowledge about fluvial processes (Leopold et al. 1964, as an example), but now treatises must be more limited in scope.

In addition to limiting treatment of fluvial geomorphology to dry environments to make the subject more manageable, a special approach to river processes is important for arid regions because many of the basic concepts in the larger field of fluvial studies were developed in humid regions. The transfer of these humid-region concepts, theories, and practices to arid regions is problematic because the magnitude and frequency of many environmental processes in general are not similar between the two regional types. Therefore, a review of fluvial processes in arid environments is in part a critique of the applicability of knowledge gained in humid regions.

Finally, geomorphology is like all sciences in that its object is to determine generalizations about known phenomena so that predictions can be made about unknown phenomena. In recent decades, however, investigations of fluvial processes in arid environments have all too often appeared piece-meal, so that it would appear that a contribution might be made by attempting the construction of integrated theory in the subject. Such formulations are in part the objective of this book, though the effort represents only a tentative first approximation. The following pages therefore represent an attempt to impose an admittedly arbitrary order on a sometimes chaotic body of research.

1.1.2 Distribution of Drylands

Theoretical definitions of dryland regions are many and varied, but their general application results in similar regions on the surface of the earth (UNESCO 1979). Most hydro-climatic definitions relevant to the problem of regionalizing fluvial processes rely on the analysis of precipitation and temperature or some combination of the two. Although the Koppen climatic classification system is a widely used approach for the regionalization of climates (see Espenshade 1978, pp. 8–9 for a simplified example), a more sophisticated system proposed by Thornthwaite and modified by Meigs (1953) is most commonly used to define drylands.

In Thornthwaite's view, the most significant aspect of climate is the interaction among rainfall and its seasonality, soil moisture, and potential evapotranspiration (reviewed by Oliver 1973, pp. 58–59). Thornthwaite and Mather (1955) proposed a climatic index defined by

$$\text{Im} = [(s-d)/PE]100, \tag{1.1}$$

where Im = the index value, s = excess of monthly rainfall over potential evapotranspiration and storage in the soil during wet months, d = deficit of rainfall plus available soil moisture below potential evapotranspiration in dry months, and PE = potential evapotranspiration (Adams et al. 1979, p. 16). Thornthwaite's index is essentially the aridity index of Budyko, the ratio of radiational energy available for evaporation to the total energy required to evaporate the actual rainfall (Sellers 1965, p. 90).

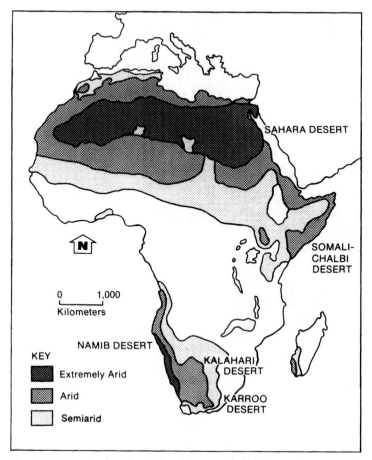

Fig. 1.1. Drylands of Africa. (After Adams et al. 1979)

Meigs (1953) established an interpretive scale of values for the Thornthwaite index, and named various classes according to specific values of the index. The maps of the regions Meigs defined as extremely arid (where 12 or more months without rainfall is the record), arid (where the index Im < −40), and semiarid (−40 < Im < −20) are the most widely used geographic definitions of drylands (Cooke et al. 1982, p. 5), and are adopted for use in this book (Figs. 1.1–1.6).

In subsequent discussions, the specific terms extremely arid, arid, and semi-arid will be used as defined by Meigs (1953). Drylands is a collective term referring to regions with any of these conditions. Generally, the term desert will be avoided, since it is more a botanical than a hydro-climatic concept.

Drylands occur in specific places because either they are isolated from ocean moisture sources or because they are located beneath semi-permanent high pressure systems (Figs. 1.1–1.6). In western North America, western South America, and Australia, interior lands are separated from oceans by some of the highest mountain ranges on the two continents. Prevailing on-shore winds

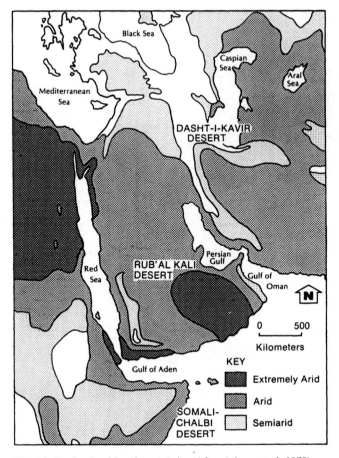

Fig. 1.2. Drylands of Southwest Asia. (After Adams et al. 1979)

cannot transport their moisture to the interiors because they are forced upward over the ranges, and concomitant cooling releases most of their moisture on the mountains (Gilman 1965). Hence, relatively rainy coastal zones contrast with arid inland zones.

Of equal importance are the belts of semi-permanent high pressure generally centered on the latitudinal lines 30 degrees north and south of the equator (Chang 1972, pp. 21–42). These high pressure zones are characterized by air decending from upper elevations as part of the global circulation system. The result is that cloud formation and precipitation mechanisms are suppressed, and rainfall is sparse irrespective of terrain conditions. Southwestern Africa, north Africa, Australia, and southwest Asia experience these atmospheric conditions and are locations of some of the most extensive arid regions of the world.

The boundaries of arid and semiarid lands are not sharply delineated on the landscape, but occur as highly mobile transition zones of uncertain location. These zones change from one year to the next as atmospheric circulation systems

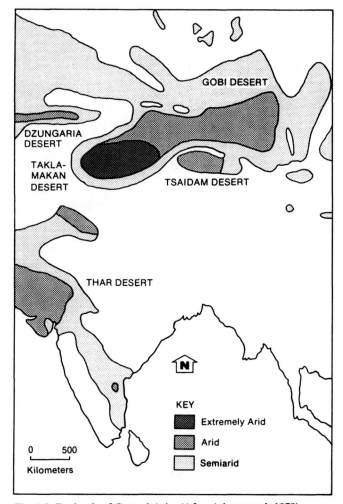

Fig. 1.3. Drylands of Central Asia. (After Adams et al. 1979)

alter moisture delivery systems. In the United States, for example, the boundary between regions with annual moisture conditions that are arid with those that are humid fluctuates several hundred km over just a few years (Fig. 1.7).

On a longer time scale, the size and distribution of drylands have changed in the recent past to include increasingly large areas through a series of processes related to desertification (reviewed by Grove 1978). In northern Africa the southern boundary of the Sahara Desert has changed remarkably over the past century, with clearly defined expansion occurring on the southern edge in the Sahel region (LeHouerou 1968). Whether the desertification is the product of overgrazing and other land management abuses or of climatic changes (e.g., Charney et al. 1975), the result is the same: an expanded area where surficial processes are more like those of water-poor areas than of water-rich areas (UNESCO 1984). Frequently,

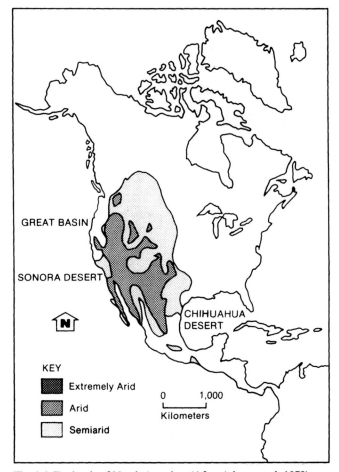

Fig. 1.4. Drylands of North America. (After Adams et al. 1979)

eolian processes become more significant in the process of desertification (Sheridan 1981)

The geographic implications of climatic classifications for geomorphologic processes are that certain climatic regions are likely to contain an identifiable suite of surface processes. The resulting area forms a morphoclimatic or morphogenetic region. In one well-known formulation, Peltier (1950) suggested that dryland assemblages of landforms and related processes relevant to drylands include savanna (a transition zone between dry and humid climates), semiarid morphogenetic regions which include commonly recognized deserts, and arid regions where extreme dryness and high temperatures occur (Fig. 1.8).

Peltier, largely following the work of Davis (1899), considered morphogenetic regions as being areas where definable landscape cycles occur. Cotton (1942, pp. 1–2) also adopted a Davisian perspective, and referred to those areas subject to processes other than humid-region fluvial as being the product of climatic acci-

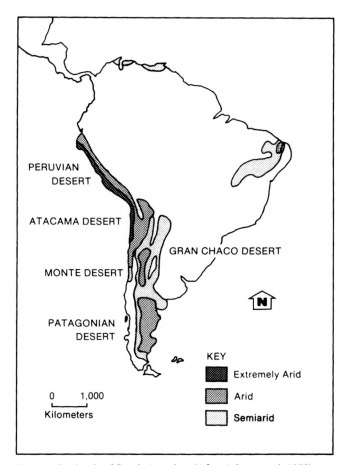

Fig. 1.5. Drylands of South America. (After Adams et al. 1979)

dents. In such regions the expected evolutionary series of landscapes was interrupted by climatic accidents such as aridity, and so demonstrated landforms and processes not accounted for by the common interpretation of the "geographic" or geomorphic cycle of Davis. The work of Peltier and Cotton demonstrates the degree to which the science of geomorphology had been nurtured by studies of humid environments and processes. When questions about drylands were encountered by the developing science they were outside the prevailing paradigm.

A more recent interpretation of the climate-surface process connection is exemplified by the French and German paradigm of climatic geomorphology wherein certain regions are dominated by one or a mixture of particular surface processes. Tricart and Cailleux (1972, pp. 244–249) envision for the "dry zone" three such regions: the subhumid steppe where fluvial processes are much in evidence, semi-arid regions where rainfall is so deficient that there is no integrated drainage and eolian processes are common, and arid regions which have no runoff.

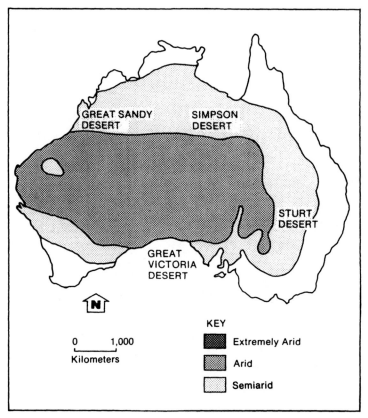

Fig. 1.6. Drylands of Australia. (After Adams et al. 1979)

In each of the these classification systems, regions of the surface of the earth are viewed as areas of process regime. In drylands, only the most arid regions lack fluvial processes, and most authors agree that on the Earth's surface areas without fluvial processes are limited, and are usually explained by highly porous lithology (Tricart and Cailleux 1972, p. 246).

1.1.3 Drylands on Other Planets

Earth is not the only planet in the solar system with arid surface conditions. As interplanetary exploration reveals more of the Earth's sister worlds, it is becoming increasing clear that drylands are probably the most common surface environment on the terrestrial-like planets (Murray et al. 1981, provide a review). Mercury probably has no water on its surface, half of which constantly faces the sun and experiences temperatures high enough to melt lead. The opposite side experiences perpetual cold so that conditions favorable to liquid moisture do not occur on the planet. The Earth's moon apparently is similar in that it is a waterless world. Venus apparently has no water on its 700° C surface, which is constantly subjected

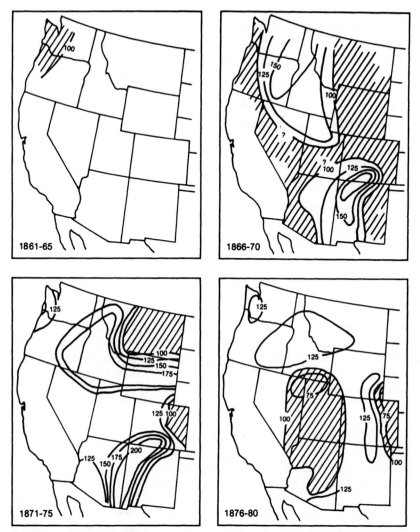

Fig. 1.7. The changing distribution of precipitation in the dryland interior of the western USA. Mapped values are the percentage of the 1951–1960 winter precipitation recorded in each time period, showing the high degree of variability during only 20 years of record. (After Bradley 1976, pp. 122–124)

to wind storms and perhaps acid rain, but it is generally without precipitation or runoff.

Mars is probably most similar to Earth, yet its surface temperatures range from $0°$ to $-100°$ C (Kopal 1979, p. 117). Although there is no evidence that moisture exists on the surface at present, numerous features that appear to be fluvial channels have been analyzed on spacecraft imagery (Fig. 1.9). These channels may have formed from fluvial processes similar to those of Earth (Baker 1982), with the possibility of debris flows and headward sapping being the most

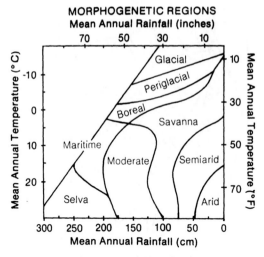

Fig. 1.8. Morphogentic regions defined by temperature and precipitation. (After Peltier 1950, p. 222)

likely commonalities (Higgins 1982). The Martian fluvial processes appear to occur on much larger scales than on Earth (one canyon system is over 2,700 km long and several km deep), but their significance is that they stimulate our otherwise earthbound imaginations to develop new explanations for familiar phenomenon (Baker and Pyne 1978).

1.2 Early Research in the American West

In the science of geomorphology, stimulation of new discoveries by spacecraft parallels the effect of earlier surveys and explorations of arid lands in the 1800s. Explorations in the western United States were especially important in theory development in fluvial geomorphology. The earliest surveys in the American West were for military purposes or were connected to the search for efficient transportation routes (Goetzmann 1959). Mapping accomplished by topographical engineers attached to the U. S. Army gave geologists and geographers their first, albeit incomplete, impressions of the continent's drylands. In most cases this work was entirely descriptive, and had little direct impact on scientific advances. The Pacific Railroad Survey involved several survey and exploration parties investigating three general alignments across the western United States and produced 17 volumes of information between 1855 and 1860. The published data included much environmental information (U. S. Topographic Corps 1855–1860), including descriptions of plant and animal life, geology, and geomorphology, but scientific generalizations were not forthcoming.

Most of the interpretative work on drylands in the western United States was left to geologists and geographers who were enquiring of mind and adventurous of spirit. The major contributors to the early development of the understanding of

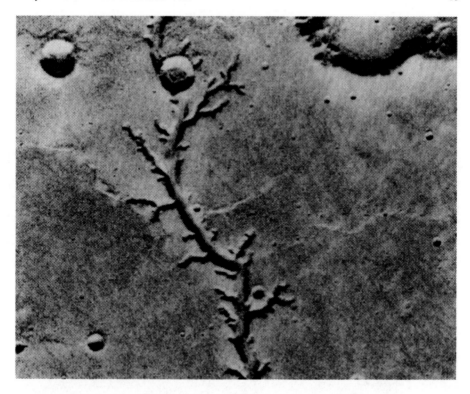

Fig. 1.9. Spacecraft image of a Martian canyon system. Vertical axis of the image spans about 50 km (30 mi). (U. S. Space Science Data Center orthographic photo 466A54)

fluvial processes in arid environments are in two groups: three workers who provided initial speculations based on reconnaissance level experiences (Newberry, King, and Hayden), and three more who conducted detailed investigations (Powell, Dutton, and Gilbert).

Each of these scientists was involved in government-sponsored surveys to provide information about unsettled lands in the arid region of the country. The major driving force behind their investigations was economic, and even though their recommendations for the wise use of the arid regions were delayed in their implementation (e.g., Powell 1879), the investigations bore immediate fruit in the development of geomorphology (reviewed by Chorley et al. 1964).

1.2.1 Newberry

John S. Newberry was a medical doctor, who early in life developed an interest in paleontology and general geology. His experiences in drylands were related to two expeditions: one along the lower Colorado River led by Lt. Joseph Christmas Ives in 1857, and the other in the central Colorado Plateau led by Capt. John N. Macomb in 1859 (Fig. 1.10). In the first expedition Newberry became the first

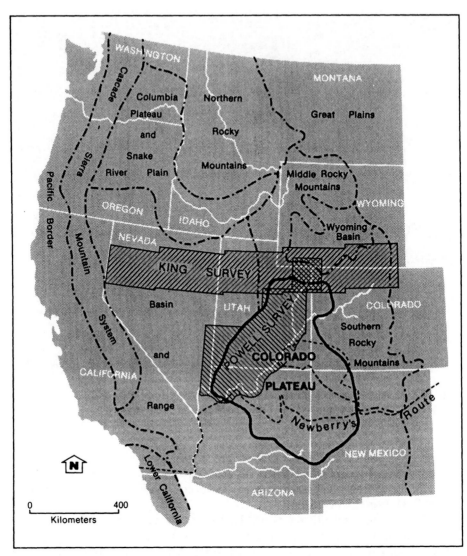

Fig. 1.10. The geomorphologic provinces of the western United States with the routes of Newberry (Chorley et al. 1964, p. 505) and the survey areas of King and Powell (Barlett 1962, p. xvii). All three efforts focused on drylands of the Colorado Plateau and Basin and Range areas

earth scientist to tread the floor of the Grand Canyon of the Colorado. His experience impressed him with the erosive potential of flowing water, so that unlike many earth scientists of his day, who ascribed major earth surface features to volcanic or tectonic processes, he stressed the role of water, even in drylands. His training did not lead him to his conclusions, but his field observations convinced him that "the broad system of valleys bounded by high and perpendicular walls belong to a vast system of erosion, and are wholly due to the action of water" (Newberry 1861, p. 398).

In the second expedition, Newberry saw parts of the Rio Grande, San Juan, Colorado, and Green River systems and their erosional impacts on the Colorado Plateau. His later observations enforced his previous opinions, and he concluded that "from this plateau, grain by grain, the sedimentary materials which once filled the broad and deep valleys of the Colorado and San Juan have been removed by the currents of these streams" (Newberry 1876, p. 84). He also recognized that the grand erosion processes must have proceeded in cycles to produce the myriad of erosional remnants he saw.

Newberry's initial impressions were supplemented by subsequent survey-or/scientists who were charged with evaluating the economic potential of public lands in the western region. Four major surveys were funded by the federal government (reviewed by Bartlett 1962). Clarence King concentrated on a survey along a probable railroad route between Salt Lake City and San Francisco, roughly along the Fortieth Parallel and including much arid terrain with internal drainage. Ferdinand V. Hayden concentrated on mountainous western Colorado, Wyoming, and Montana, with little emphasis on dryland processes. John Wesley Powell focused on the Colorado Plateau, while Capt. George M. Wheeler attempted to map nearly the entire arid Southwest.

1.2.2 King, Hayden, and Marvine

King was trained as a geologist at Yale University by the noted geologist James Dana, and by the time he undertook his survey duties in 1867 he had considerable experience in the California Geological Survey (Wilkens 1958). King was mainly interested in the mineral resources along his route across the Basin and Range Province, and his observations of fluvial process, especially of the Humbolt River and its termination in a closed basin, had little impact on the development of geomorphology. Perhaps of greater importance were the photographs produced by his associate, Timothy O'Sullivan, which documented a terrain that was radically different from that contemplated by earth scientists from humid regions (Horan 1966).

Hayden, like Newberry, was a medical doctor by training who was a naturalist first by avocation and later by profession (Bartlett 1962, pp. 3–9). Hayden's writings showed an appreciation for structural landscapes that were prominent in the mountain fringe zones. He clearly elucidated the uplift and deformation of the mountain ranges, followed by degradation by erosion over long time periods. He and his geologist companion, A. R. Marvine, also pointed out that terrace deposits were important indicators of climatic change, that "they date back to a period when there was much more water in the streams than at the present time, and in consequence, the aqueous forces were much more marked than they are now" (Hayden 1876, p. 200). It would be almost a century before this tantalizing observation was followed by detailed research.

Marvine was an innovative geologist who taught at the Harvard Mining School before joining the western surveys (Powell 1876). He was mostly preoccupied with questions of stratigraphy, but he recognized the significance of fluvial processes in the drylands where he was working. He envisioned the uplift of mountains followed by the erosion of glaciers and rivers resulting in a relatively level surface (Marvine 1874, pp. 144–145). He presaged Davis in postulating a final surface that beveled geologic structure, had accordant summits, and could contain a superimposed drainage system. His intellectual influence on G. K. Gilbert is unknown, but he and Gilbert were close friends and were related by marriage (Davis 1922, p. 67).

1.2.3 Powell

John Wesley Powell, born in 1834, is a significant figure in the development of fluvial geomorphology in drylands because of his exploratory efforts, ability to interpret the landscapes he saw, and skill in managing the scientific work of others. He had little formal schooling, and though he was once a school teacher, he never completed a degree in any of several colleges he attended (following bibliographical information from Darrah 1951; Stegner 1954; Chorley et al. 1964). His largely self-taught appreciation and understanding of the natural sciences was gained by direct experience, especially during boat trips made during the late 1850s down the lengths of the Mississippi, Ohio, Illinois, and Des Moines rivers.

His career was interrupted by the American Civil War, where in service with the Federal Army he was wounded at the battle of Shilo and lost his right arm. He later rose to the rank of major, a title that he informally carried for the rest of his life. After the war he became a professor of geology at Illinois Wesleyan College and at Illinois Normal College.

Powell's interest in the scientific exploration of rivers led him to begin a series of expeditions to the western territories. In 1867 he directed an exploration party in the Pikes Peak region and the central Colorado Rockies, and in 1868 he penetrated the region near the headwaters of the Green River. These forays prepared him for his most famous expeditions in 1869 and 1871, the exploration by boat of the then unknown canyons of the Green and Colorado Rivers. These river trips made Powell a popular public figure and led directly to governmental support for an extensive survey of the northern and central Colorado Plateau. The result was the Geographic and Geological Survey of the Rocky Mountain Region, unofficially known as the Powell Survey (Fig. 1.10).

The importance of the Powell Survey for fluvial geomorphology in drylands is that it emphasized the development of scientific explanations for the features observed. Survey members prepared the first topographic maps of the region, but intellectually they went much further than mere description and attempted explanation and generalization. Powell's effectiveness at securing governmental support insured that he and his researchers enjoyed adequate financing for their development of the study of landforms and landscapes in the dryland region.

Powell's partially romanticized account of the river expeditions (for literary convenience he recounted the story as a single trip rather than the two that actually occurred) makes an exciting adventure tale, but it also contains the first insights into the large-scale behavior of dryland rivers (Powell 1875). It was Powell's speculations about the development of drainage across geologic structures that brought to the attention of Davis and others the concepts of consequ-

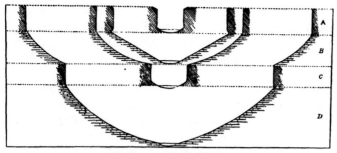

Development of Cañon Profiles.

Fig. 1.11. Dutton's rendition of the recession of canyon walls and cliffs in strata of varying resistances based on his experiences in the Grand Canyon region. (Dutton 1880, Plate XL). *A−D* represent individual rock units

ent, antecedent, and superimposed drainage. In the arid Colorado Plateau with its easily deduced geologic structure, the geological history of river systems seemed obvious. Subsequent interpretations have added details and reveal some errors of interpretation (e.g., Hunt 1969; Young 1987), but Powell's original statements have mostly stood the test of time and they have performed a major foundational function in the general science of geomorphology.

In 1879 the Powell Survey was combined with the surveys of Hayden, King, and remnants of the Wheeler Survey (an army effort), into a new organization, the U. S. Geological Survey (Rabbitt 1980). After a year of indecisive leadership by King, Powell directed the new survey through its formative years and established its mission as a research organization. In his bureaucratic capacity he contributed little directly to the development of geomorphologic theory, but he made it possible for others to expand knowledge and theory beyond his original contributions. His primary associates were Clarence E. Dutton and Grove Karl Gilbert.

1.2.4 Dutton

Dutton was born in 1841 and was educated at Yale University, where he developed effective literary skills (Diller 1911). Like Powell, he served in the Federal Army during the Civil War, but he remained a career officer. He joined the Powell Survey as a field leader in 1875, and after the formation of the Geological Survey he was placed in charge of the entire Colorado Division. Bureaucratic disagreements with Powell led to the eventual estrangement of the two (Stegner 1954, p. 330), and Dutton returned to active duty with the Army in 1890.

Dutton's major efforts focused on the Grand Canyon region, where his studies made several significant contributions (Fig. 1.11). First, he described the dryland landscape more effectively than any other scientific writer. He appreciated the scientific problems posed by the landscapes and their processes, but he also appreciated the inherent beauty of the landscapes. His description of a portion of the central Colorado Plateau as seen from the heights of a volcanic ridge is typical. The plateau country "is a maze of cliffs and terraces lined off with stratification, of crumbling buttes, red and white domes, rock platforms gashed with profound canyons, burning plains barren even of sage – all glowing with bright color and flooded with blazing sunlight. Everything visible tells of ruin and decay. It is the extreme of desolation, the blankest of solitude..." (Dutton 1880, pp. 286–287).

Dutton's second significant contribution was to expand upon Powell's interpretation of the Colorado River system as antecedent, that is, inherited from a previous episode of landscape stability (Dutton 1882b). In the geologic record revealed in the walls of the canyon he saw evidence of ancient erosion surfaces that were nearly planar, and he likened these surfaces to a sort of end product of erosion which subsequently would be uplifted and subjected to new erosion (Dutton 1882a, p. 120). He recognized the probability that previous climatic episodes in the now arid country had alternated between relatively moist and relatively arid conditons.

Dutton's third major contribution was not limited to arid regions alone. He postulated that river systems tended to erode their associated landscapes to a final base level, below which no further erosion could occur. He carried the concept to new lengths, pointing out that base level need not be synonymous with sea level, but that it might be locally introduced by closed basins or structural barriers (Dutton 1882a, pp. 76–77).

Finally, his deductions concerning the development of alluvial fans (which Dutton termed "cones") represented the first sophisticated analysis of these features typically found in arid environments. He pointed out the typical fan shape and profile, the sorting of particles, the role of diminished transport capacity of the stream that created the fan, and the instability of distributary stream courses on fan surfaces (Dutton 1880, pp. 219–220). He drew upon a well-reasoned process explanation for fan characteristics that was based on fluvial processes at a time when the term "geomorphology" was not even part of the vocabulary. Dutton's development of process-based explanations was parallel to the work of his now better-known co-worker, G. K. Gilbert.

1.2.5 Gilbert

Gilbert was born in Rochester, New York, in 1843, and graduated from Rochester University. His first professional experiences in the earth sciences were as a post-graduate geological assistant at the university (general biographical information from Davis 1918, 1922; Mendenhall 1920; Birot 1958; Chorley et al. 1964; Pyne 1980). In 1869–1870 he worked as an assistant with John Newberry in the Ohio Survey, and from that association apparently was stimulated to see the dryland western terrains. From 1871 to 1874 he was a geologist with the Wheeler Survey in Arizona, Nevada, and Utah, but he was dissatisfied with the conduct of the survey. Gilbert was a scientist bent on accurate and detailed description followed by contemplative explanation, while the survey's mission was one of rapid topographic survey and description. The rigid military character of Wheeler and restrictions on publication of results led Gilbert to accept the offer of a position in the Powell Survey.

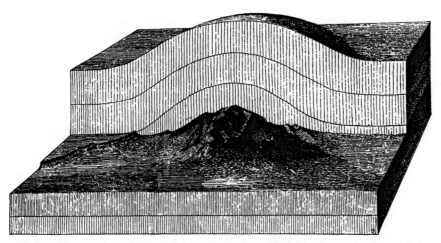

Fig. 1.12. The frontispiece of Gilbert's (1877) work on the geology of the Henry Mountains. The distant half of the diagram shows the deformation of strata by laccolithic intrusion, while the product of fluvial erosion is shown at the front half of the diagram. The mountain is an accurate rendition of Mount Ellsworth, one of the smaller masses of the Henry Mountains. This illustrative technique was later adopted frequently by W. M. Davis

In association with the stimulating minds of Powell, Dutton, and Marvine, Gilbert's scientific expertise blossomed. In 1875–1876 he investigated the Henry Mountains region of Utah and wrote one of the most enduring pieces of geomorphologic literature in only 2 months (Tinkler 1985, p. 142). In the subsequent 2 years he worked with Powell on evaluating the potential of the survey area for irrigation (Gilbert 1879), and with Powell's appointment he became the most important assistant in the formulation of the new U. S. Geological Survey. The bureaucratic work limited his scientific efforts for more than a decade. It may be that Powell's demise as director was a most positive influence on Gilbert's career, because thereafter he began a series of investigations into the action of flowing water, tectonic activity, and planetary geology that provide data for researchers more than 75 years later. He died in 1918.

Of all of Gilbert's publications, three are most important to an analysis of fluvial processes in dryland environments. First, his work describing forms and processes in the Henry Mountains represents a simple, direct explanation of geomorphic processes of flowing water in channels and on hillslopes (Fig. 1.12; Gilbert 1877). The chapter of the report on "Land Sculpture" provided the first complete outline of basic principles of geomorphic dynamics, though some of the material had been published in a little-cited work of the previous year (Gilbert 1876). The major contributions of the Henry Mountains work were the development of a mechanical explanation of landform processes, the application of the concept of an interdependent system of definable subcomponents, and an outline of dynamic equilibrium in geomorphic applications. In Gilbert's view, each portion of the earth surface in a given locale adopted a form that represented a balance between its resistance and the forces of erosion exerted upon it.

He viewed components of hillslopes and channels as being interrelated, with each component responding to changes in input of water and materials from above. "The disturbance which has been transferred from one member of the series to the two which adjoin it, is by them transmitted to others, and does not cease until it has reached the confines of the drainage basin. For in each basin all lines of drainage unite in a main line, and disturbance upon any line is communicated through it to the main line and thence to every tributary. And as any member of the system may influence all the others, so each member is influenced by every other. There is an interdependence throughout the system" (Gilbert 1877, p. 118). Schumm (1973, p. 307) would later echo the same concept under the appropriate title of complex responses: "Within a complex natural system, one event can trigger a complex reaction (morphologic and/or stratigraphic) as the components of the system respond progressively to change".

Gilbert's second major work reflected further applications of equilibrium concepts. His investigation of Lake Bonneville, the precursor to the modern Great Salt Lake, dealt with isostatic rebound of the crust after the evaporation of most of the lake's waters (Gilbert 1890). His analyses of shoreline erosion, transportation and deposition, and of the interaction between lake currents and discharge of tributary river waters in the construction and modification of bars and spits were logical extensions of his work on rivers and tributary slopes.

Finally, Gilbert's investigation of the transportation and deposition of mining debris in the semiarid foothills of the Sierra Nevada of California represents a level of sophistication in measurement and analysis that was half a century ahead of its time (Gilbert 1917). The field studies of the California rivers were extensions of flume experiments in which he had attempted to mathematically explain the transportation of sediment particles by flowing water (Gilbert 1914). In the laboratory it had been his objective to achieve complete description of the river-sediment system by a set of unified equations, but he found that the goal was impossible because the system was too complex and there were too many unknowns (Pyne 1980, pp. 241–242). He concluded that "the actual nature of the relation is too involved for disentanglement by empiric methods" (Gilbert 1914, p. 109), but his application of the general principles proved useful in explaining the impact of mining waste on semiarid mountain streams.

The impact of Gilbert's work on modern geomorphology is difficult to overestimate (reviewed by Yochelson 1980). His direct intellectual contributions included the application of sound scientific method and the judicious use of mathematics along with clear visions of the concepts of equilibrium and interconnectivity of systems. The reverence with which his work is held in the 1980s is reflected in the fact that the Association of American Geographers and the Geological Society of America have instituted research awards given in his honor.

Taken together, the work of Powell, Dutton, and Gilbert represents the beginnings of modern geomorphology. Individually each was a powerful scientist, but because they worked closely with each other their effectiveness was magnified. No matter whose name appeared on the Powell Survey reports, they represented works of collaboration among Powell, Dutton, and Gilbert. In many cases, the researchers themselves did not know where their own ideas ended and those of their colleagues began (Stegner 1977, p. vii). The historical irony is that their theories developed in drylands led directly to W. M. Davis' development of an evolutionary theory of landscape development applied mostly in humid regions. The majority of fluvial geomorphologic research then remained in humid regions and only recently have the drylands become subjects of interest in their own right.

1.3 Early Developments from Other Drylands

The exploration of the arid southwestern United States came at a fortuitous time in the development of the science of geomorphology. Geology and geography were developing as support sciences, researchers were involved in the survey activities, financial support was available to pursue and publish the work, and the workers were numerous enough and in close enough contact to share ideas and intellectually support each other. With regard to the development of fluvial geomorphology in drylands, this set of circumstances never developed elsewhere, so that the contributions of other drylands of the world to the development of the science have not been of similar magnitude. Research experiences on all the continents (except Antarctica) have made substantial contributions, however. The following section briefly outlines the major examples. The objective is to provide a sampling of early work rather than an exhaustive literature review, which has been provided by Lustig (1968).

1.3.1 Africa

Early research interest by Europeans in the landscapes of Africa was mainly a function of colonial or political interests and the resulting resource survey activities. Expeditions to the Sahara produced descriptions of landscapes hardly perceived by thinkers in the developing science of geomorphology (e.g., Flammond 1899; Gautier and Chudeau 1909; Tilho 1911; Newbold 1924; Bagnold 1931, 1933). Although these reports made important contributions to the development of geomorphology in areas other than fluvial (e.g., Bagnold 1941), they did little to further the understanding of rivers because modern processes in the Sahara do not include conspicuous fluvial activity. The notable exception is the Nile River which attracted geomorphologic interest (e.g. Lyons 1905) and provided extensive input for hydrologic studies (e.g., Hurst and Phillips 1931; summarized by Hurst 1952).

French efforts in north Africa, though mostly descriptive, provided important inputs to the developing theme of regional climatic geomorphology. The observations of workers such as Gautier and Chudeau (1909), Gautier (1935), and Urvoy (1936, 1942) led directly into later generalizations by Birot (1968), much in the same way that the work of Powell, Dutton, and Gilbert had led to the generalizations of Davis in the United States. Although the French African work did not focus on fluvial processes, it clearly outlined the problem of climatic change and its implications for surface hydrology.

The Kalahari Desert of southwest Africa also provided valuable input into the French and German theme of climatic-geomorphologic regions. Passarge's

(1904a) monumental description of the region brought to light river systems quite unlike those of Europe, rivers that had experienced significant hydroclimatic changes not directly related to the advance and retreat of glaciers. Additional work in southern Africa by Passarge (1904a, b, c) provided important input to Davis' (1905) development of a specialized version of his geographic cycle theory for arid regions. He also contributed observations from his travels in northern Africa (Passarge 1930). Descriptions of geographic and geologic phenomena were common in the early twentieth century (Herman 1908; Range 1910, 1912, 1914), but only Dixey (1939) devoted special attention to questions of geomorphology. Jaeger (1921a, b) was also a widely traveled scientific observer who contributed to the slowly developing European theories from his experiences in southwest Africa.

. Early work in the Namib Desert was similar to that in the Kalahari. An early period of broad description based on general exploration and travels (Gaulton 1852; Bockemeyer 1890; Vageler 1920; Jessen 1936) was followed by more intensive investigations focused on geology (Stromer von Reichenbach 1896). Investigations of geomorphologic questions did not appear until well into the twentieth century (Kaiser 1921, 1923; Jaeger and Waibel 1920; Jaeger 1921a, b, 1923, 1927). Thus, the timing of the work on southwest African drylands was such that it was unlikely to contribute to the foundation of the science of geomorphology because the investigations came too late. Instead, the basic theories had already been developed, and in many cases the southwest African experience was used as one more example rather than as a source of new ideas.

The Arabian Desert is largely an eolian surface, so that its contributions to theory development in fluvial geomorphology have been relatively limited. Even general geomorphologic descriptions are recent, with those by Little (1925) and Kaselau (1928) being among the first. Petroleum exploration in more recent decades stimulated the interest of earth scientists in the region, and subsequent economic development turned some attention to surface processes. Vajda and Smallwood's (1953) work on the Wadi Jisa' is representative, as is that by Schattner (1962) on the Jordon Valley.

1.3.2 Asia

In Asia, contributions to the early development of fluvial geomorphology in arid regions were frequently the byproduct of endeavors undertaken for other purposes. Geographers were searching for support for the then popular but doomed theories of environmental determinism, but their work provided information from previously little known areas (e.g., Huntington 1914a). Geologists used the form of the land not as an intellectual puzzle in itself, but as an aid to interpretation of the geologic environment (e.g., Blanford 1877). Archeological investigations also provided the impetus for some of the earliest Asian investigations in geomorphology and environmental change (e.g., Raverty 1878).

The literature on three Asian deserts illustrates a range of interest in fluvial processes in a variety of publication periods. The Iranian Desert was the subject of early descriptions and surveys at about the same time as the deserts of the western

United States, but survey parties were rarely accompanied by research scientists (Todd 1844; McMahon 1897). An exception is the description by Blanford (1873) of basin deposits, but none of the work seems to have influenced the generally insular American theoretians.

The Thar Desert of West Pakistan, on the other hand, was the subject of much early descriptive work and some analytic discussions of geomorphic features. Many of the early descriptions were at a reconnaissance level (e.g., Masson 1843), but many also provided insights into the geology and geomorphology of the region (Carter 1861; Blanford 1872) and addressed specific questions of geomorphic processes, such as the work of Stiffe (1873) on mud volcanoes. Fluvial geomorphic research was dominated by investigations of the economically critical Indus River Basin, of which the work of Drew (1873) on alluvial materials and forms was probably the earliest example. Elsewhere in the Thar Desert in India, the dominant research interest in early years was the problem of channel change (Oldham 1886). This early fluvial work (most of it in English) did not seem to influence American writers in the nineteenth century, but it later found utility in archeological associations (Ghosh 1952) and investigations of climatic changes that were reflected in surficial processes (Bharadwaj 1961).

The Asian experience at the dawn of the twentieth century had limited influence on the course of events in the development of geomorphology for three primary reasons. First, many of the areas studied (though not all) had features other than fluvial ones that were of primary interest. Second, because the objectives of many of the descriptions and studies were not geomorphological, they had less of an impact on the science than the work in the American West. Finally, by the time the majority of Asian work reached print, the Davisian approach dominated American geomorphology, while work by Penck (1953) claimed much attention in Europe.

1.3.3 South America and Australia

Unlike the work in the Thar Desert, early research in the arid portions of Australia did not greatly impact developments elsewhere. Almost all the early publications from the southern continent were journal descriptions of travel and expeditions, without attempts at broadly applicable principles. For central Australia see Sturt (1849), Tietkens (1891), and Spencer (1896); for western Australia see Warburton (1875a,b), Lindsay (1889, 1893), Wells (1902), and Barclay (1916). Once scientific investigations began in Australia, they emphasized landscape evolution and weathering, logical interests given the great age of the continental surfaces. Jutson's (1934) comprehensive work on the Western Australia Desert represents one of the most integrated physical geographic descriptions of any dryland. Fluvial topics appear in the Australian literature much later, but they then had great influence on theory development for channel change (e.g., Schumm 1965), and the understanding of misfit streams (Dury 1964a, b, 1965).

Except for general comments by Darwin (1846), scientific observations in the arid portions of South America came about too late to be generally included in the early development of fluvial studies in drylands. In the Monte-Patagonian Desert,

Kuhn's (1922) physiographic summary was the first generally available. Saline deposits in closed lake basins attracted some early attention (Frenguelli 1928). In the Atacama Desert, one of the earliest physiographic descriptions was by Bosworth (1922), who had some interest in river terraces, alluvial fans, and closed lake basins. Other general geomorphic efforts were by Mortensen's (1927) exploration and Iddings and Olsson (1928). It was not until much later that drainage features were the subject of intensive investigations (e.g., Garner 1959). Observations by Penck (1920, 1953) in the Patagonian and Atacama deserts found their way into European efforts at theory building during the early decades of the twentieth century.

2 Theoretical Perspectives

The basis of successful theories for geomorphic phenomenon is an understanding of the spatial and temporal distributions of force, resistance, and work as the outcome of the combination of force and resistance (Graf 1979b, p. 266). All too often geomorphologic research has been conducted, reported, and synthesized without a view toward the construction of such theories and broad generalizations. Chorley (1978, p. 1) has lamented that "whenever anyone mentions theory to a geomorphologist, he instinctively reaches for his soil auger" and presumably departs for the field. Perhaps this trend reflects the dissatisfaction with the first widely accepted theory in the science, the "geographic cycle" of W. M. Davis (Davis 1899), but in recent years this disdain for theory building has prevented the science from achieving its full potential. Thornes (1978) contended that the science is at the threshold of an era of theory construction, and it appears that fluvial studies in drylands may be at a similar stage; hence the interest in this volume in theory building in a preliminary fashion.

An entire generation of scientists has developed who have little or no appreciation for the philosophy of science and little understanding of its applicability to geomorphology. The purpose of the following chapter is to outline a variety of scientific methods that have been applied (sometimes knowingly, sometimes not) in studies of dryland fluvial processes with attending examples. The chapter also provides a review of concepts or perspectives that have significant application in subsequent analyses: general concepts that in themselves may constitute general theories, temporal concepts largely contributed by geology, and spatial concepts mostly contributed by geography.

Before these subjects are addressed, however, the specification of commonly used but sometimes confusing terms is useful. According to Newton, an hypothesis represents an educated guess about a phenomena – it is an unproven explanation (for an original discussion from Newton's work and others, see Tweney et al. 1981). Occasionally, several hypotheses may be advanced simultaneously as multiple working hypotheses (Chamberlin 1897). Often in geomorphology hypotheses are expressed as expected relationships, such as increased water discharge in a channel causing increased sediment transport.

There are two generally recognized processes of hypothesis verification and law establishment: induction and deduction. In induction the researcher begins with numerous particular cases and tries to generalize them into a universal statement. The familiar Manning Equation relating discharge of water in a channel to the hydraulic roughness, channel dimensions, and an empirical constant is an example of a law-like statement verified by induction. The equation is the product of observations in canals and later natural rivers (Barnes 1967). Other hypotheses may be confirmed by deduction, which proceeds from some universal statement to particular sets of events. A geomorphic example is provided by the analysis of stream power deduced from the established principles of physics by Bagnold (1966).

Once "proven" by induction or deduction, hypotheses may graduate to the status of laws, which are usually relationships between variables that have been verified. Several laws may be associated with each other and assembled into a structure that can then be labeled a theory (Harvey 1969, p. 87–129). Hempel (1965, p. 182) refers to a scientific theory as "a set of sentences expressed in terms of a specific vocabulary". Increasingly in geomorphology, the vocabulary is either statistical or mathematical.

Models in science represent analogies to the real world (Chorley and Haggett 1967, pp. 23–24), and in geomorphology they are often in the form of quantitative statements or equations that express the relationships among two or more variables. A model may therefore be in the form of an hypothesis, law, or theory, depending on its degree of confirmation. Considerable debate may attend this latter point. The relationship between rainfall and runoff, for example, may be in its exact nature an hypothesis to a field researcher, while to the engineer it may be accepted as a well-defined law, and the scientist may use it as a building block in a more complex theoretical structure.

2.1 Scientific Methods

The objective of a science such as geomorphology is "to establish general laws covering the behavior of empirical events or objects . . . to enable us to connect together our knowledge of separately known events, and to make reliable predictions of events as yet unknown" (Braithwaite 1960, p. 1). These general laws are the products of standardized methodologies commonly referred to as scientific methods. Thus, if geomorphology has matured as a science in recent years it is not because it has adopted quantitative methods which allow increasingly precise measurement, but rather it is because the field has come to increasingly common use of scientific methods. Three commonly identified scientific methods are of interest to geomorphologists: logical positivism, fallsification, and critical rationalism. The question of whether or not these methods have been adopted in the field will be addressed later in this volume.

2.1.1 Logical Positivism

Before the advent of scientific methods, natural studies were conducted using free speculation and systematic doubt, individualistic approaches that led to unique explanations for each observed phenomenon. In the 1830s Auguste Comte, a Frenchman, established the concept of positivism which defined science for more than a century (Holt-Jensen 1981). Positivism depended on the identification of a question, formulation of empirical hypotheses, and verification of these hypotheses by comparison with observable phenomena. The approach is positivist be-

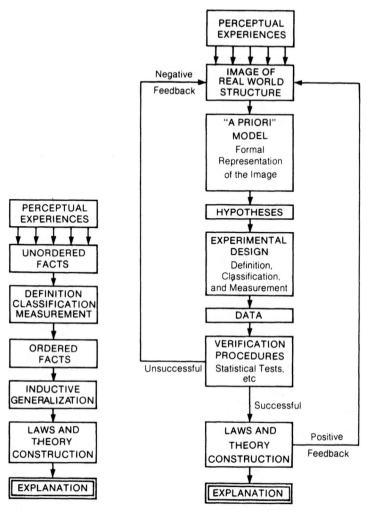

Fig. 2.1. Two routes to scientific explanation. The original positivist approach on the *left* espoused by Comte and Francis Bacon, and a revised version by Harvey on the *right*. (After Harvey 1969, p. 34)

cause it depends on positive tests that permit the elevation of the proven hypotheses to the level of laws. Hypotheses are repeatedly tested until they can be characterized as universally correct, at which point they become laws. These laws can then be assembled into an integrated system or theory (Johnston 1983).

Two issues develop when natural sciences such as geomorphology deal with the positivist approach of Comte. First, in the natural world it is unlikely that any hypothesis will be found to be universally correct because it is impossible to test the entire universe for possible cases (Mahoney 1976). It is unlikely, for example, that all dryland rivers can be analyzed to determine if each one can be correctly classified as braided or meandering according to its slope and discharge properties

(Leopold and Wolman 1957). Even in the most basic of physical sciences, physics, universal laws appear to be quite rare (Newtonian geometry having limited application to celestial mechanics, for example), so it is unlikely that the more complex sciences of geomorphology and hydrology can achieve laws. A compromise position suggested by (Ackoff 1964, p. 1, among other authors) is that natural (and social) sciences develop "law-like" statements which are hypotheses that have been extensively but not universally tested. It is likely that any theory building in geomorphology will depend on law-like statements rather than true laws.

A second issue with the positivist approach is that in its initial stages of hypothesis formation it is not entirely free of theory. The observations from which hypotheses develop are disorganized until ordered in some classification system, for example, and measurements imply some reasonable starting point which has a theoretical basis (Churchman 1961, p. 71). Harvey (1969, pp. 34–36) has

Table 2.1. An example scientific method as it might be applied in geomorphology

---

1. Observations
   a. Empirical observations of real-world phenomena
   b. Reasoning by formal logic or mathematical formulations
   c. Theoretical imputs from previous research

2. Creation of research questions

3. Generation of general hypotheses as speculative answers to the research questions

4. Identification of the significance of the research
   a. Basic theory development
   b. Applied problem solving

5. Definition of the components of the experimental design
   a. Geographic study area
   b. Basic working definition of terms
   c. Formal assumptions
   d. Constants
   e. Dependent variables
   f. Independent variables
   g. Models to link variables with each other in formal statements
   h. Specific testable hypotheses stated in the form of expected model arrangements
   i. Creation of a timetable for planning purposes

6. Collection of data
   a. Establishment of a sampling scheme
   b. Definition of measurement techniques
   c. Execution of data collection

7. Analysis of data
   a. Specification of temporal and spatial scales of analysis
   b. Execution of statistical and mathematical tests

8. Definition of results of testing – acceptance of hypotheses with positive feedback to number 1 above or rejection of hypotheses with negative feedback to number 2 above

9. Eventual acceptance of successful hypotheses as laws

10. Use laws to construct theory

---

proposed a revision of the positivist approach which accounts for the role of theory in the early part of the scientific process (Fig. 2.1). Almost all of the research discussed in subsequent chapters of this volume appears to have been generated along the lines of this revised methodology, though in most cases the original authors did not expressly admit it. As originally conceived by Comte, the positivist approach began by generating testable hypotheses by empirical evidence. In the 1920s several researchers in Vienna proposed that positivist principles could be extended to include hypotheses that grew out of logical or mathematical observations. Thus, "the scientific method" became known as a logical positivist approach (Gregory 1985, p. 25). A more detailed outline of the approach is given in Table 2.1.

2.1.2 Falsification

Given that it is impossible to verify an hypothesis through repeated testing because all possible cases cannot be analyzed, Popper (1963) proposed that scientific testing be conducted through falsification. In Popper's approach, testable hypotheses are stated in such a way that the objective of the research is to find the case which falsifies the hypothetical statement. The argument is that if the statement must be universally proven to be accepted, all cases in the universe must be tested. If the false case is the object of the search, only one case needs to be uncovered in the testing process. The falsification approach is probably the most powerful one available to scientists (Mahoney 1976, p. 131), but it does not appear to be widely accepted among geomorphologists at this time.

2.1.3 Critical Rationalism

One possible resolution to the problem of scientific verification is to dispense with the question of "proof" altogether, and to allow hypotheses, laws, and theories to compete with each other in critical rationalism. Bartley (1962, 1964) points out that justification approaches which require universal proof are irrational, and proposes instead that beliefs (that is hypotheses, laws, theories) are always open to criticism, and that there are no absolutely true statements, only those that fare well in critical competition with other statements. The criticism of competing views might take place using logic, data, previously accepted theories, and problem-solving adequacy. Bartley's approach seems well suited for geomorphology (Haines-Young and Petch 1980), but it represents such a radical departure from justificational methods with their emphasis on familiar "proof" that critical rationalism seems relegated to obscurity for the time being.

Irrespective of the philosophical approach, when is the time for a researcher to graduate an hypothesis to the level of a law-like statement and to include it in a body of theory? This decision is a sociological one, because the researcher must convince other workers in the field of the veracity of a supposed relationship. It is a decision of collective judgment rather than of complete proof. In analyzing the progress of sciences, Kuhn (1970) points out that paradigms develop when groups of researchers agree on valid problems and acceptable solutions. When the accepted theories generated by such paradigms fail to answer large numbers of important questions, a scientific revolution occurs and a new paradigm is established.

Although there have been some philosophical objections to this generalization about science (Shapere 1964; Masterman 1970), it usefully describes at least parts of the history of geomorphology (Graf 1983a). Geomorphology is a relatively small science (Costa and Graf 1984), and its community of researchers enjoys extensive personal as well as professional linkages. Social pressure to accept the prevailing views of the ruling paradigm is substantial. Even so dominant an author as S. A. Schumm must convince his fellow travelers of the veracity of his generalizations. Regarding the newly published work of Schumm et al. (1984) on incised channels, Lewin (1985, p. 471) noted that the "theory and methods have been subjected to little rigorous field testing and it would be prudent to follow the spirit of the approach rather than expect too wide an application of the design equations until such testing is further carried out." The conduct of the science is therefore not nearly as objective as is frequently stated, and at some point the acceptance of a theory is less a question of proof and more an act of collective faith.

2.2 General Systems Concepts

Theory development in fluvial geomorphology as applied to drylands requires the use of a series of fundamental concepts that are not restricted solely to the matter at hand. It is useful to review these concepts in the abstract before their application to specific issues in the following chapters. In some cases, the specific definition of these concepts may avoid confusion since the same terms are occasionally applied in the literature with different meanings (the term "dynamic equilibrium" is a notorious offender in this regard).

The most broadly applicable concept in fluvial geomorphology is general systems theory (Hugget 1985), and most subsequent concepts are in some way related to or use general systems approaches. A general system is a structured set of objects or attributes (Chorley and Kennedy 1971, p. 1). The objects in a system are frequently variables which have measurable magnitudes, so they become part of the scientific methods outlined above because hypotheses frequently take the form of relationships among variables and those relationships can also be expressed as part of the system. To view a geomorphic system in this way, as a set of elements with a definable structure and interrelationships, forces the researcher to account for the complexity often encountered in natural systems. General systems theory is a body of knowledge and conventions that formalize the way in which we view systems and can aid in accommodating the complexity.

Natural systems that are the subject of geomorphologic research occur in a field defined by degree of complexity and randomness of behavior (Fig. 2.2; Weinberg 1975, pp. 17–19). Highly organized simplistic systems are machines or engineered works, and are usually in the purvey of engineers who can analyze these machines using deterministic mathematical functions. It is not often that

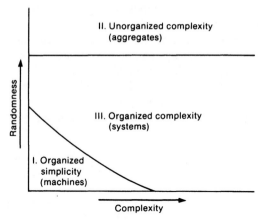

Fig. 2.2. A field representing the randomness and complexity of systems. (After Weinberg 1975, p. 18)

the geomorphologist can afford the luxury of such approaches with more complex structures. At the opposite end of the spectrum of randomness are those systems which exhibit a high degree of random behavior so that their behavior may be described by stochastic methods. Large data sets frequently encountered in long-term hydrologic records might fall in this region of Fig. 2.2 and are susceptible to statistical analysis with relatively high degrees of predictability.

Unfortunately, most systems with which specialists in geomorphology must deal fall into the third portion of Fig. 2.2., between the one extreme where deterministic mathematics are most useful and the other extreme where random-based statistical methods are most successful. In this third portion of the diagram we deal with systems that are moderately complex and moderately random. Often we deal with such systems by applying combinations of mathematical and statistical methods, but we are often near the limits of such methods. Statistical inferences from small samples, for instance, become important, or if we use mathematical formulations, some of the underlying assumptions are violated. This is not an argument to avoid these intermediate systems, but there are limits to the strength of the theories built from their analysis.

There are four commonly recognized types of general systems: morphologic, cascading, process-response, and control systems (Chorley and Kennedy 1971, pp. 5–10). Morphologic systems are made up of instantaneous physical properties connected by functional correlations. The strength and direction of causality in the connections may be tested by correlation and path analysis. Many morphologic systems are not so conveniently simple, however, and include bi-directional causality between elements or feedback loops that suppress (negative) or enhance (positive) change once initiated. An example of such a morphologic system is the set of relationships among channel characteristics as envisioned by Schumm (1977, p. 134) expressed by the generalization

$$Q \simeq f(b, d, l)/S, \tag{2.1}$$

where Q = discharge, b = channel width, d = channel depth, l = meander wavelength, and S = channel gradient (Fig. 2.3).

In cascading systems input and output are identified and system elements are connected to each other by flows of mass or energy. The major interest in such systems from the research perspective is in characterizing the relationship between input and output. In "white box" cases the internal connections between the two are understood, in "black box" cases they are not considered, and in "grey box" cases they are only partly known. Canonical correlation techniques and

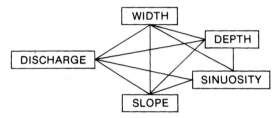

Fig. 2.3. The basic physical characteristics of a stream channel represented as a simple morphologic system. Some of the connections are stronger than others. (From concepts by Schumm 1977, p. 134)

transfer functions are commonly applied methods of analysis for cascading systems. Melton's (1962) suggested analysis of energy in a stream channel provides an example (Fig. 2.4).

Process-response systems represent a combination whereby cascading arrangements of energy or mass pass through elements in morphologic systems. The systems mutually adjust to changes in inputs, frequently with negative feedback arrangements which cause the systems to be self-regulated. In such systems, research efforts are usually focused on specifying the relationship between processes (the cascades) and forms resulting from them (the morphologic systems) through correlation analysis. A simple example is provided by an internally drained basin with a pluvial lake. The volume of water stored, depth, and surface area of the lake might be considered as components of a morphologic system, while inputs through runoff and precipitation and outputs from evaporation consititute the cascading portion of the process-response system. Analyses of pluvial lake systems following this approach have permitted estimates of likely temperature and precipitation conditions during the Pleistocene in New Mexico (Leopold 1951) and Nevada (Broecker and Orr 1958; Snyder and Langbein 1962).

Finally, control systems are process-response systems in which key elements influencing the cascade of energy or mass are controlled by human activities. Such influences can be intentional, as in urban storm-water management systems, or they may be unintentional as in the case of the hydrologic effects of overgrazing. Lusby et al. (1971), for example, showed that the runoff (output) from a small semiarid basin may increase by 43% when overgrazing influences the important system element of vegetation coverage and density.

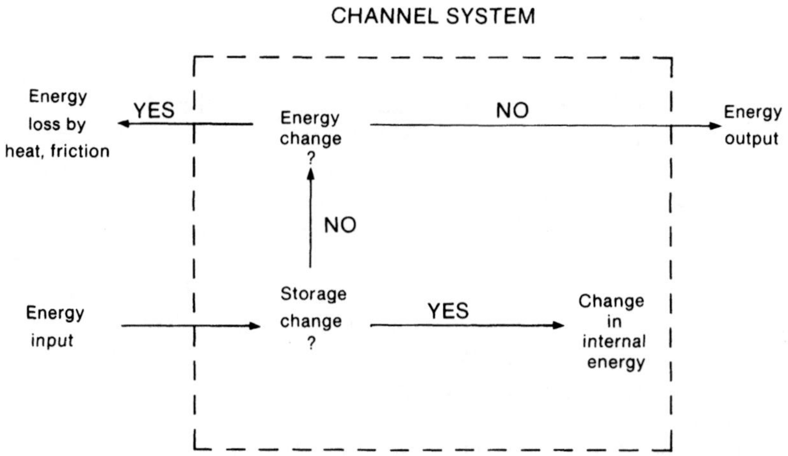

CHANNEL SYSTEM

Fig. 2.4. The basic flow of energy through a channel reach represented by a simple cascading system. (After Chorley and Kennedy 1971, p. 101; from concepts suggested by Melton 1962)

Earth surface systems are obviously open systems, with the transport of energy and mass across the system boundaries. Although these boundaries are usually artificial theoretical constructs designed to meet the needs of the researcher, they are porous, and input-output relationships must almost always be considered. Closed systems which are sealed from the surrounding environment are not generally successful as models of geomorphic processes.

Formal general systems theory was developed by von Bertalanffy (1950) and was explicitly introduced to geomorphologic applications by Chorley (1962). As a concept it became bound up with debates about the various forms of equilibrium observable in the geomorphic environment (Howard 1965; Ollier 1968) and interpretations of allometric change (Woldenberg 1968). For a time it seemed to capture the fancy of the science, but perhaps it was over-represented and could not make up for a lack of theoretical underpinnings (Smalley and Vita-Finzi 1969). It does, however, provide an implicit framework for analysis in geomorphology that is frequently used but little recognized (Kennedy 1985b). Its formal recognition in geomorphologic theory building may increase its utility and permit the researcher to draw on its many powerful derivative concepts such as entropy, equilibrium states, self-regulation, and hierarchial analysis.

2.3 Temporal Concepts

The construction of integrated theories for fluvial geomorphology in arid environments might begin with four families of basic concepts regarding system change through time: uniformitarianism, evolution, equilibrium, and rate laws. These components might later be combined with spatial concepts to provide relatively complete explanations of processes and forms.

2.3.1 Uniformitarianism

The first fundamental debate in the earth sciences with regard to systems change through time concerned the question of the rate of change. In the 1700s religious views dominated the interpretation of landscapes, and strict interpretations of Biblical accounts of earth history led to the assumption that the age of the earth was only a little over 4,000 years (Davies 1969). Rapid cataclysmic changes in the surface of the earth through floods seemed the only reasonable explanation for geomorphic features and gave rise to the concept of catastrophism. As the apparent slow rates of change in life forms became better known in the late 1700s, earth history took on much longer dimensions. James Hutton (1788) and James Playfair (1802) proposed an alternative theory, uniformitarianism, which indicated that surface processes operated in the past much as they do in the present.

Uniformitarianism has had a clear and dominant effect on geomorphology, and it has been a major assumption underlying most theory development in the science. However, the concept may not be strictly applicable in the sense that rates of change, especially in drylands, are subject to adjustment through climatic change. Processes may rapidly reverse themselves, so that erosion may cease and deposition begin within a short period over an entire stream network. The role of human activities is also an important modifying factor, so that erosion, for example, proceeds under present circumstances at rates unlike those in the past (Sherlock 1922). These reservations aside, however, uniformitarianism in the general sense is nearly universally accepted in describing geomorphic system change. Uniformitarianism was a part of the earth sciences before the development of specialized studies in drylands, and it was largely incorporated in such studies automatically.

Uniformitarianism recently has taken a different meaning in recognition that its strict application to slow rates of system change may not be appropriate. In some formulations, the term merely expresses a scientific research strategy in which the simplest explanation consistent with known evidence is adopted (Kennedy 1985, p. 448). In this revised form the concept is less rigorously tied to the concept of rate of change; it is more of an ethic of scientific behavior and less of an explanation of the processes of natural systems (Shea 1982).

2.3.2 Evolution

Evolution is a conceptualization of change through time characterized by a series of successive forms. The concept has seen its greatest development in the life sciences, where Charles Lyell and Jean-Baptiste de Lamarck used it in the 1830s to describe embryological development (Armstrong 1985, pp. 174–175). Charles Darwin and Alfred Russel Wallace later used and publicized the concept, though they did not often use the term. The concept of change through a successive series of identifiable forms consitituted a scientific revolution in the middle 1800s, and the idea was imported by many fields other than biology (Oldroyd 1980).

The infusion of the concept of evolution into science was most remarkable at precisely the time that geomorphology was becoming organized, so it is not surprising that the new science should take up the concept. Gregory (1985, p. 18) claims that Darwin's work exerted strong influence on Davis' formulation of the geographic cycle or cycle of erosion, but it appears that this influence was indirect. At the time Davis was developing his theory for landscape change, the most prominent textbook concerning evolution was probably that by Huxley (1877), but there is no evidence that Davis was influenced by the work. Instead, he appears to have adopted the evolutionary perspective on landscapes promulgated by T. C. Chamberlin in his descriptions of the "driftless area" of southwestern Wisconsin (Chorley et al. 1973). Chamberlin (1883) used the terms "youth" and "old age" to describe the terrain, and emphasized the role of base level. There is some evidence that Davis used these ideas as starting points for his concepts of the geographic cycle (Martin 1950, p. 178).

Soon after the appearance of the geographic cycle, modifications began. Davis (1905) proposed a specific revision for drylands in which a youthful uplifted landscape exhibited much relief between rugged mountains and intervening basins. His examples included the Basin and Range Province of the American Southwest as well as the Iranian and Tibetan deserts. Continued erosion produced mature landscapes in Davis' scheme, with basins filled with sediment and mountains deeply dissected and reduced in size by the expansion of pediments. His examples of mature arid landscapes were limited to some of the bolsons areas of New Mexico. In old-age arid lands, according to Davis, pediments dominate the landscape with a few erosional remnants where high mountains once stood. Old-age landscapes were exemplified by those in South Africa. In his 1905 paper he specifically borrowed concepts from publications on evolution, and he even characterized the progressive replacement of high base levels with low ones as "inorganic natural selection" (Johnson 1909, p. 302).

The Davisian theory of landscape evolution eventually lost much of its support among geomorphologists for two major reasons. First, from the scientific perspective, it could not be verified. No one has ever seen a real landscape evolve through the predicted stages because of the long time periods required. The theory cannot be tested, so according to philosophers of science, it is not scientific but rather mere speculation. Second, the theory failed in an operational sense because it did not explain observations. Although Davis relied on the observations of the surveys in the western United States to formulate his ideas (he frequently acknowledged the work of Powell, Gilbert, Dutton, and others (Daly 1945, p. 272), he seems not to have internalized their emphasis on processes. His geographic cycle consists of a parade of successive forms with no explanation of the processes that cause the observed changes.

Evolution as a potentially useful concept in the explanation of fluvial processes in drylands has not generally developed as an avenue of explanation. In some cases river channel changes have been viewed as an ordered series of events (for example, by Schumm and Lichty 1963; Burkham 1972), but the emphasis has been on the processes causing the transitions from one equilibrium form to another.

2.3.3 Equilibrium

One of the most pervasive concepts in geomorphologic theory building is that geomorphic systems tend toward equilibrium conditions where the input of mass and energy to a specific subsystem is equal to the outputs from the same subsystem. A corollary of this condition is that the internal forms of the system remain unchanged during the transfer process. The concept was developed in geomorphology in the arid environment of the Colorado Plateau by G. K. Gilbert (1876). He used the term grade to denote the condition of a channel that had a gradient such that a perfect balance occurred among corrasion, resistance, and transportation. In the engineering literature a stream exhibiting this equilibrium condition was referred to as being "in regime", a concept developed by French and Italian civil engineers dealing with river management problems (Baulig 1950).

In the development of his theory of the geographic cycle, Davis (1899) adopted the idea of grade and expanded its definition beyond the concept of equilibrium among processes. He maintained that a graded stream had a "balance between erosion and deposition" (Davis 1902, p. 86), but he also suggested that a graded stream was one with a particular profile, steepest in the headwaters and nearly flat at its junction with the base-level control (normally the ocean; in drylands at a more localized level, the floor of an internally drained basin). Thus, as with many Davisian terms, the word "grade" soon accumulated considerable theoretical baggage, and the application of this one word carried enormous connotations of interpretation.

Mackin (1948, p. 471) emphasized the role of slope and the utlity of profile analysis in the concept of the graded river which he considered to be "one in which, over a period of years, slope is delicately adjusted to provide, with available discharge and with prevailing channel characteristics, just the velocity required for the transportation of the load supplied from the drainage basin. The graded stream is a system in equilibrium." Mackin concluded that if present river channels were in a graded condition, many prehistoric channels might also obtain the same condition. The result was that it would be possible to map portions of terraces left by the paleorivers and mathematically extend them over long reaches of the river system.

In general application of the concept of grade, subsequent workers usually equated grade and equilibrium, but it was soon realized that most natural rivers never achieve a perfect equilibrium, so that the longitudinal profiles of their channels rarely coincide with the expected perfect model. Nickpoints related to bedrock outcrops may interrupt the profile which in every other way fits a broad definition of equilibrium (Wolman 1955). Additional work showed that many rivers can adjust their profiles in a relatively short period on a geologic time scale (Wolman and Miller 1960; Hack 1960), in many cases in response to climatic changes (Knox 1975). The implications of these facts are serious for the concepts of both Davis and Mackin (Dury 1966a, pp. 228–229). For Davisian theory they imply that graded stream profiles are not limited to one particular portion of the geographic cycle, and that they may occur even in "youthful" landscapes. For Mackin's theory of grade, they imply that mathematical extension of limited segments of terraces may not be valid because of nickpoints which may be part of an equilibrium arrangement.

Because the concept of grade is unfortunately strongly associated with gradient and base level, because it does not make exclusive statements concerning the connection between morphology and process, and because it appears to be a transient condition, the concept is not useful in analysis of equilibrium. Dury (1966a, p. 231) argued for the complete abandonment of the term, while Knox (1975, p. 179) presented a case for substantial modification of its usage. The literature contains more clearly defined concepts concerning system equilibrium that the geomorphologist may unambiguously employ. The concept of grade is best regarded as a model against which actual conditions are measured rather than as something that actually exists.

In recognition that equilibrium in its perfect definition does not exist in natural rivers, Leopold and Maddock (1953, pp. 50–51) suggested the use of the term quasi-equilibrium to characterize natural systems that tend toward a true equilibrium state, but do not realize it. They viewed natural rivers establishing a relationship between channel geometry on one hand and the water-sediment load on the other as an approximate equilibrium. On the Rio Grande River in New Mexico, Leopold and Wolman (1956) showed that scour and fill of the channel during the annual spring flood returns the channel system each year to approximately the same geometry. Leopold and Maddock made no restrictions on the precise nature of the equilibirum that was approached, and subsequent authors appear to have extended the meaning of the term beyond its original intent. Chorley and Kennedy (1971, p. 203) and Knighton (1984, p. 92), for example, equate quasi-equilibrium with a tendency toward a particular type of equilibrium, "steady state." As originally proposed by Leopold and Maddock, quasi-equilibrium could include either steady state or dynamic equilibrium as defined below.

Quasi-equilibrium is also not equivalent to grade because, as originally defined, channels in headwaters areas that are actively downcutting might have consistent geometry and thus be in quasi-equilibrium (Dury 1966a, p. 226). Leopold and Maddock (1953, p. 47) indicated that "graded reaches of a river are shown to have width, depth, velocity, and discharge relations similar to those of reaches not known to be graded." Quasi-equilibrium is a term commonly found in the literature of the 1950s and 1960s, but it seems to bear no direct correlation to earlier concepts of grade and only a general relationship to later-published reviews on equilibrium as applied in general systems theory. The phrase approximate equilibrium would be less confusing and would carry no particular connotations.

Chorley and Kennedy (1971, pp. 201–203) provide relatively unambiguous definitions for a variety of types of equilibrium likely to be experienced by geomorphic systems. These terms are adopted in the remainder of this volume with the following meanings to describe the observed changes in a variable over time (Fig. 2.5).

Static equilibrium – no change over time.

Stable equilibrium – the tendency of the variable to return to its original value through internal feedback operations within the system following a disturbance.

Unstable equilibrium – the tendency of the variable to respond to system disturbance by adjustment to a new value.

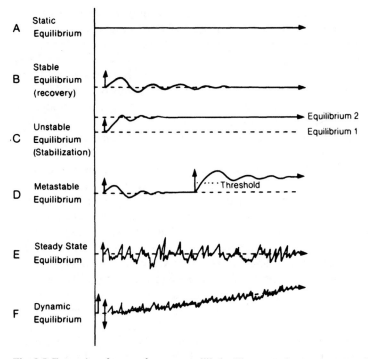

Fig. 2.5. Examples of types of system equilibria. The *vertical axis* represents the value of a typical system variable such as stream channel width. The *horizontal axis* represents time. *Dashed lines* are mean conditions, *solid lines* are observed conditons. (After Chorley and Kennedy 1971, p. 202)

Metastable equilibrium – a combination of stable and unstable equilibrium except that the variable settles on a new value only after having crossed some threshold value--otherwise it returns to original value.

Steady state equilibrium – the variable has shorter-term fluctuations with a longer-term constant mean value.

Dynamic equilibrium – the variable has shorter-term fluctuations with a longer-term mean value that is changing.

The final two concepts in this list are the most widely used in geomorphology, but their use has been obfuscated by careless authors who fail to adhere to established definitions and do not provide their readers with substitute definiti- ons. Unfortunately, steady state and dynamic equilibrium are frequently used as though they were synonymous (Hugget 1980). The difference between the two is intellectually signficant and must be taken into account by theory builders. As defined above (and by Chroley and Kennedy) steady state equilibrium is a special case of dynamic equilibrium when the mean value of the variable is unchanging. The difference between the terms is critical because one of the issues facing analysts of dryland fluvial processes is the choice of a suitable time period over which observations may define a state of equilibrium. If the selected time period is too short, a true steady state may not be recognized and instead be labeled dynamic.

The concept of equilibrium in stream channels reached a sophistication in engineering long before Gilbert's (1877) introduction of the concept into geomorphology. Guglielmini, an Italian engineer (Rouse and Ince 1957), and Surell, a Frenchman, made the connection between equilibrium and gradient in 1841 (Kesseli 1941, pp. 567–568). Eventually the field of hydraulic engineering focused these concepts of equilibrium into the "regime theory". An equilibrium stream is referred to as being "in regime", though the concept originally applied strictly to designed channels and canals (Mahmood and Shen 1971).

Early work by Kennedy (1895) empirically defined an equilibrium canal form based on his measurements (at 30 sites) of a stable canal, Upper Bari Doab in Pakistan:

$$V = 0.84 \text{ m } D^{0.64}, \tag{2.2}$$

where V = mean velocity (ft s^{-1}), m = a critical velocity ratio (1.1–1.2 for coarse sand, 0.8–0.9 for fine sand), and D = depth of flow (ft) (Shonemann 1914, 1916). The function is actually a regression equation with statistically derived constants rather than a deterministic equation derived from first principles and with balanced units.

Lindley (1919) modified the concept of regime by including significant additional factors. Again based on empirical data (786 sites from the Lower Chenab Canal, Pakistan), he generated the following statistical formulae:

$$W = 3.8 \ D^{1.61}, \tag{2.3}$$

and

$$V = 0.95 \ D^{0.57}, \tag{2.4}$$

where W = channel bottom width (ft), D = depth (ft), and V = discharge (ft^3 s^{-1}). Lindley added an additional function for channel gradient with a Manning Roughness Coefficient of 0.0225.

The most complete statement of regime equations is that by Lacey (1958). Although the family of equations is extensive and underwent a series of changes after its initial publication (Lacey 1930), its most basic components were three relationships (Scheidegger 1970, pp. 208–209):

$$v^2/h = B, \tag{2.5}$$

$$v^3/b = s, \tag{2.6}$$

$$(v^2)/(g \ h \ S) = c \ (vb/v)^{0.25}, \tag{2.7}$$

where v = velocity (ft s^{-1}), h = depth of flow (ft), B = a bed factor related to type of bed load, b = channel width (ft), s = a side factor related to type of bank material, g = acceleration of gravity (ft s^{-2}), h = depth of flow (ft), v = kinematic viscosity (ft^2 s^{-1}), c = a dimensionless constant.

From the standpoint of the geomorphologist dealing with drylands, the significance of the regime equations is that they were not originally designed to be applied to rivers in their natural conditions and that they assume a stable flow with low Froude numbers that are rarely, if ever, observed in dryland rivers. This assumption is pervasive in engineering approaches (see Shen 1971, for example). For these reasons, equilibrium as envisioned in regime theory is not directly applicable to natural rivers except in the sense of providing a standard for comparison with actual conditions. In analyzing channel change on the Gila and Salt rivers in central Arizona, Stevens et al. (1975) and Graf (1981) reached similar conclusions: that these dryland rivers could not be considered to be systems in equilibrium.

Much research in fluvial geomorphology has emphasized the search for and the analysis of equilibrium conditions. In humid regions with continuous operation of the fluvial system and constant feedback processes operating among morphology, energy, and mass in the system, equilibrium is a useful explanatory concept. In dryland rivers, however, discontinuous operation (through time and across space) of the channel systems makes the achievement of even approximate equilibrium unlikely in many cases. The tendency of systems to operate toward an equilibrium state may be useful in such circumstances, but more emphasis is needed on the disequilibrium behavior of dryland rivers.

2.3.4 Rate Laws

One aspect of disequilibrium analysis is the investigation of observed system changes between defined states of equilibrium. Because streams in drylands are perhaps more often than not observed in such transition states, the interpretation of inter-equilibrium conditions is especially important. There are three aspects of system behavior in the transition from one equilibrium state to another: a reaction time, a relaxation period, and a characteristic path for system change.

Reaction time is that time during which system operation absorbs and transmits the stimulus for change, but before the system variable under analysis evidences the change (Fig. 2.6). The disruption that initiates the process of change in a system variable from one state to another might be external, such as climatic perturbations or human management strategies, or the disruption may be internally derived by intrinsic adjustments (Schumm and Hadley 1957). For example, Fig. 2.6 might represent a variable in a sample reach of river such as channel width, which during some initial period is in a steady state condition with minor fluctuations about a stable mean. The disruption might take the form of urbanization of the watershed upstream from the sample reach (for specific examples, see Hammer 1972, Leopold 1973). After the urbanization occurs, the downstream channel width may not respond for a considerable reaction time until the water and sediment components of the system absorb and transmit the disruption through the fluvial system.

Once the impacts of the disruption affect the width of the sample reach, however, change occurs until a new steady state equilibrium is achieved with the width variable fluctuating about a new stable mean. The relaxation time is the period between the beginning of change and the establishment of the new equilibrium state. The relaxation time is an indicator of the responsiveness of a

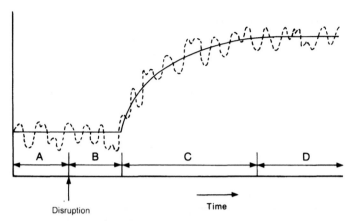

Disruption   Time

Fig. 2.6. Example of system change between two states of equilibria. The *vertical axis* represents the value of a typical system variable such as stream channel width. The *horizontal axis* represents time. The *solid line* is the mean condition of the system, while the *dashed line* represents observed values. This particular example represents a system adjusting from one steady state to another because the mean conditons in the periods of equilibrium are stable. *A* and *D* are steady states; *B* is reaction time; *C* is relaxation time. (After Graf 1977c, p. 179; later modified by Gregory 1985, p. 176)

system to imposed changes, and if it can be defined, accurate predictions of the time to stability might be possible.

The path of system change over time during the relaxation period is likely to be characterized by rapid change at first, followed by decreasing rates of change (Fig 2.7). This arrangement has been indicated by Knox (1972) for general environmental systems and for variables that measure fluvial processes; it has been demonstrated experimentally by R. S. Parker (1976; see Schumm 1977, p. 73). Rapid rates of change followed by decreasing rates is an arrangement that characterizes many physical and chemical adjustments (for example, Feynman et al. 1965), and is best represented by an exponential function (Sumner 1978, pp. 85–87):

$$Y_t = Y_0 e^{-bt}, \tag{2.8}$$

where Y_t = the value of a system variable (such as channel width) at time t, Y_0 the value of the same variable at some initial time 0, e = base of the natural logarithm system, and b = an empirically derived rate constant. The rate constant is more easily interpreted if it is used in the form of a half-life, much in the manner of the description of half-lives for radioactive materials (Laidler 1965):

$$T = (\ln 2)/b, \tag{2.9}$$

where T = half-life, ln 2 = logarithm to the base e of 2, and b is the rate constant derived in (2.8).

Graf (1977a) proposed that the rate law defined in (2.8) be applied to geomorphic processes and that the half-life concept defined in (2.9) could be used to characterize eroding channel networks. If Y_0 were equal to the maximum poten-

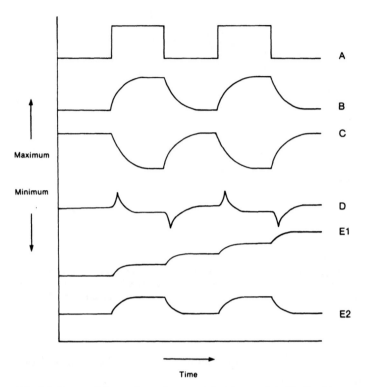

Maximum

Minimum

Time

Fig. 2.7. Types of change in environmental systems related to fluvial processes. The *vertical axis* represents parameters of the systems (*Y*), and the *horizontal axis* represents time. Curve (*A*) represents a climatic variable such as precipitation. The other curves represent system responses to the climatic changes: (*B*) vegetation cover, (*C*) hillslope potential for fluvial erosion, (*D*) geomorphic work such as sediment production, (*E1*) dimensional or spatial characteristics of geomorphic systems that do not have recurring states, (*E2*) dimensional or spatial characteristics of geomorphic systems that have recurring states. Curves *A* and *B* modified from Bryson and Wendland (1967), *C* and *D* from Knox (1972), *E1* and *E2* from Graf (1977c)

tial equilibrium length of an entrenched channel, for example, then various values of Y_t could be observed at times t, and equation (2.8) could be solved for b. In a test case involving three gully systems in Colorado, Graf (1977a, p. 189) found that the half-life of the process of gully erosion was about 17 years. In about 17 years the gullies eroded to about half their ultimate length, in 34 years to three-quarters of the ultimate length, seven eighths in 51 years, and so on.

The rate law offers the prospect of a useful characterization of change in geomorphologic systems in drylands (Gregory 1985, p. 177), but its major drawback is that it has yet to be widely tested (Graf 1977a, p. 189; Knighton 1984, p. 181). Time-dependent data are required to determine the values of the rate constant, and such data is difficult to obtain, especially in field investigations. The primary advantage of the rate law is that it is well established in the physical sciences and that it is likely to be a useful, time-related tool in the construction of theory.

2.4 Spatial Concepts

Time-related components of geomorphologic theory are complemented by space-related components. Spatial concepts provide information and explanation about system operation because processes take place within a spatial or geographic framework. Concepts related to spatial structure can be combined with process concepts for explanations of fluvial systems, and to ignore either the structure or the process results in incomplete explanations. Three families of basic concepts that address spatial structures are available for inclusion in geomorphologic theory: regions, networks, and distance.

2.4.1 Regions

A region is a geographic area with similar characteristics throughout its extent and capable of differentiation from surrounding geographic areas with other characteristics. Regions may be separated from each other by sharply defined boundaries, but more often in the natural world they are separated by zones of transition. The use of the concept of region as a building block for generalizations was a major paradigm of American and British geography in the first half of the twentieth century (Johnston 1983b, pp. 42–49). Hartshorne (1939, p. 462) argued that the fundamental purpose of geography was areal differentiation where the earth's surface was divided into distinct regions.

Regional approaches continue to form a foundation for the science of geography (Pattison 1964, Taaffe 1974), but its importance has declined because as it was originally practiced it offered little explanation of phenomena. Greater emphasis on functional relationships among system elements within regions, however, makes the concept of region useful in modern research (Guelke 1977), and any integrative theory of fluvial processes in drylands is likely to include regional elements. At the very least the definition of the drylands is inherently regional in that it sets off some areas of the earth's surface from others, based on climatic conditions.

Within dry regions, further regionalization of processes is likely. H. Faulkner, for example, has analyzed runoff and erosion processes in semiarid basins in western Colorado, and found that in some parts of the basin snow-melt processes dominate. Other areas in the same basins are little influenced by snow melt. In order to understand the hydro-geomorphic systems of the basins, a regional division is required.

On a similar scale, the variable source area concept is also a regional approach. The drainage basin is a functional region which feeds runoff to a particular stream. All parts or subregions of a given drainage basin do not contribute equally to the runoff process, and during a given precipitation event only limited areas provide water for streamflow (Branson et al. 1981, p. 77–79). The number and sizes of subregions of the basin supplying runoff change through time in a single storm (Fig. 2.8, Hewlett and Nutter 1970), and are different for different sizes of storms. The implications of the areal differentiation for the processes of erosion and sedimentation have not been fully analyzed, but must be considerable.

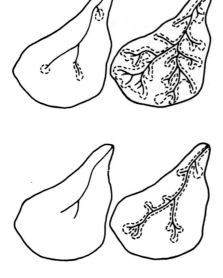

Fig. 2.8. An example of the variable source area concept showing the slope areas contributing runoff to a channel network as time progresses during a single storm. For larger or longer storms larger areas contribute. (After Hewlett and Nutter 1970)

It is on scales larger than the single drainage basin that the concept of region has been most explicitly used in geomorphology. Davis (1899, 1905) recognized that his theoretical geographic cycle operated differently in regions with different climatic characteristics, and although the theory is not now completely accepted, the idea that processes operate in different areas with different intensities is widely acknowledged (Peltier 1950). A. Penck (1910) may not have agreed with Davis in questions of detail of operation, but he, too, recognized process variation on a regional basis and proposed a physiographic regional classification system. He proposed that from the geomorphological perspective there are three principal regions: nival, humid, and arid, each with its distinct climatically influenced landscape (Penck's work along with that of other German writers on the subject of climatic geomorphology has been translated into English by Derbyshire 1973).

The linkage between climate and earth surface processes later became the heart of a European paradigm in geomorphology. Budel (1948) outlined the fundamentals of this perspective by pointing out that landforms were the products of endogenic processes, exogenic processes, and local geologic materials. Since the exogenic processes are direct functions of climate, and since climate is highly regional on earth, geomorphic processes and the landforms they produce are also highly regional. Budel (1944) took the regional approach a step further by proposing that the three major physiographic regions defined by Penck (1910) were insufficient, and that there were further regional divisions controlled by the other factors influencing surface processes. Following the suggestion of Mortensen (1930) Budel conducted an inductive investigation of surface landforms rather than deductively deriving a classification, and found some nine regions defined by geomorphic processes.

Budel's final refinement was to suggest that through the Tertiary the size and distribution of climatogenetic regions changed (Budel 1963). The concept of region in this application becomes dynamic, with changing areas of process domination as a response to large-scale climatic changes. The precise timing of changes through the Tertiary has changed as more evidence has come to light, but the fundamental concept of process regions which change their areal coverage over time remains. The significance of the concept for the study of fluvial processes in drylands is that present climatic conditions may not be a satisfactory guide to the conditions which dominated the formation of the landscape now visible.

As climatic data became more readily available, refinements in the concept of climatic-geomorphic regions were inevitable. Tanner (1961) adopted a deductive scheme, contrary to Mortensen's (1930) suggestion, and suggested that four general process regions could be identified from climatic station data. Tanner placed emphasis on moisture availability, and so defined the regions based on evaporation and precipitation information. He, too, defined arid regions as distinct from others.

Climatic geomorphology in the form of applied regionalism probably reached its major expression in the work of Tricart and Cailleux (1972). They envisioned a series of regions (including an arid region) in which landforms had achieved some equilibrium with the prevailing climate. They termed this state morphoclimatic equilibrium, and they recognized that there were many exceptions due in part to large-scale changes as outlined by Budel (1963). In Tricart and Cailleux's final map of the morphoclimatic regions of the earth, semiarid and arid regions appear to be more finely subdivided than other regions, a testament to the ease of application of the regional concept to drylands.

The concept of region is a powerful tool for the organization of observations and data, and it can serve a useful function in formulating research questions and defining the domain of resulting theory. The region in the form of the drainage basin is a most fundamental concept in fluvial analysis, but regional concepts in the form of subdivision (as in Schumm's 1977 division of the drainage basin into source, transportation, and depositonal areas) or in the form of aggregation (as in process regions or morphoclimatic regions) are also useful. Regionalism cannot stand alone in fluvial research because, like the Davisian geographic cycle, it does not provide direct explanation. Regional definition demonstrates that functional relationships exist and that their areal extent can be mapped, but it does not explain how those functions operate. A geomorphic theory without regional components is therefore incomplete, but a theory that consists only of regionalization is likely to be relatively uninformative.

2.4.2 Networks

Explanation of fluvial processes is in part dependent on generalizations about the stream networks within which they occur. Generalizations about stream networks are advanced relative to some other areas of geomorphologic inquiry and provide useful components of integrated theory for fluvial geomorphology. Early researchers into network properties were not concerned with rivers, but rather biological systems including bifurcations of the arms of starfish (Winthrop

1670; Lyman 1878) and size components of the human arterial network (Keill 1708; Young 1809).

The first application of network concepts in geomorphology was probably by Playfair (1802, p. 102), who noted that each river consisted of a main trunk with a variety of branches, and that the branches joined the trunk in a concordant manner. It is possible that Playfair developed his perception of river systems in conjunction with Young's perceptions of the human arterial system because they knew each other and were at Edinburgh University simultaneously (Jarvis and Woldenberg 1984, p. 10). Playfair's network-based interpretation of stream systems attracted little attention, a fate that was also accorded to the work of Gravelius (1914) who deduced that the branching characteristics of stream networks would be useful in the analysis of geomorphic processes.

Horton (1932, 1945) introduced rigorous network analysis in geomorphology by the concept of an ordering system which identified the finger-tip tributaries as first order. A second order stream resulted from the junction of two first order streams and so forth (Fig. 2.9). Then the highest order stream was extended back up the network to the furthest extremity. Strahler (1952) eliminated this latter provision. Strahler orders can be derived on a strictly mathematical basis without reference to geomorphic or hydrologic process (Melton 1959). For two decades the Horton system and the Strahler modification dominated the methodology of network analysis.

Stream ordering accomplished more than mere description. It provided the basis for the construction of a series of law-like statements about the spatial properties of stream systems. For example, statements were generated relating the numbers and lengths of tributaries to their specific order. Schumm (1956a) added to the list of law-like statements by identifying specific relationships between basin areas and order as well as between channel lengths and basin areas across successive orders. Work by Schumm was firmly grounded in geomorphic processes, but subsequent efforts began to show that the various law-like statements developed by geomorphologists were not geomorphic. They were the logical consequences of applying the ordering system to any network (Woldenberg 1969).

Shreve (1966, 1967) introduced the perspective of the topologically random network which he expected to find in the absence of geologic controls. He began by proposing a new ordering system in which a given network was assigned a magnitude dependent on the number of sources in the network (a link being the connection between two junctions or between a beginning point and a junction). He pointed out that there were a limited number of network arrangements for networks with the same magnitude and that within networks of a given magnitude there was a most likely number of streams of particular Strahler orders. He found that by using this probability-based approach he could approximate the laws earlier defined by Horton from hydro-physical approaches. More elaborate models followed with the assumption of topological random networks (e.g., Smart 1978; Smart and Werner 1976; Shreve 1975).

With a randomly generated topologic network for comparison with real stream networks, the influence of geologic materials or geomorphic characteristics of the basins can be measured as deviations from the random model. Howard

A. Horton   B. Strahler   C. Shreve

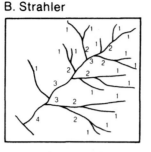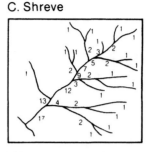

Fig. 2.9. The stream ordering systems of Horton, Strahler, and Shreve applied to the same network. (After Doornkamp and King 1971)

(1971) found that none of the networks he studied were topologically random, an observation at least partially explained by Abrahams (1977) and Smart (1978) as network responses to the influence of relative relief or slope in the basins. Abrahams and Flint (1983) showed conclusively that geologic structure also influences the topologic characteristics of channel networks.

An alternative approach to network analysis has been to focus on the areas that contribute runoff to the network segments. Woldenberg (1969, 1971) proposed that the division of the physical landscape into hierarchical and nested hexagons would inevitably result in networks similar to those observed by other workers. Werritty (1972, pp. 191–192) challenged the basic assumptions of Woldenberg's (1969) approach, but analyses deriving from a space-filling approach have continued in other ways. Woldenberg (1972) found geometric and topologic importance in networks of drainage divide lines rather than channel lines. Jarvis (1976) showed that the lengths of segments in a drainage network are constrained by the available space in areas near drainage divides, and Marcus (1980) showed the importance of location within the basin as a control on segment length.

A combined area and length approach to the analysis of network links and their associated drainage areas was introduced by Smart (1972). The approach has important hydrologic consequences because it can be used to interpret the geometric connections between runoff areas and channels that collect the runoff. Graf (1977c) found that suburban development introduced measurable changes in the area-topologic characteristics of drainage networks using Smart's approach.

The highly developed theories (reviewed by Abrahams 1984; Jarvis and Woldenberg 1984) in network analysis hold some promise for analysis of dryland fluvial systems, though there has been little effort at combination of network theories and process studies (Fig. 2.10). The arrangement of network links has important but as yet undefined implications for the movement and storage of sediment in dryland systems which have the additional intellectual problem of accounting for spatially discontinuous operation. The distribution of link lengths has implications for transmission losses of water from channels. Theories which explain dryland fluvial processes may be significantly strengthened by the addition of the statistically derived concepts for networks in general.

NETWORK A   NETWORK B

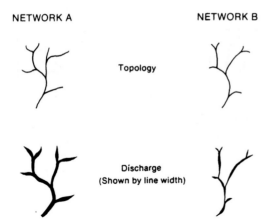

Topology

Discharge
(Shown by line width)

Fig. 2.10. Hypothetical topologic and process networks in an arid region. Discharge attenuates over the long segments of one network but not the other. The two networks are the same topologically (Shreve 1966), but different in terms of processes related to discharge

2.4.3 Distance Decay

Distance is the most fundamental measure of space and is part of the most basic geographic law: things that are close together are closely associated with each other, and as distance between things increases, the degree of their association decreases. In the social sciences, this generalization is characterized as a spatial interaction law because it may be used to describe the social or economic interactions among selected places that are located at varying distances from each other (Chapman 1979, pp. 107–138). The generalization is sometimes referred to as the gravity law in recognition of its application in physics (for review of use in social science, see Haynes and Fartheringham 1984).

Although the analogy between gravity systems and geomorphic systems is an imperfect one, a review of the application of the concepts of distance in gravitation is a useful introduction to basic principles that may then be translated and modified for use with fluvial systems. In Newton's Law of Universal Gravitation, the gravitational attraction between two bodies is directly related to a gravitational constant, the masses of the two bodies and the square of the distance separating them (for reviews, see Feynman et al. 1965, vol. 1, pp. 13/3–13/9; Resnick and Halliday 1977, pp. 338–342). The mathematical form is

$$F = G \, [(m_1 \, m_2)d^{-2}], \qquad\qquad (2.10)$$

where F = attractive force, G = constant of universal gravitation (empirically determined to be $6.67 \times 10^{-11} \, N \, m^2 \, kg^{-2}$), m = masses of the bodies, and d = distance between the two bodies.

Most gravitational systems are more complex than two bodies, so that the force acting upon any one entity is the combination of the influences of all the other bodies in the system. This combined set of forces is gravitational potential, represented by

$$P = \Sigma \{G[(m_i\, m_j)d_{ij}^{-2}]\}, \tag{2.11}$$

where i, j = serial identifiers.

Graf (1982b) has argued that the gravity law can be adopted for the analysis of fluvial geomorphic systems by redefining some of the concepts of the physical application. For example, the concept of force of attraction might be replaced by a variable such as amount of work accomplished, the amount of downcutting in a disrupted stream, for example. The gravitational constant in a spatial application for a fluvial system would be an interaction constant that, like its physical counterpart, would be empirically defined. The masses of the physical application would be replaced in the fluvial application by dimensional characteristics of two different channel segments in a system, but the distance variable would remain a measure of physical separation.

In an example application, Graf (1982b) derived the following function to describe arroyo development triggered in a tributary stream by the lowering of base level in the trunk stream:

$$x_j = K\,(Q_j)^{b_1}\,(d_{ij})^{-b_2}, \tag{2.12}$$

where x_j = amount of downcutting at location j in the tributary, K = an empirical constant that includes the magnitude of the disruption, Q_j = a discharge of some standard return interval as a measure of the available energy at site j, d_{ij} = distance from the location of base level lowering i and the site j, and $b_{1,2}$ = empirical constants. If there were several locations of disruption in the system, they might be included in the function (as Q_j) as measures of the discharge (read available energy) at various sites i. In applying (2.12) in several basins with differing materials (Fig. 2.11), the value of the distance decay constant, b_2 varied from 0.0 for sandy channels (suggesting no influence by the structural considera- tion of distance) to almost 2.0 (suggesting strong distance influences).

Other applications or analyses of the structural role of distance in fluvial processes include a distance-related decline in the heights of nickpoints as they migrate upstream (Brush and Wolman 1957). Eventually the nickpoints decline in height to zero, and they "wash out." Begin et al. (1981) showed how the effects of lowering of base level are propagated upstream through a network, with the magnitude of impacts declining with increasing distance.

All fluvial processes in drylands do not necessarily operate with simple distance decay influences in their spatial structures. The transportation of sedi- ment in pulses, for example, might result in sediment distributions that follow functions more complex than those presented above. For many aspects of fluvial systems, however, a combination of the rate law to account for variation through time with a distance decay function to account for variation across space can provide useful descriptions of forms and processes.

2.5 Integrative Concepts

Fundamental perspectives on spatial and temporal changes can be combined into more complex components of theoretical statements about fluvial processes in

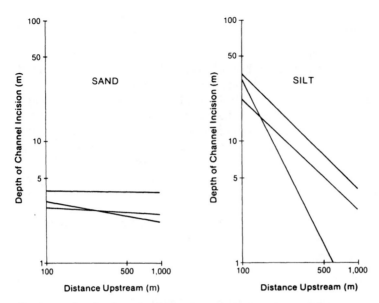

Fig. 2.11. Distance decay in depth of arroyo incision in the Henry Mountains region, Utah. Channels incised in sand have depths that are relatively unresponsive to the spatial structure, but those in silt show dramatic changes over relatively short distances. (Data from Graf 1982)

drylands. These more complex integrative concepts permit the construction of frameworks for accurate description of spatial and temporal phenomena that include some explanation. In geomorphology physical and chemical processes can be explained from first principles, but those processes operate within space and time structures which can to a certain degree be generalized. The integrative concepts in the following paragraphs include the notions of thresholds, complex responses, scale, magnitude-frequency, entropy, allometric changes, and catastrophe theory.

2.5.1 Thresholds and Complex Responses

If geomorphic processes can be characterized as the interplay between force and resistance (Graf 1979b, p. 266), then the place and time at which the two are equal is a threshold. When force exceeds resistance, erosion and transportation occur, and when resistance is greater, deposition or stability result. Schumm and Hadley (1957) showed that channel slope could represent the threshold concept in small gullies of the American Southwest. They found that gullies developed on those valley floors with steeper gradients, while those with lesser gradients were stable. Brice (1966) found in Nebraska that it was possible to define the threshold slope for channel reaches with a given drainage area upstream (Fig. 2.12). Patton and Schumm (1975) later refined the concept using data from western Colorado (Fig. 2.13).

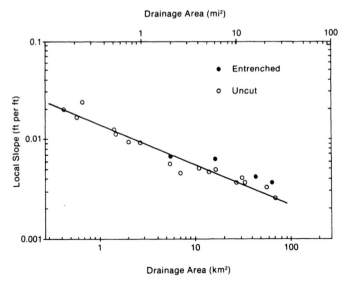

Drainage Area (mi²)

Drainage Area (km²)

Fig. 2.12. Valley-floor slopes and gully development in small streams in Nebraska. Those valleys above the discriminating line are gullied in the larger basins. (After Brice 1966, p. 296)

Further refinements of the threshold approach appeared in the late 1970s. Bull (1979, 1980) suggested that the threshold concept in geomorphology could be applied in the same fashion as it is in hydraulic transport problems. The resistance to motion can be defined in the same terms as the available force in flowing water, and the threshold can be defined as the value of force at which the two are equal:

$$\text{(stream power)/(critical power).} \tag{2.13}$$

Graf (1979b) showed that in mountain valleys of central Colorado the resistance offered by vegetation on the alluvial surface defined the critical threshold. When tractive force or shear stress exceeded this threshold, erosion and entrenchment occurred (Fig. 2.14). The data for this work were from an area subjected to mining activity, and the vegetation changes associated with human use of the landscape played an important role in destabilizing the stream systems.

Howard (1980) demonstrated that fluvial systems operate within a complex set of thresholds rather than at an single threshold. Velocity thresholds control the nature of bedforms, thresholds for sediment transport have varying relationships with gradient and velocity of flow, depending on particle size, and there are differences in behavior between alluvial and bedrock channels. It is unlikely that system operation can be characterized by a single function that identifies a threshold, but simplification may be helpful to the researcher as long as there is recognition that much unexplained variation is to be expected because of those parts of the system which are ignored in the analysis (see Coates and Vitek 1980, for a general review).

The concept of "complex response" is dependent on the identification of system elements and their associated thresholds. Gilbert (1877, pp. 117–118) suggested that in arid (and presumably other) regions slope units are connected

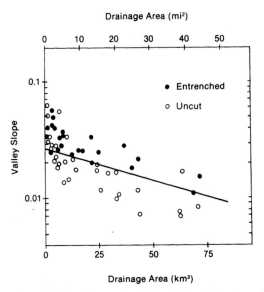

Fig. 2.13. Valley-floor slopes and gully development in streams of northwest Colorado. Those valleys plotting above the discriminating line are gullied except in those basins with less than about 5 square miles, where other influences such as vegetation or particle size become more important than drainage basin area. (After Schumm 1973, who used data by P. C. Patton)

together in an integrated series of input and output subsystems. When a given segment of a slope erodes, it contributes material to the next segment downslope, but it also receives material coming down from the next segment above. Changes are therefore propagated throughout the slope system. Schumm (1973, p. 305) pointed out that the same complexity exists in stream systems, where channel segments are connnected to each other. If a threshold is exceeded and a process change occurs in one part of the system, the effects may eventually extend throughout the entire network. The series of responses may be varied (Schumm 1977, pp. 13–14): if base level is lowered, erosion that began in the trunk stream of a basin may propagate itself into headward tributaries. As the erosion extends into lower order streams, the eroded materials may be deposited in the trunk stream, so that process adjustments that began as erosion may be spatially and temporally converted to sedimentation. This complex response is a logical outcome of the operation of a general system.

The importance of thresholds and complex response perspectives in considering dryland rivers is that precipitation events of the same magnitude will not always elicit the same response from the same drainage basin. The available energy from a given rainfall will be the same no matter when it occurs, but the resistance to that rainfall varies over time. Major amounts of erosion will occur only when the event occurs at a time when the resistance factor represented by channel gradient, particle sizes, or vegetation, is lower than the available force. Thresholds also vary across space, so that the same precipitation event may cause widespread erosion in the channels of one basin while no similar response is stimulated in a similar sized basin nearby. System complexity insures that once

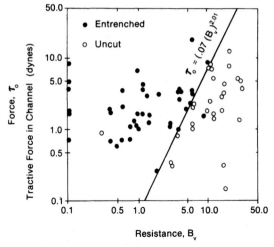

Fig. 2.14. Force versus resistance as a discriminating relationship to define gullied and uncut valley floors in the Front Range of Colorado. The calculated force is for the ten-year flood, and resistance was defined empirically. (After Graf 1979a, p. 7)

initiated, adjustments are likely to be propagated throughout the drainage network, but because fluvial processes are intermittent in drylands the responses are also discontinuous. At any given time the system may present evidence of only having partially adjusted to external or internal forces for change.

2.5.2 Scale

In fluvial systems every variable has both a cause and effect relationship with every other variable, making simplified generalizations about causality impossible if the entire system operation is considered (Mackin 1953, p. 149). It is therefore necessary to divide time and space into compartments for purposes of understanding the interactions among system elements. The size of these intellectual compartments in part determines the outcome of research efforts. Theory construction in geomorphology is therefore scale-dependent, and problems associated with varying scale become significant. For example, what we learn about processes in restricted time and space is not necessarily true for larger units of analysis, so that the way in which data is collected and aggregated strongly influences conclusions. Geologists have emphasized the problem of scale in a temporal sense, while geographers have emphasized the spatial aspects. The conclusions in each case are similar: explanations vary according to the magnitude of the system to which they are applied.

Schumm and Lichty (1965) illustrated the importance of time scales in assessing causality for fluvial systems (Table 2.2). They organized geomorphic time into three hierarchical scales: cyclic, graded, and steady state time. Cyclic time consists of those periods up to several millions of years in length when the many characteristics of fluvial systems are dependent variables. A fluvial system might be viewed as progressing through a cycle of erosion as envisioned by Davis (1899) during a period of cyclic time. Graded time corresponds to periods of several hundreds or a few thousands of years, when the fluvial system might be characterized by Mackin (1948) to be in grade, with general overall adjustment about a mean condition. Steady time refers to relatively brief periods when none of the system variables change with time (Schumm and Lichty 1965, p. 115). Although Schumm and Lichty referred to this condition as "steady state," it is better characterized as "static equilibrium" as defined by Chorley and Kennedy (1971, p. 201).

The importance of considering the different time scales is that variables describing the fluvial system assume different causal relationships with each other depending on the scale of analysis. In cyclic time (very long periods), only time, initial relief, geologic structure and lithology, and climate are independent variables. At the other extreme in the very short periods of steady time, all the variables are independent except for discharges of water and sediment. The implication of this arrangement is that in order to explain system behavior, data must be collected over the relevant time scale.

Space, like time, is also arranged in a hierarchical fashion that influences questions of causality. Small drainage basins are nested within larger drainage basins, and depending on the scale of analysis, system variables take on different causal relationships with each other. Fenneman (1916) generated a series of landform divisions for the United States based on scale-dependent criteria. Un-

Table 2.2. The effect of time scales on causality for drainage basin variables

Variable	Time scale		
	Cyclic	Graded Spatial scale	Steady
	Continent	Province	Tract
Time	Independent	Not Relevant	Not Relevant
Initial relief	Independent	*Not Relevant*	*Not Relevant*
Geology (lithology, structure)	Independent	Independent	Independent
Climate	*Independent*	Independent	Independent
Vegetation (type and density)	Dependent	Independent	Independent
Relief	Dependent	Independent	Independent
Hydrology (water and sediment yield per unit area)	Dependent	*Independent*	Independent
Drainage Network	Dependent	Dependent	Independent
Hillslope Morphology	Dependent	Dependent	*Independent*
Hydrology (water and sediment yield from the system)	Dependent	Dependent	Dependent

Spatial scales as defined by Linton (1948) added to table from Schumm and Lichty (1965, p. 112).

stead (1933) continued Fenneman's work by subdividing the larger regions defined previously, with the resulting landform units reflecting process-form regions. Further detail was added by Linton (1948). Whittlesey (1956) operationalized the issue of scale in a representative landform study in southern Rhodesia by attaching map scales to the various sized units in the hierarchy (Table 2.3).

A table representing the variation in causal relationships among variables in a fluvial system can be constructed for spatial considerations similar to that constructed by Schumm and Lichty for temporal considerations. If the variables considered are the same and the scales are roughly borrowed from Fenneman's (1916) and Unstead's (1933) proposals, the results are similar to the temporal perspective, with increasingly large scales encompassing more dependent variables (Table 2.2). At the continental scale with areas of 10^{10} km^2, only geologic time, initial relief, geologic structure and lithology, and climate are causally independent. At the opposite extreme, relatively small drainage basins about 1 to 1,000 km^2 in extent have only their sediment and water yields as dependent variables.

Tables 2.2 and 2.3 show a linkage between time and space with regard to scale. "Most processes produce variability in areal or temporal series which is best identified at particular intervals" (Thornes and Brunsden 1977, p. 183). Slow processes are likely to be manifest to the researcher at the largest spatial scales over long periods of observation. Over short time periods and limited spatial scales, these slow processes are not separable from the "noise" of the system. Hence, it makes little sense to attempt to deduce continental denudation rates from observations of small or even medium-sized watershed which are subject to the influences of short-term human activities and internal storage (Trimble 1977).

The importance of scale can be illustrated in a quantitative sense by considering the spatial distribution of values of a given variable as represented by a trend surface. The value at any point in space of the surface is the product of at least three influences: the regional trend, local covariation, and specific variation such

Table 2.3. Spatial hierarchies in geomorphic regions

Approx. size mi^2	Fenneman 1916	Unstead 1933	Linton 1949	Whittlesey 1954
10^0			Site	
10^1		Stow	Stow	Locality
10^2	District		Tract	
		Tract		District
10^3	Section		Section	
		Sub-Region		Province
10^4	Province		Province	
		Minor Region		
10^5	Major Division		Major Division	Realm
10^6		Major Region	Continent	

Haggett et al. (1977, p. 454).

as measurement error (Haggett et al. 1977, p. 383). The regional trend is usually the influence of major interest to the geomorphologist because it permits regional generalizations. The local covariation represents the local influences on point values and represents spatial autocorrelation--the values of data points geographically close to each other exert a structural influence on each other. This spatial autocorrelation frequently obeys laws of distance decay, and might be as informative as regional variations, but it would provide explanation at a scale smaller than regional.

A dryland example of this regional versus local influence as a manifestation of scale issues is the study by Lustig (1969) of the mountain ranges of the Basin and Range Province of the southwestern United States and northwest Mexico. Lustig fitted a trend surface to dimensional characteristics of the individual ranges and identified regional trends in these values. The result was a generalization on a subcontinental scale. Deviations from the general surface resulting from local tectonic influences were not investigated, however, so that the regional generalizations are not completely successful in explaining the observed values.

Scale poses two problems in experimental design because of its influence on observations and conclusions. First is the problem of the "ecological fallacy" which Haggett et al. (1977, p. 10) identify as a sampling problem. As the spatial resolution of the sampling scheme changes, the results change. Robinson (1956) found that results of a statistical analysis involving map data changed radically as the size of the sampling units changed. Therefore, findings at one scale of analysis may be misleading if applied at finer or coarser scales of analysis.

The second problem is the limiting effect that scale has on sample sizes. If individual units of time or space are small, the number of such samples is likely to be large and susceptible to confident statistical analysis. If the sampling bases are large areas or long periods they are likely to be so few in number that cumulative generalizations are not possible.

Scale has important implications for the analysis of dryland rivers. What is learned of processes in small basins is not likely to be informative with regard to the operation of large river systems, and what is learned of the large rivers will have limited application to small channel behavior. Small basins of a few km^2 may respond to individual rainfall events triggered by thunderstorm cells of about the same size. The behavior of large rivers may not be explicable by reference to individual precipitation events, but rather be more closely attuned to longer-term climatic changes in drought indices. The scale of the explanatory variable must match the scale of the responding variable.

2.5.3 Magnitude and Frequency

One of the most important concepts that integrates time and dimensional ideas is that of magnitude and frequency. In natural systems, small events occur often and large events are rare. This arrangement sets up a conundrum whereby it is uncertain which events account for more geomorphic work, small ones that occur so often that their cumulative effort is greatest, or the more rare events that may be so large that their impact is most important.

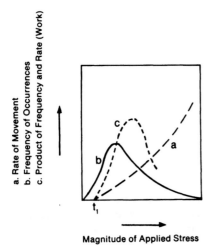

a. Rate of Movement
b. Frequency of Occurrences
c. Product of Frequency and Rate (Work)

Magnitude of Applied Stress

Fig. 2.15. The original formulation of magnitude-frequency relationships among *A* rate of movement as a measure of magnitude, *B* frequency of events, and *C* work accomplished as a product of *A* and *B*. (After Wolman and Miller 1960, p. 56)

From the fluvial geomorphological perspective, the magnitude of an event is measured by the rate of water or sediment transport through a channel cross section in a given time period. The frequency of an event is measured as an annual frequency of occurrence for events of the same magnitude. In determining which events are most likely to have the greatest geomorphological significance, the magnitude is multiplied by its related frequency over a standardized time period as a measure of total work accomplished, such as total amount of water or sediment moved.

Wolman and Miller (1960) proposed that the magnitude and frequency concepts could be quantified and represented by mathematically derived curves (Fig. 2.15). The magnitude of the events may be described by a power function:

$$q = a\,x^n, \tag{2.14}$$

where q = the rate of movement, x = a variable positively related to the magnitude of the event (e.g., shear stress), and a, n = parameters of the function. Figure 2.16 shows various forms of the power function representing magnitude.

The function describing the frequency of fluvial events is similar to the functions describing eolian or marine events and is a probability density function in the form of a log-normal distribution (Chow 1954; Krumbein 1955):

$$f(w) = \{1/[wB(2\,pi)^{0.5}]\}\exp\{-[1/2(B)^2][\ln w - a]^2\}, \tag{2.15}$$

where $f(w)$ = the probability density function, w = a measure of the observation magnitude, B = the empirically determined variance of the distribution, pi = the value of pi (about 3.14), and a = the empirically determined mean of the distribution (Koch and Link 1970; Krumbein and Graybill 1965). Fig. 2.17 shows various forms of the log-normal distribution.

In their original presentation of the concept of magnitude and frequency, Wolman and Miller (1960, p. 56) used as examples power functions and log-normal distributions that represented conditions for perennial streams. When they multiplied the two functions to derive the amount of work accomplished, they

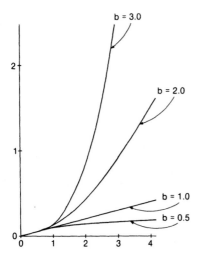

Fig. 2.16. Various forms of the power function [Eq (2.13)] representing rate of movement as a measure of magnitude

concluded that under perennial flow conditions the maximum work was accomplished by the moderate-sized events, as shown in Fig. 2.15. Leopold et al. (1964, pp. 67–80) provided an extended discussion of the problem of defining the relative geomorphic significance of events using data from stream gauging records. They concluded that the magnitude-frequency concept had several different implications depending on the system involved:

1) For water yield in perennial streams the intermediate-size events contribute most of the water.
2) For sediment yield the major work is accomplished by modest-sized, relatively frequent floods.
3) For dissolved load the major work is done by low, frequent flows because dissolved concentrations decline in high flows.
4) The more variable the flow of the streams, the larger the percentage of total load that is carried in a few large events.

The magnitude-frequency concept is climatically dependent (Graf 1985, p. 273), though this is rarely stated in discussions of the idea. In some subsequent reports of the Wolman and Miller interpretation, the concept of the importance of the moderate events was repeated without the qualification that such a conclusion rests on a special arrangement of the functions or that in drylands the largest events become relatively more important (e.g., Thornes and Brunsden 1977, p. 9; Richards 1982, pp. 122–124). The magnitude of the geomorphologically most important event in dryland rivers is likely to be large compared to the moderate events significant in humid systems and those with perennial flow. Neff (1967) showed that because of the importance of the high flows in low rainfall areas dryland rivers transport as much as 60% of their sediment in events with return intervals of 10 years or greater. Humid-region rivers transport only about 10% of their sediment in similar sized events.

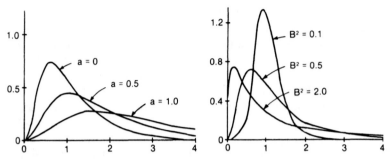

Fig. 2.17. Various forms of the log-normal function [Eq. (2.14)] representing frequency of events

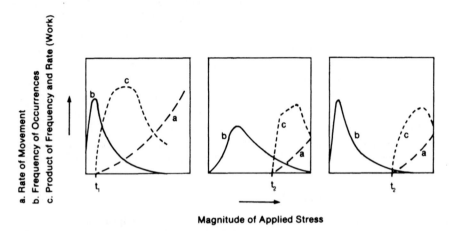

Magnitude of Applied Stress

Fig. 2.18. Modifications of the original formulation of magnitude-frequency relationships among *A* rate of movement as a measure of magnitude, *B* frequency of events, and *C* work accomplished as a product of *A* and *B*. (After Baker 1977, p. 1057, 1059)

Baker (1977, pp. 1057–1059) introduced the concept of thresholds to the analysis of magnitude and frequency by pointing out that the transportation of sediment will not always begin automatically with increasingly large discharges. Transportation will not begin until a threshold of motion for particles is passed (2.13), so that if the particles are large, no transport and no work will occur until the largest events (Fig. 2.18). This result is likely in many dryland rivers that have gravel or larger bed materials. Because of the threshold phenomena and the highly skewed frequency distributions related to discharge events, the larger event are likely to be geomorphologically the most significant in dryland rivers.

2.5.4 Allometric Change

Allometry refers to the "study of proportional changes correlated with variation in size of either the total organism or the part under consideration" (Gould 1966, p.

629). The general principle of proportional rates of change between elements in a system was first formulated in biology by Huxley (1924), and first applied to river systems by Woldenberg (1966). According to Woldenberg (1968, p. 776) the law of allometric growth specifies that the rate of growth of a system element is a constant fraction of the rate of growth of the entire system. Bull (1975b, p. 113) suggested that the term change should be substituted for "growth" since many geomorphic systems undergo positive and negative change. Mosley and Parker (1972) illustrated that there are two modes of allometric analysis: (1) static allometry, concerned with dimensional changes among elements at the same time, and (2) dynamic allometry, concerned with dimensional changes of a single element through time.

In allometric changes, as size increases or decreases, the geometry of form changes. As an example, islands and bars in the Green River, Utah, change shape as they grow larger (Graf 1978, pp. 1498–1499). Small islands and bars are relatively short and wide, while large ones are relatively narrow and long. Channel width restricts the width of the islands and bars which can otherwise expand in the downstream direction through accretion. The allometric changes might also be explained by dynamic streamlining of the features as they adjust to the competing influences of reduced friction (favoring compact forms with little perimeter per unit surface area) and reduced turbulence (favoring long, narrow forms).

Allometry has a relatively simple mathematical expression. If the rate of change of one element (a dimensional measure, for example) is a constant fraction of the rate of change of another element (or of a dimensional measure of the entire system), then

$$(dy/y)(1/dt) = b(dx/x)(1/dt), \qquad (2.16)$$

where dy/y = proportional change in dimension y, dx/x = proportional change in dimension x, dt = time increment, and b = a constant of proportionality, a constant fraction. If both sides of (2.16) are multiplied by dt and integrated, the result is

$$\int (dy/y) = b \int (dx/x), \qquad (2.17)$$

$$\log y = \log a + b \log x. \qquad (2.18)$$

Taking antilogs in (2.18) produces the simplified statement of allometric change

$$y = a x^b, \qquad (2.19)$$

where b remains the constant of proportionality. Bull (1975a) proposed that all relationships of the form of (2.19) are rooted in allometric priniciples, but Kennedy (1985d, p. 13) argued that such a position obscures the underlying theoretical philosophy of the concept.

A special case of allometric change is isometry. Isometric relationships have a constant of proportionality (b) of 1.0 if the variables have the same units of measure. In isometric change, as size changes the form remains the same. If the variables have different dimensional units, in isometric relationships the constant of proportionality balances (2.19) (von Bertalanffy 1960). If x is in the dimension area (length²), y is in the dimension length, and isometry prevails, then b = 0.5.

Church and Mark (1980) have proposed that isometric relationships are completely specified in terms of physical processes, and that allometric relationships are incomplete specifications including unidentified scale-dependent controls.

Whatever the interpretation of the allometric-isometric relationships, they form useful law-like statements that can assist in theory construction. Allometry is merely descriptive, and provides no explanation, but it can provide a convenient tool for analyzing landform changes over time and space. Woldenberg (1966) and Graf (1978, 1979a) have shown the utility of allometry in the analysis of fluvial features.

2.5.5 Entropy

In physics, entropy represents the amount of heat liberated or absorbed by a perfect engine divided by the temperature of the engine (Feynman et al. 1965, p. 44/10). In perfectly reversible processes the entropy generated by one engine is absorbed by another engine operating in reverse, so that the total amount of entropy in the overall system is unchanging. Physical processes are generally not perfectly reversible, and therefore entropy accumulates in a system as time progresses. As the irreversible engine operates and gives up amounts of heat divided by temperature, the available energy in the system declines and the amount of expended energy increases until the system decays to a state of no available energy, a maximum of entropy, and no further operation.

An alternative interpretation of entropy relates to order and disorder (Chorley and Kennedy 1971, pp. 219–225). At its original definition, a closed system may be assumed to be highly organized, but as it operates, the expenditure of energy is required to maintain organization. Eventually the system decays into disorder, as progressively less energy is available for work and more energy has been expended. The expended energy is referred to as entropy in this approach, so disorganization is associated with high amounts of entropy. In an open system energy may be transported into the system to increase the amount of available energy and reduce the relative amount of entropy (Denbigh 1951, p. 40).

In the thermodynamic example, energy is transferred from a warm body to its surroundings until the body and its surroundings contain the same amount of heat (Lewis and Randall 1961). This simple system has then made an adjustment from a low entropy case where heat was concentrated in one place (the body) and not the other (the surroundings) to a high entropy case where the heat distribution is equal throughout the system. In the thermodynamic example, the level of heat energy is measured on a scale of absolute temperature which has a base level of absolute zero where there is no molecular motion.

The statistical interpretation of entropy is the portion of the concept most likely to be useful to geomorphologists. From the statistical perspective, a highly organized system in a particular configuration (low entropy) has a low probability of occurrence (Wilson 1970; Gould 1972). A disorganized system with a random configuration (high entropy) has a high probability of occurrence. This arrangement provides a direct link among the distribution of energy, the state of a system, and the probability of occurrence (Bell 1956, p. 159).

In the thermodynamic case (following Leopold and Langbein 1962, pp. 3–4), the change in thermal energy per unit mass in a system (dE) at an absolute temperature (T) is a function of specific thermal energy of the substance involved (C):

$$dE = C \, dT. \tag{2.20}$$

Entropy (S) is the sum of expended thermal energy

$$S = \Sigma \, (dE/T). \tag{2.21}$$

Entropy defined per unit of mass is

$$S = C \Sigma \, (dT/T). \tag{2.22}$$

Temperature (T) may also be thought of as a probability (p) that the energy in the system exists in a given state above absolute zero. In this interpretation $0 < p < 1$, so p does not equal T, but the two are related to each other, and (2.22) can be written

$$S = C' \Sigma \, (dp/p), \tag{2.23}$$

where C' = specific heat, or

$$S = C' \ln p + \text{constant}, \tag{2.24}$$

where ln = logarithm in the natural logarithm system.

Because the concept of entropy is bound up with the entire history of a system, in the statistical sense entropy is the sum of the individual probabilities attached to each possible state (that is, $p_1, p_2, \ldots p_n$). The sum of these probabilities constitutes p in equation 2.24, so that the general statement for entropy is

$$S = C' \Sigma \ln p. \tag{2.25}$$

Because logarithms of numbers less than 1.0 are involved, entropy S is in negative units. Entropy is at the minimum of zero when one state has a probability of 1.0 and all others have zero probability and do not occur ($\sigma p = 1.0$, $\ln 1.0 = 0.0 = S$). Such a system would be highly organized. At the opposite extreme, a system with 100 possible states, all with equal probabilities (0.01 each), would be a system with maximum entropy ($\ln 0.01 = -4.61$; $-4.61 \times 100 = -461 = S$). Such a system would have little organization and much randomness: it would be statistically indeterminant.

In the fluvial geomorphologic application of entropy, a scale of measure analogous to that for temperature is available in the form of elevation which is directly related to the amount of potential energy available for work in the system. Absolute zero in the geomorphologic application is base level for the stream in question. As erosion processes approach the base level, entropy increases, the system becomes progressively less well organized and progressively less energy is available for geomorphic work.

Entropy concepts also lead to the principle of least work in geomorphic systems. Prigogine (1955, p. 84) contends that natural open systems tend to operate such that the rate of production of entropy per unit volume corresponds to a minimum possible. Although sometimes applied to geomorphic systems

(Rubey 1952, p. 135), the principle of least work is not a complete explanation because geologic controls and climatic changes interrupt the system adjustments (Leopold and Langbein 1962).

The analogy between basic physics and complicated geomorphologic systems in terms of entropy has been criticized as an inappropriate connection (Kennedy 1985a, p. 155). The statistical implications of entropy have more relevance to geomorphology than the physical interpretations. The state of maximum entropy is a special case because it is the most likely one. The most likely distribution of individual state probabilities is that they are all equal (Leopold and Langbein 1962, p. 4), resulting in a maximum value for entropy (S) in (2.25). This statistically most probable case rarely occurs in nature because constraints on system operation prevent perfect adjustment.

Analysis of geomorphic systems using entropy concepts results in a research strategy different from the usual approach (Thornes and Brunsden 1977, pp. 172–173). Instead of constructing a model with many variables and assumptions identified a priori, the researcher makes no assumptions and has no preconceived model in the entropy approach.

A potentially useful application of entropy concepts in the analysis of dryland rivers is to use the concept as a standard against which actual conditions can be compared. Dryland rivers adjust through a series of states; a probability of occurrence might be empirically assigned to each state. This actual probability can then be compared with the maximum entropy case, with the difference between the two a measure of the organization of the channel system's operation and a measure of the amount of energy added to the system to prevent the development of maximum entropy.

2.5.6 Catastrophe Theory

Catastrophe theory is a language that describes time-space changes in general systems. As a language it accounts for smooth transitions from one state to another as well as abrupt changes. Catastrophe theory is not a "theory" in the sense of the word as used by geomorphologists because it provides no formal explanation. The language of calculus-based mathematics is suited to relatively gradual changes over the range of possible values of given variables, so catastrophe theory is a possible substitute for calculus in some applications. Thom (1975) provided the original outline and proof of catastrophe theory as an exercise in differential topology. Zeeman (1976), Woodcock and Davis (1978), and Woodcock and Poston (1974) provided non-mathematical treatments of the concepts. Recent detailed reviews including examples of applications include those by Gilmore (1981) and Wilson (1981). Bennett and Chorley (1978) review the underlying mathematics.

The basis of Thom's theory is that a series of topologic singularities describes the interactions of control and response variables. In topology, a singularity is a phenomenon that occurs when points are projected from one surface to another surface that has a different shape. In the transition through this geometric distortion, the singularity may change in size or magnitude, but it retains its

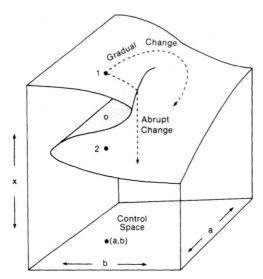

Fig. 2.19. A cusp catastrophe surface defined by the equilibrium values of (a, b, x). In the fold area, the control factors (a, b) produce two possible equilibrium values for x, shown as system states *1* and *2*. The sequence of changes in (a, b) influence the path of change on the surface and may result in gradual or abrupt changes depending on the relationship between the path and the fold

topologic integrity. Thom suggested that system changes can be described by topologic singularities which he called catastrophes. He mathematically and topologically proved that for a given number of control factors and responding variables there is only one catastrophe (that is, singularity) that describes system changes in time-space dimensions.

If it is true that geomorphologic problems can be reduced to an analysis of two control factors, force and resistance, and a responding variable, then the systems involved may be represented by the catastrophe referred to by Thom as the cusp (Fig. 2.19).

The cusp is a three-dimensional surface defined by three variables: the two control factors (a and b in Fig. 2.19) and the responding variable (x in Fig. 2.19). The shape of the cusp catastrophe with its internal fold is a product of Thom's topologic proofs. The state of the system at any particular time is represented by a point which represents one combination of a, b, and x. If the system is in equilibrium, the point plots on the surface of the catastrophe. If it is not in equilibrium, the point plots off the surface and is not accounted for by the catastrophe theory.

At different times, different combinations of a, b, and x occur in system operation, so that moving the point about on the catastrophe represents the change of the system through time. Some paths that the point representing system conditions might take as it moves on the surface result in slow changes in the responding variable. Other paths might cause the point to move across the fold and move vertically to another level, resulting in abrupt change in the responding variable even though the control factors changed only slightly. Because the surface of the cusp represents equilibrium conditions and because in part of the

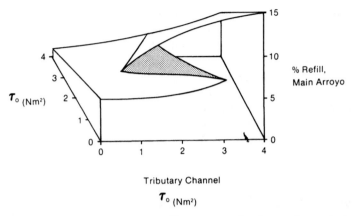

Fig. 2.20. A catastrophe theory representation of sedimentation at the junctions of channels in an arroyo network. The control factors are shear stress in the tributary channel and shear stress in the main channel, while the responding variable is the amount of sedimentation measured by the per cent of the arroyo refilled by sediment. The surface is smoothed from 28 data points from streams in the Henry Mountains region, Utah. (After Graf 1982, p. 211)

catastrophe two surfaces overlap each other, there are some values of (a, b) where two equilibrium conditions are possible.

Graf (1979c) found that the applicability of catastrophe theory in fluvial geomorphologic research depended on the successful definition of measures of force and resistance, and on the selection of a measure of system response that reflected a variety of system states. He found that if stream power and resistance offered by vegetation were control variables, the sizes of arroyo (for definition see section 5.8, this volume) cross sections served as a responding variable that generated a cusp catastrophe. Catastrophe theory was not useful in describing response variables expressed as probabilities of occurrence. Thornes (1980) used catastrophe theory to model the spatial and temporal variation of sediment transport in semiarid Spanish streams, and Graf (1982a, p. 211) used the catastrophe theory concepts to model deposition at stream junctions (Fig. 2.20). Other applications in geomorphology include that by Richards (1982, p. 215) for channel patterns and by Hutter (1982, p. 31) for glacial processes.

Problems associated with catastrophe theory include the issue of defining the control factors. In fluvial systems the definition of force and resistance measures for use as control factors is common, but whether or not enough of the total system operation can be described by using only these two measures is debatable. The tendency toward equilibrium is a prerequisite for the application of catastrophe theory because the catastrophe represents an equilibrium surface (in the case of the cusp), but all aspects of dryland rivers do not necessarily tend toward equilibrium (Bull 1975b). Catastrophe theory is also mainly qualitative, and though it can include quantitative aspects, its major contribution is the knowledge of the shape of the surface. The supporting mathematics are nearly intractable for normal geomorphologic applications except in the definition of the location of the fold.

The advantages of catastrophe theory include its ability to accommodate a variety of types of change through time and across space. It represents a more complicated view of thresholds because instead of defining dividing lines (Schumm 1973), it defines zones of transition where two states of equilibrium are possible. Movement of points representing system states on the surface of the catastrophe allows description of most types of system behavior observed in geomorphology, including divergence, abrupt and smooth changes, bimodal behavior, and hysteresis. Even if the entire catastrophe defies simple mathematical definition, specification of the fold area provides information on the most interesting and important part of the surface. Finally, catastrophe theory provides a significant world view: if there are two control factors (such as force and resistance) and one responding variable (such as channel width), then the system invariably will exhibit behavior described by the cusp catastrophe. In this circumstance, occasional rapid adjustments are to be expected and are not unusual outliers of system behavior.

Although Thom's topological proof of catastrophe theory has not been successfully challenged, applications of the approach have been limited. Croll (1976), Zahler and Sussman (1977), and Kolata (1977) questioned the general utility of the concepts, as did Wagstaff (1976) for geographic applications, and Gregory (1985, p. 181) feared that it might be a methodological fad. Zeeman (1977) showed the wide applicability of catastrophe theory, and Cubitt and Shaw (1976) and Henley (1976) were more optimistic for its use in geology. Huggett (1980, p. 193) proposed an important role for it in geography. Because of its recent publication, catastrophe theory has yet to be adequately tested, but it offers the possibility of efficiently describing a variety of system behaviors with a single model.

Part II   Processes and Forms

3 Surface Water in Drylands

Fluvial processes in drylands are driven by water that falls on the earth surface as precipitation, so that an understanding of the temporal and spatial variability of precipitation is a prerequisite to understanding the variation of river behavior. The purpose of this chapter is to explore briefly some aspects of precipitation, followed by an analysis of the behavior of surface waters in the forms of runoff and streamflow. Because fluvial systems operate at a variety of scales, the chapter includes a discussion of climatic change and its hydrologic implications.

A thoroughly developed integrated theory for rivers in drylands does not yet exist, but the construction of such a theory must include at least the fundamental principles outlined in this chapter. In many cases, the fundamental principles are manifestations of the application of the abstract concepts reviewed in Chapter 2.

3.1 Dryland Precipitation

Precipitation in drylands results from several distinct atmospheric processes, each one operating at a particular scale and producing a different temporal and spatial pattern. Of special interest in drylands are thunderstorms, which strongly influence small (< 10 km^2) drainage basins. In some drylands, however, precipitation events in other nearby regions are important because the dryland fluvial system may be dominated by through-flowing rivers that head outside the dryland area. The Colorado River in North America, the Nile River in North Africa, and the Indus River in Asia are examples. Most of the following discussion focuses on processes within dryland regions and excludes, except for the analysis of paleo-climatology, the through-flowing streams from external regions.

3.1.1 Mechanisms

Precipitation in drylands is the product of four general atmospheric processes: frontal activity, tropical storms, orographic effects, and convection. Most dryland areas of the world are located beneath the semi-permanent high pressure systems of the general global circulation system and are therefore air-mass source regions (for a specific example, see Mitchell 1976). The passage of frontal systems through these air-mass source regions is relatively unusual compared to the more humid mid-latitudes. When frontal activity occurs in drylands, it brings about rare but important fluvial events because the frontal processes may generate long-term rainfalls. Saturation of surfaces, rapid runoff, and river floods result.

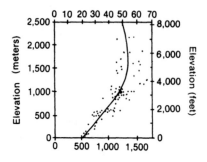

Mean Annual Precipitation (inches)

Fig. 3.1. The relationship between elevation and rainfall for stations in the Sierra Nevada, California, showing the initial increase of precipitation with elevation and then a reversal of the relationship. (After Armstrong and Stidd 1967)

Tropical storms are even more rare in dryland areas than frontal activity, but these storms (including cyclones, hurricanes, or typhoons) occasionally drift from their tropical ocean sources to invade arid and semiarid landmasses. Large floods from incursions of tropical storms may form what appear to be outliers in the hydrologic record, but they must be taken into account in the explanation of dryland precipitation regimes. An analysis by Eidemiller (1978, p. 26) of 22 years of record for tropical storms in the east-central Pacific Ocean showed that the drylands of northwest Mexico and southwest United States had a probability of experiencing a tropical storm of about 1% per year. Similar storms move into the drylands of most of the world, exemplified by southwest Asia (documented by Schick 1971a, b) and the interior of Australia (Boughton 1980, reviews the problems of statistical analysis).

As moisture-bearing air masses lift over mountain barriers, the reduced pressures and cooler temperatures associated with increased elevation induce condensation and precipitation. This orographic effect operates in dryland areas even though the atmosphere may have relatively low levels of moisture. The result is a distinct association between elevation and precipitation and a pronounced concentration of rainfall on some upland surfaces. Above a certain elevation most of the available moisture has precipitated out of the air mass, and dry conditions prevail (Fig. 3.1). The high plateaus of the southwestern United States, portions of India, western China, and the high deserts of western South America are examples. As air descends, it also heats, producing a rainshadow effect whereby frontal and tropical storms are rare in the lee of mountains. Orographic influences in Israel and Jordan are obvious in data collated by Schick (1971, p. 112): at nine stations not influenced by orographic effects the mean annual rainfall was 37 mm, while at two stations with orographic effects the mean was 220 mm.

Finally, thunderstorms produce local precipitation in drylands. Usually the product of convection of heated air, most thunderstorm cells are only a few kilometers in diameter, so they affect small drainage basins of about the same size by inducing flash floods. When observed in the field, dryland thunderstorms take on an awesome appearance, perhaps accentuated by the general aridity. They do not produce heavier precipitation than those generated in other regions, as

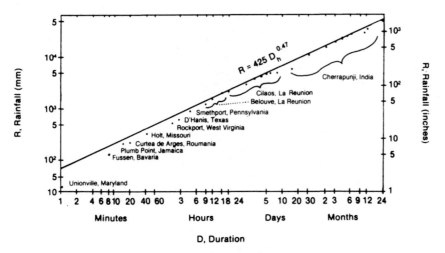

Fig. 3.2. World record rainfall events showing the lack of high intensity events in drylands. (After Paulhus 1965 and Shaw 1983)

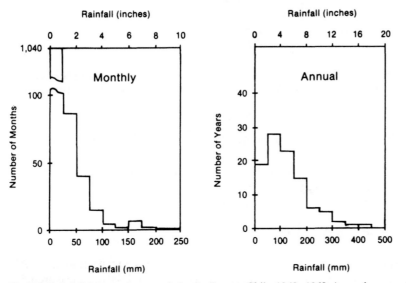

Fig. 3.3. Rainfall frequencies recorded at La Serena, Chile, 1869–1968. Annual average is 126 mm. (After Shaw 1983, p. 231)

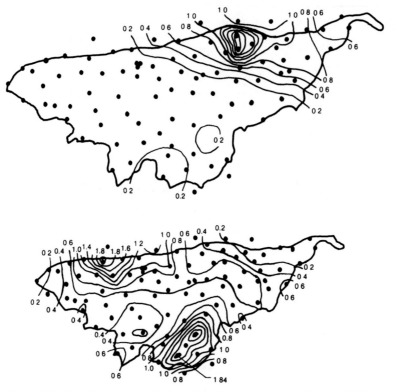

Fig. 3.4. The distribution of rainfall from thunderstorm cells over Walnut Gulch, Arizona: Above, a single cell event; Below, a multiple cell event. Dots represent rain gauges. (After Renard 1970)

illustrated by a review of world-record rainfalls (Fig. 3.2). Figure 3.3 presents precipitation data from a representative dryland station (mean annual precipitation = 126 mm) with an exceptionally long record (a century), and shows the dominance of relatively light rainfalls.

3.1.2 Spatial and Temporal Characteristics

The importance of convective precipitation in drylands means that available records from these regions are of questionable value unless the measurement period is especially long. Rainfall patterns from single thunderstorm cells are roughly circular, though if prevailing winds are pervasive enough to move the cell an elongated ellipse may describe the wetted area (Stall and Huff 1971). Experimental watersheds with exceptionally dense raingauge networks show that steep gradients occur between areas of little rain and the peak rainfall area (Fig. 3.4). Renard and Brakensiek (1976) have shown that the scattered nature of thunderstorm events and the limited spatial extent of individual cells make the distribution of the observation network critical if a regional picture of thunderstorm

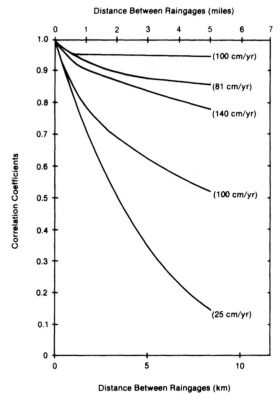

Fig. 3.5. The correlation between rain gauges and distance for stations representative of several climatic regions. Note that the dryland example has the lowest correlation by a large degree. (After Hershfield 1967 and Sharon 1972)

activity is to emerge from the data. In most dryland areas of the world, reliable weather stations are widely separated and observations cover only a tiny fraction of the total arid and semiarid region. Extrapolations of rainfall data even a short distance into ungauged areas in drylands is a hazardous business (Fig. 3.5).

Presently available records represent only a sample of questionable reliability, and the connection between thunderstorm activity and fluvial response remains largely unexplored. Branson et al. (1981) concluded from the limited available data in the southwestern United States that fluvial system behavior may respond to the following known aspects of thunderstorms:

(1) Convective storms are multi-cellular, random, short duration (less than one hour), of low rainfall production (less than 2 cm), and limited in area (groups of moving cells generally affect areas of less than 70 km²).
(2) Rainfall peak occurs in the first ten minutes of the storm, but there may be up to four bursts of precipitation.
(3) Few storms produce channel flow.

(4) Runoff-producing storms are high-intensity events.

These results suggest that basins larger than about 70 km^2 will rarely be complet-ely covered by a thunderstorm event. Flash flooding from such storms and asso-ciated geomorphic activities are likely to be most pronounced in basins equal to or smaller than 70 km^2.

3.1.3 Data Availability

Until recently, rainfall data were typically recorded on standardized paper forms, with monthly or annual summaries published in many countries. Historical records in many parts of the world are still stored in hand-written form, so that the user must be prepared to sift through voluminous accumulations of records to abstract useful materials. Since the late 1970s, some countries have converted their meteorologic records to computerized forms with data stored on magnetic tapes. This more efficient storage form does not allow the user the freedom to browse the data as was possible with the older paper forms, and one must know specifically what is needed before a request is made. Reitan and Green (1968) provide a useful (and not yet outdated) guide to obtaining climatic data in drylands.

   The availability of climatic data is variable among nations administering dryland areas. Generally, more developed countries have longer-term records. Israel has the densest raingauge network of any dryland nation. Table 3.1 reviews the status of climatic data for dryland nations.

Table 3.1. Published climatic data for dryland nations (m = 20 years or more of monthly data; a = 20 years or more of annual data; x = less than 20 years of data)

Aden – x	Iran – m,a	West Pakistan – m,a
Afghanistan – m	Iraq – m	Peru – m,a
Algeria – m	Israel – m,a	Qatar – x
Angola – m,a	Jordan – m	Saudi Arabia – m,a
Argentina – m	Kenya – m,a	Somalia – x
Australia – m	Kuwait – m,a	South Africa – m,a
Bahrain – x	Libya – m	Sudan – m,a
Botswana – m,a	Mali – m	Syria – m,a
Bolivia – a	Mauritania – x	Tunisia – m,a
Cameroon – m,a	Mexico – m,a	U.S.S.R – m
Chad – m	Morocco – m,a	United Arab Emirates – x
Chile – a	Muscat and Oman – x	U.S.A. – m,a
China – x	Namibia – m,a	Upper Volta – x
Djibouti – m	Niger – x	Western Sahara – x
Ethiopia – m	Nigeria – m	Yeman – x
India – m,a	Outer Mongolia – x	

Note: Unpublished records exist in most countries in this list. Basic data from Reitan and Green (1968).

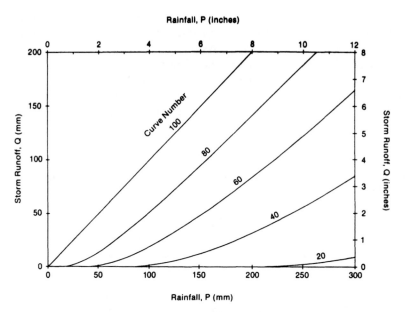

Fig. 3.6. Nomogram for estimating runoff from rainfall according to the U. S. Soil Conservation Service method for rainfalls of less than 30 cm (12 in). Curve numbers represent hydrologic and soil conditions as outlined in Tables 3.2, 3.3, and 3.4. (After U. S. Soil Conservation Service 1972)

3.2 Runoff

Upon arrival on the surface, water from precipitation seeps into the subsurface until water fills the interstices of the soil particles causing a saturated condition. Once this infiltration capacity is met, water ponds on the surface in small depressions. Eventually, the amount of available water exceeds this depression storage, the natural reservoirs overflow, and the water moves downslope in the form of runoff. The runoff represents the contribution of mass and energy to the channel system and fuels the fluvial processes of dryland rivers. Runoff on slopes performs geomorphologic work, but because this volume concerns river processes, the following discussion is limited to the quantitative contribution of water to channels by runoff. For more complete reviews of hillslope runoff processes, see Carson and Kirkby (1972), Dunne and Leopold (1978), and Branson et al. (1981).

3.2.1 Soil Conservation Service Method

The prediction of the amount of runoff expected from the slopes of a drainage basin could be accomplished by complex numerical modeling of the infiltration, storage, and overland flow processes, but in dryland geomorphologic applications such approaches require too much data and are not widely proven. Empirical methods may be less precise, but are likely to be more accurate than a deterministic approach. One widely used empirical method developed by the U. S. Soil Conservation Service is the "curve number method" (see Ralliston 1980, for a discussion of the evolution of the method). As published by the U. S. Soil Conservation Service (1972), the user of the method begins with the estimated rainfall from a storm and traces the relationship between rainfall and expected storm runoff on a nomogram (Fig. 3.6). The relationship varies according to soil and vegetation conditions which are represented by different curves in the nomogram.

Table 3.2. Runoff curve numbers for use in the Soil Conservation Service Method for estimating runoff (see Fig. 3.6)

Land use or cover	Hydrologic	Hydrologic soil group			
	Condition	A	B	C	D
Range or pasture	Poor	68	79	86	89
	Fair	49	69	79	84
	Good	39	61	74	80
Meadow	Good	30	58	71	78
Woodland	Poor	45	66	77	83
	Fair	36	60	73	79
	Good	25	55	70	77
Dirt roads		72	82	87	89
Paved roads		74	84	90	92

Source: U. S. Soil Conservation Service (1972). See Tables 3.3 and 3.4 for information on hydrologic condition and hydrologic soil groups.

Selection of a particular curve, identified by a number between 0 and 100, is the critical point in the process. Originally, guidelines specified the selection of curve numbers on a generally subjective basis (Table 3.2), with evaluations of soils (Table 3.3) and hydrologic conditions (Table 3.4) as inputs. The original Soil Conservation Service efforts were for humid agricultural regions and the application of the results in drylands were questionable.

Additional work by Jencsok (1968), Malone (1972), and Simanton et al. (1973) extended the method to applications in dryland conditions. The general method forced emphasis on the selection of a particular curve number, so that these attempts at adapting the method to drylands consisted of formulating ways of identifying the appropriate curve number. Jencsok (1968) provided a nomogram that combined vegetation type and percent ground cover in a prediction of the curve number (Fig. 3.7). Increasing the percent cover of grasses and desert brush resulted in gradual reduction of the curve number and brought about an associated gradual decline in the amount of runoff expected from a storm of a particular magnitude. Increasing the percent cover of mountain brush (chaparral in the

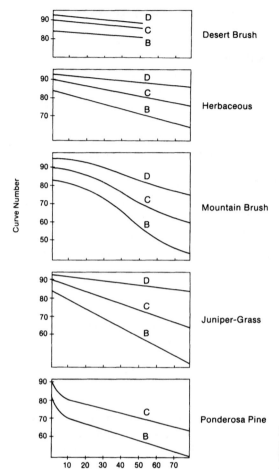

Fig. 3.7. Nomogram for estimating curve numbers in drylands for use in the U. S. Soil Conservation Service Method for runoff as depicted in Fig 3.6. For letters, see Table 3.3. (After Jencsok 1968)

Table 3.3. Classification of hydrologic soil groups for use with the Soil Conservation Service Method for runoff estimation (see Table 3.2 and Fig 3.6)

Group	Description
A	High infiltration capacity – sands, gravels, deeply weathered, well-drained
B	Moderate infiltration capacity, moderately to deeply weathered moderately to well-drained, moderately fine to moderately coarse texture
C	Low infiltration capacity, moderately fine to fine texture, usually with a horizon that impedes drainage
D	Very low infiltration capacity, swelling clays, soils with permanent high water tables, soils with clay lenses, shallow soils over impervious material

Source: U. S. Soil Conservation Service (1972).

Table 3.4. Classification of hydrologic conditions for use with the Soil Conservation Service Method for runoff estimation (see Table 3.2 and Fig 3.6)

Vegetation Cover	Condition	Description
Range or Pasture	Poor	Heavily grazed or having plant cover on less than 50% of the area
	Fair	Moderately grazed or having plant cover on 50–75% of the area
	Good	Lightly grazed or having plant cover on more than 75% of the area
Meadow		100% dense grass cover
Woodland	Poor	Heavily grazed or regularly burned, eliminating small trees and brush
	Fair	Grazed, not burned, with some litter
	Good	Not grazed, surface covered with litter and shrubs

Source: U. S. Soil Conservation Service (1972).

United States and Mexico) and tree species resulted in a much more rapid reduction in curve number and runoff.

Simanton et al. (1973) used empirical data to solve the function

$$CN = k\, x^e, \tag{3.1}$$

where CN = curve number for use with the nomogram shown in Fig. 3.6, x = drainage area in acres (area in m^2 divided by 4047), and k, e = empirical constants. Data from semiarid watersheds in southern Arizona produced the following results:

$$\text{desert shrub } CN = 85.75\, x^{-0.0087}, \tag{3.2}$$

$$\text{grass } CN = 88.00\, x^{-0.0085}, \tag{3.3}$$

$$\text{grass and shrub } CN = 86.74\, x^{-0.0088}. \tag{3.4}$$

Standard errors of the estimates of the curve numbers ranged from 1.6 to 3.5 in text calculations for instrumented basins up to 2.27 km^2.

Despite these adaptations of the curve method to dryland conditions, important reservations attend its use. The selection of the curve number is critical to successful use, yet many of the judgements required for the choice are subjective. In the United States, published government soil surveys include recommended curve numbers for mapped units, lending some standardization to the process. Users in areas outside the United States must rely on interpretations of the descriptions and discussions by the U. S. Soil Conservation Service (1972) which may not account for the full range of conditions on a global basis. The second major limitation of the curve method is spatial: it is most effective for drainage areas less than 1 km², though it may be extended to areas up to about 25 km² with care. Modifications of the method for areas greater than about 25 km² are required because single storms are not likely to cover the entire basin. Division of large basins into subcomponents may provide reasonable estimates for the largest amount of expected runoff, but the resulting value would be a limit rather than an accurate prediction.

Finally, transmission losses are important in dryland areas but are not accounted for by the method. Boughton and Stone (1985) recommend that the curve number be reduced by about 5 when the area of the watershed increases by two orders of magnitude. Lane (1982) recommended the use of a distributed model wherein the curve method predicts runoff for small subunits of large basins followed by calculations of transmission losses to modify the final expected runoff.

3.2.2 Rational Method

An alternative to runoff estimation by the Soil Conservation Service Method is the so-called rational method, termed rational because the units of the function are balanced. The method assumes that rainfall of a uniform intensity covers the entire basin, and that after a time of concentration the entire basin contributes a consistent amount of runoff to define

$$Q_{pk} = C\,I\,A, \tag{3.5}$$

where Q_{pk} = peak runoff (ft³s⁻¹), C = a dimensionless coefficient, I = rainfall intensity (in h⁻¹), and A = drainage area (acres). The metric version is

$$Q_{pk} = 0.278\,C\,I\,A, \tag{3.6}$$

where Q_{pk} is in m³s⁻¹, I in mm h⁻¹, and A in km² (Dunne and Leopold (1978, p. 299).

The coefficient C accounts for soil type, surface roughness, vegetation, and land use. Users have most often applied the method to urban areas, so that accepted values for C are best defined for developed landscapes (Table 3.5). In small basins in these settings, almost all of the basin surface probably contributes runoff. In nonurban settings, variable source area considerations (see Sect. 2.4.1 above) suggest that only a fraction of the basin surface contributes to the runoff peak, and the role of the coefficient C is further complicated. The coefficient in the nonurban setting describes the ratio of rainfall to runoff only in saturated areas which must be partitioned in the analysis of the drainage basin (Dunne et al. 1975).

Research on extremely arid hillslopes in the Negev Desert of Israel suggests that the variable source area concept applies to small hillslope areas as well as to whole basins. Yair and Lavee (1982) found that variation in rainstorm duration, slope aspect, and compaction explained spatial variation of runoff from small hillslope plots. In these extremely arid regions rainfall is so infrequent that artificial rainfall must be supplied for the measurements of runoff (Yair and Lavee 1974). In badland areas of extremely arid areas, deep surface cracks also exert strong influence on runoff processes by abstracting water that might otherwise contribute to the flow in a slope version of channel transmission losses (Yair et al.

Table 3.5. Values of coefficient C in the rational method

Drainage area characteristics	C
Urban areas	
Central business districts	0.70–0.95
Neighborhood commercial	0.50–0.70
Heavy industrial	0.60–0.90
Light industrial	0.50–0.80
Single-family residential	0.30–0.50
Detached multiple-family residential	0.40–0.60
Attached multiple-family residential	0.60–0.75
General suburban	0.25–0.40
Apartments	0.50–0.70
Parks	0.10–0.25
Rail yards	0.20–0.40
Undeveloped areas	0.10–0.30
Paved streets	0.70–0.90
Brick streets	0.70–0.85
Drives and walkways	0.75–0.85
Roofs	0.75–0.95
Non-urban areas	
Clay and shallow soils: Cultivated	0.50
Pasture	0.45
Wooded	0.40
Loam soils: Cultivated	0.40
Pasture	0.35
Wooded	0.30
Sandy and gravelly soils: Cultivated	0.20
Pasture	0.15
Wooded	0.10

Sources: Chow (1964a, p. 14/8) and Dunne and Leopold (1978, p. 300).

1980b). In limestone areas with extremely arid climatic conditions in the Negev, Yair et al. (1980a) found that even under extreme rainfall conditions little runoff makes the complete journey from hilltop to channel below because of rapid infiltration.

The time of concentration is critical to the application of the rational method because it signals the beginning of runoff and the start of accounting for the total amount of runoff. Empirical studies of small agricultural basins by U. S. Soil Conservation Service (1972) defined the function

$$t_c = (0.00013)(L^{1.15})(H^{-0.38}), \tag{3.7}$$

where t_c = time of concentration (h), L = length of the catchment along the main stream (ft), and H = elevation difference between the outlet and the most distant divide (ft). Kirpich (1940) found from a different sample the function

$$t_c = (0.00013)(L^{0.77})(S^{-0.385}), \tag{3.8}$$

where t_c is in h, L = length of the catchment measured in a straight line from mouth to farthest divide (mi), and S = basin slope defined as the length divided by the elevational difference between the mouth and the top of the farthest divide (dimensionless).

The rational method has an appealing simplicity which has led to its wide application, especially in urban areas. Several reservations make its application to dryland areas problematical. Krimgold (1946) reviewed the underlying assumptions of the method, providing a basis for the following points. First, the method assumes that the selected rainfall intensity lasts as long as the time of concentration or longer. In dryland areas precipitation events are highly variable in terms of the delivery of precipitation to the surface, and except for rare tropical storms or slow-moving frontal events, arid and semiarid storms probably violate the assumption.

Second, the method assumes a simple linear relationship between precipitation intensity and runoff with coincidental ranges. In arid and semiarid regions with highly porous soils commonly associated with fans and valley fills interspaced with relatively impervious rock surfaces and occasional fine soils, the relationship is likely to be curvilinear and highly complex. As a simplification, the assumption of linearity is reasonable, but distortion of predictions is the likely price.

Third, the method assumes that the frequency of peak discharges is the same as the frequency of the rainfall intensity for the given time of concentration. Nonurban runoff events do not meet this assumption because the recurrence interval of a rainfall event is less than that of the resulting flood (Dunne and Leopold 1978, p. 300). Fourth, the method does not account for the scale effects of basin area and the concept of variable sources for runoff at variable frequencies. The method is strictly limited to applications for small basins of a few km² at most.

Finally, the coefficient C is the same for storms of different intensities and frequencies. The method does not account for antecedent moisture conditions which have critical influences on the runoff from dryland surfaces. A relatively frequent, small storm may not generate runoff at a time when soils are dry, but a storm of the same size might result in flood conditions if it were to occur when soils are saturated.

It appears that the rational method is not especially accurate, even in urban applications where it might be expected to perform best. Jens and McPherson (1964) found that when they compared predictions from the rational method with measured runoff from 66 measured precipitation events in American urban drainage basins, the method produced values with a 34% mean deviation from the observed values. They found that in 34 of the 66 cases predictions were greater than 20% in error.

These reservations concerning the rational method suggest caution in its application to dryland problems. The advantages of simplicity and wide acceptance of the method may outweigh the reservations. In application for a large complex basin, the general area might be partitioned into small subunits, with predictions from each subunit by the rational method decreased to account for transmission losses (a solution pursued by Lane 1972, in applying the Soil Conservation Service Method). Fig. 3.8 demonstrates the importance of account-

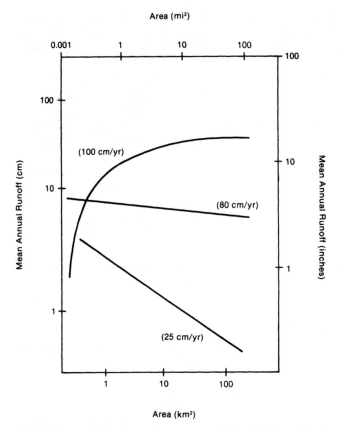

Fig. 3.8. Relationship between area and runoff showing the relative importance of the scale factor for dryland areas. In the dryland example runoff decreases with area because of transmission losses and because storms rarely cover entire basins in the larger size ranges. Neither of these conditions generally prevails in humid regions. (After Glymph and Holtan 1969)

ing for transmission losses and the impact of increasing the scale of analysis in dryland areas.

3.2.3 Impact of Human Activities

Predictions of runoff from dryland slopes must take into account the influence of human structures which introduce changes into the permeability of natural materials. Structures that concentrate overland flow have impacts on the amount of runoff expected from otherwise natural slopes. If the drainage area under consideration is subdivided, modern human influences can be taken into account.

Intentional modification of slope runoff processes are common in some dryland regions. Geddes (1963) coined the term "water harvesting" for the process of enhancing, collecting, and storing runoff. In the United States, modern water harvesting techniques are referred to as rain traps, paved basins, trick tanks, and guzzlers, while in Australia terms include roaded catchments and flat batter tanks (Branson et al. 1981, p. 101). Although in 1975 there were about 3,000 water harvesting systems, most were located in the western United States and western Australia. They are probably not significant for geomorphologic researchers who could avoid them in a search for natural slope systems.

Subtle human influences of past cultures, however, may be more difficult to account for despite their importance in influencing modern runoff processes. Shanan et al. (1969) found extensive evidence of a sophisticated water harvesting system on the slopes of the Negev Desert of Israel. As early as 3,000 BP these systems were constructed over large areas of the Negev Highlands. Extensive engineering works included surface modification and stone walls and mounds on slopes that diverted and conducted the runoff to irrigated fields below (Yair 1983). These ancient works significantly influenced infiltration and runoff patterns to the advantage of the ancient farmers, and continue to influence modern conditions, though the systems are no longer maintained (Shanan and Schick 1980).

In the southwestern United States, the Anasazi Culture constructed less extensive systems for runoff collection in the semiarid Colorado Plateau about 1200 AD. In southern Arizona the Papago Indians developed successful water harvesting systems that influenced runoff processes from the seventeenth to the twentieth centuries (Nabhan 1986, provides a complete review). It may well be that ancient cultures influenced runoff processes in most dryland areas. The application of predictive models and the construction of explanatory theories for modern processes therefore must account for these influences.

3.3 Streamflow and Floods

By their location, dryland rivers are usually ephemeral, flowing only occasionally and remaining dry for most of the year. Large through-flowing streams with their origins outside the dryland zone and small spring-fed streams are the only exceptions. The analysis of streamflow in dryland rivers therefore emphasizes an analysis of flood events. The hydrologic literature generally defines floods in relationship to experiences in humid regions: "A flood is a relatively high flow which overtaxes the natural channel provided for the runoff" (Chow 1964). Ward's (1978, p. 5) definition is more useful: "A flood is a body of water which rises to overflow land which is not normally submerged." In dryland streams, a flood event occurs whenever there is water in the normally dry channel, irrespective of the amount. The following discussion of dryland streamflows and floods emphasizes the particular characteristics of the processes in arid and semiarid regions. The discussion begins with a review of the application of standard streamflow concepts to dryland rivers followed by a review of types of floods (Riggs 1985 offers more details). The section continues with the important characteristics of transmission losses, magnitude and frequency, and the relationship between flows and channel characteristics. Concluding remarks address sheetfloods which may be considered as special cases of channel floods.

3.3.1 Streamflow

Discharge in dryland rivers has several properties of importance to the geomorphologist: the magnitude representing total mass transfer, width as the most diagnostic characteristic, velocity that is important but difficult to measure or estimate, depth that is less useful than in humid-region streams, and interaction with roughness elements which exert substantial control on processes. The discharge also controls the shear stress and power of the stream, characteristics directly related to the processes of sediment transport.

The equation of continuity describes the magnitude of discharge that passes through a given channel cross section:

$$Q = W D V, \tag{3.9}$$

where Q = discharge (m^3s^{-1}), W = width (m), D = depth (m), and V = velocity (m s^{-1}). To the geomorphologic or hydrologic researcher the equation of continuity is probably the simplist of functions and most basic of concepts, but researchers have not been effective in using it to communicate with planners and decision makers. Proposals to develop dryland rivers in the United States and Israel, for example, include narrowing the natural channel to a more manageable width. Many of the proposers are unfortunately unaware of the equation of continuity and its implications: that if width is reduced, either velocity or depth of flow must increase, with concomitant changes in channel stability that endanger planned near-channel structures.

Of the components of the equation of continuity, width is most easily measured and is most diagnostic of the channel system (Dunne and Leopold 1978, pp. 643–646). Width is a reasonably dependable predictor of discharge and in dryland settings where gauging information is scarce it may be the only consistent measure available. From a research perspective, width is a valuable measure because it changes more rapidly at a station or in the downstream direction than either depth or velocity, so width is more sensitive to discharge changes than the other measures.

Mechanical measurements or numerical calculations may estimate velocity. Current meters are available that consist of propellers or vanes mounted on the front of streamlined bodies that are inserted directly into the flow. The number of revolutions made by the propeller shaft records the rate of flow by sound or on recording paper or magnetic tape (Fig. 3.9). Measurements from several depths at several points across the cross section provide data for averaging purposes. Direct measurement of velocity is difficult or impossible during most peak flows in dryland rivers because of bed and bank instability. Flows with depths of 0.5 m (1.5 ft) cause particles up to gravel sizes to become unstable on the beds of channels, so that without cable systems across the channel there can be no direct measurement of velocity. A common alternative approach is to calculate an estimated velocity from other data.

The two most frequently used methods of velocity calculation are the Chezy and Manning equations. Antoine Chezy, a French engineer, proposed his formula in 1769 based on experiments on the Courpalet Canal and the Seine River (Herschel 1897). The function form is

Fig. 3.9. Field measurement of stream-flow velocity using a hand-held current meter. View in the upper photo includes a cable-way across the channel for measurements during high flows and an instrument shelter and stilling well on the bank. (U. S. Geological Survey photographs, Stacey photos numbers 500 and 501)

$$V = C (R S)^{0.5},\tag{3.10}$$

where V = velocity (m s^{-1}), C is a constant, R = hydraulic radius (m), and S = slope of the energy gradient, approximated in field applications by the gradient of the channel bed (Chow 1959, pp. 93–94 provides the derivation of the function). The constant C is related to channel roughness and must be estimated or calculated. A common estimator is that by Bazin (1897):

$$C = 157.6 / [1 + m/(R^{0.5})],\tag{3.11}$$

where C = Chezy's C for use with English units, m = a coefficient of roughness ranging from 0.11 for smooth cement to 3.17 for earth channels in rough condition, and R = hydraulic radius (ft).

Scott and Culbertson (1971) also provided a predictor function for Chezy's C using mean particle size on the channel floor (D_{50}, in mm)

$$C = 10.2 \, D_{50}^{-1/2}.\tag{3.12}$$

The obvious problem of applying the Chezy function in dryland rivers is that the function is a regression rather than a deterministic formula (the units are not compatible on either side of the equal sign), and the function was derived under conditions unlike those in dryland river channels.

The Irish engineer Robert Manning offered a different formulation of Chezy's approach in 1889. After later modification, the Manning Equation became known as

$$V = 1.49 \, n^{-1} \, R^{2/3} \, S^{1/2},\tag{3.13}$$

where V = velocity (ft s^{-1}), n = a coefficient of roughness, R = hydraulic radius (ft), and S = slope of the energy gradient (Manning 1895). In metric units, the constant 1.49 is 1.0 and the function becomes

$$V = n^{-1} \, R^{2/3} \, S^{1/2}.\tag{3.14}$$

The value of the Manning n as a roughness coefficient has a direct impact on the predicted velocity. The value of n for natural dryland rivers commonly ranges from about 0.020 for smooth sandy channels to about 0.15 for channels with phreatophyte vegetation on the channel floor. Many dryland rivers with sand to boulder beds have n values between 0.030 and 0.045. The estimation of the Manning n is a matter of experience, though some guidelines are available. Chow (1959, pp. 109–123) and Barnes (1967) offered extensive tables of verbal descriptions and photographs to define representative values of n. Ree and Palmer (1949) offered guidelines useful to dryland researchers. Bedforms also introduce roughness into flows, sometimes to a greater degree than that introduced by grain roughness (Shen 1979, p. 1/15; see Simons and Senturk 1976, p. 225 for recommended values).

There have been several attempts to provide estimator functions for Manning's n, including efforts by Scobey (1939), Cowan (1956), and Limerinos (1969). Possibly the most useful function for researchers in drylands is that by Strickler (1923):

$$n = 0.0151 \, D_{50}^{1/6},\tag{3.15}$$

where n = Manning roughness and D_{50} = mean particle size (mm). Strickler's function may be useful in dryland streams because it represents data on gravel bed streams in pro-glacial environments similar to dryland streams with coarse bed materials.

Further issues in using the Manning Equation and its roughness factor in dryland settings include changes in the roughness value with changing discharges and changes across the channel cross section. As discharge increases at a cross section, the value of n declines because the roughness elements become progressively smaller relative to the depth of flow. Because the roughness value figures directly in the equation, small changes in n propagate themselves through the calculation of velocity and ultimately to estimations of total discharge through the equation of continuity. Spatial variation of particle sizes or of vegetation cover imply that the value of n is not constant across a single cross section, and when flows exceed bankfull capacity, near-channel surfaces are not likely to have roughness characteristics similar to the channel floor. Finally, Manning's n accounts for two types of roughness, skin roughness determined by particle sizes and vegetation, and form roughness controlled by the bed conditions. Bedforms may be highly variable and range from low roughness conditions on plane beds to very rough conditions represented by dunes and antidunes.

Application of the Manning Equation to dryland rivers is difficult because, like the Chezy Equation, it is not a deterministic function containing physical explanation but rather it represents a regression used by Manning to summarize empirical data from canals and humid-region streams. The constants commonly used in the function may not be appropriate for dryland streams, but until proven replacements are published and widely accepted, the Manning Equation will see continued (though admittedly flawed) application.

Depth of flow in dryland streams also presents problems of measurement or estimation. Depth is difficult to measure during peak flow conditions because of the discontinuous processes and because at peak flow on many dryland streams bed materials are in motion. It may be difficult to determine on the vertical dimension where the bedload stops and where the bed begins. Estimates of depth of flow are possible using known or estimated discharges with estimates of velocity, width, and roughness. One estimator equation for depth is a rearranged combination of the equation of continuity (3.9) and the Manning estimate of velocity (3.14):

$$D = [(Q\ n)/(W\ S^{0.5})]^{0.6}, \qquad\qquad\qquad (3.16)$$

in SI units with symbols as before. The reservations concerning this approach are that depth substitutes for hydraulic radius (acceptable in many dryland rivers which are wide and shallow), and the rearrangement of a regression without recourse to the original data (Williams 1983). Graf (1979d, 1983a) used the function for dryland rivers, and Tinkler (1983) recommended refinements to improve accuracy, including the use of Newton/Raphson root finding algorithm to determine depth from a combined equation of continuity and Manning Equation.

3.3.2 Force, Stress, and Power

From the perspective of the geomorphologist, the most important aspect of discharge in a river is its ability to move sediment, which in turn is related to the force, stress, or power of the flow. Rhoads (1987) provided a review of the connections among the various measures. The shear force of the flow on the wetted perimeter of a channel segment is

$$F = \gamma \, W \, D \, X \, S, \tag{3.17}$$

where F = shear force (N), γ = specific gravity of water (N m^{-3}), specific gravity of water is the density of the water (1,000 kg m^{-3}) times acceleration of gravity (9.807 m s^{-2}), W = width (m), D = depth (m), X = length of the channel segment (m), and S = gradient of the bed.

A word about units of measure may be appropriate at this point. Rhoads (1987) and many others use as units for force the dimension of mass-length-time^{-2} (for example, kg m s^{-2}). In the International System of Units (SI System) the unit of measure for force is the newton (Stelczer 1981). A newton is the force required to accelerate a mass of 1 kg in 1 s to a velocity of 1 m s^{-1}. Therefore 1 N = 1 kg m s^{-2}. Use of newton as a measure of force is simpler than dealing with units expressed with kilograms. The latter causes confusion because kilograms measure mass rather than force. The following discussion departs from that of Rhoads by using newtons. The concepts and conclusions are the same, however, in applying the results to geomorphologic problems.

Shear force is not commonly used in geomorphologic applications, but a modified measure, shear stress, commonly occurs in fluvial theory. Shear stress refers to the amount of stress or drag applied to a unit *area* of the bed of the channel:

$$\tau_o = \gamma \, R \, S, \tag{3.18}$$

where τ_o = shear stress (N m^{-2}), γ = specific gravity of water (N m^{-3}), R = hydraulic radius (m), and S = slope. The shear stress is the drag exerted on a unit area of the channel bed and is directly related to the competence of the stream. The competence refers to the maximum size of bed materials that the stream can transport.

Shear stress in SI units is expressed as N m^{-2}, while in the English system the measure is in kg m^{-1} s^{-2}. These two sets of units are exactly equivalent. Many authors refer to (3.17) or a modified form substituting depth for hydraulic radius as tractive force but technically the label is incorrect because the units of measure are in force per unit area or stress (Bogardi 1974, pp. 80–84).

Bagnold (1966, 1977) introduced the concept of unit stream power which he related to the capacity of the stream. Capacity refers to the total amount of sediment that the stream is capable of transporting. Unit stream power represents the power exerted on a unit segment of the channel cross section and equal to shear stress times velocity or

$$\omega = \gamma \, D \, S \, V, \tag{3.19}$$

where ω = unit stream power (N m^{-1} s^{-1}, convertible to Watts per unit length), γ = unit weight of water (N m^{-3}), D = depth (m), S = slope, and V = mean velocity (m s^{-1}).

The total amount of power exerted on the entire channel cross section represents the power available for work and is not yet frequently used in geomorphology except as a general index. Stream power is unit stream power times the width of the cross section or

$$\Omega = \gamma\, D\, S\, V\, W = \gamma\, Q\, S \qquad (3.20)$$

where Ω = stream power ($N\, s^{-1}$, convertible to watts) and the other symbols are as before.

This discussion of force, stress, and power generally follows the precepts of Bagnold (1966, 1977) with SI units. Other definitions of the term stream power with different interpretations and different units have empirically defined relationships with the amount of sediment transported (e.g., Chang 1979; Chang and Hill 1977). Bagnold's approach has the advantage for geomorphologic researchers that it is based on deductive arguements grounded in physical explanation.

3.3.3 Types of Dryland Floods

A limited set of climatologic and surface factors control the magnitude and duration of flood flows in dryland areas. In an extensive statistical analysis in the southwestern United States, Benson (1964) found that in those areas where snowmelt floods were not significant, most of the variability in peak discharges could be statistically explained by seven factors: drainage-basin area, rainfall

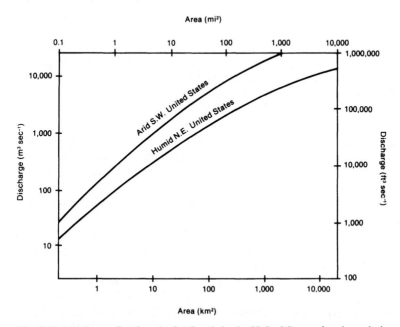

Fig. 3.10. Maximum flood peaks for floods in the United States showing relative magnitudes of floods from dryland and humid regions. (After Chippen and Bue 1977, pp. 7, 15)

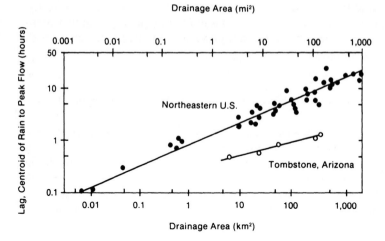

Fig. 3.11. Lag time between the centroid of rainfall and peak flow for dryland and humid region examples showing the relatively short lag times for drylands. (After Dunne and Leopold 1978, p. 337)

intensity for a given duration and frequency, main-channel slope, basin length, surface area of lakes and ponds, the ratio of runoff to rainfall during the months of annual peak discharge, and the annual number of thunderstorm days. The last two factors played minor roles. These results are probably typical of dryland areas, and demonstrate the importance of geomorphology in determining the character of floods.

Because dryland regions generally have poor soil development and relatively little vegetation, peak discharges for maximum floods of record are larger for dryland basins than similar sized humid basins. Chippen and Bue (1977) show that maximum floods of record over the United States portion of North America vary from one climatic zone to another, but that the highest maximums are generally from the arid and semiarid zone (Fig. 3.10). This generalization appears valid for basins less than about 2,600 km² (1,000 mi²), but in basins larger than this limit, the larger amounts of rainfall generate larger floods in humid regions. Lack of vegetation and thin soils also lead to shorter lag times between rainfall and discharge events in drylands (Fig. 3.11).

Floods in dryland rivers are generally of four types: flash floods, single peak events, multiple peak events, and seasonal floods. These types are not mutually exclusive and are partly scale dependent, with flash floods occurring on small streams and seasonal floods characterizing the large throughflowing streams originating outside the dryland areas.

Flash floods are almost always associated with convectional precipitation and thunderstorms (International Association of Hydrological Sciences 1974). Because most thunderstorm cells are relatively small with diameters of about 8 km (5 mi) (Morgan 1966), flash floods are limited in drylands to basins of 100 km² or less. Flash floods represent streamflows which increase from zero to a maximum within a few minutes or at most a few hours, and are often associated with an advancing front of water or bore. The advancing wave of water (frequently reported by lay observers as a "wall of water") is usually turbulent, has large amounts of air creating foam, and may push debris before its main mass (Fig. 3.12).

The rapid rise in water level in the channel produces hydrographs with vertical rising limbs (Fig. 3.13), a feature common to the largest examples (Fig. 3.14). Woolley (1946) using data from Utah, and Schick (1970, 1971) using data from southern Israel have shown that the characteristics of dryland flash floods are strongly influenced by watershed characteristics and antecedent moisture conditions. Because in drylands flash floods occur in previously dry channels, and because the flow declines to zero again after the event, simplified models of the flood hydrograph are possible. Triangular representations whereby the apex of the triangle is the peak discharge and the other corners represent the times of zero discharge are probably adequate for geomorphological research. Multiple stacked triangles also have proven useful in some dryland applications (e.g., Lane et al. 1985).

Single-peak flood events are longer than flash floods and range from a few hours to many days in duration (Ward 1978, p. 19). In dryland areas they are the products of tropical storms or frontal precipitation and may affect basins of several thousand km^2. In 1980, for example, the Salt River (Arizona) rose from no discharge to a single peak flow of 5,000 m^{3}s^{-1} (180,000 ft^3 s^{-1}) over the course of a few days. The flood resulted from warm frontal rains falling on saturated snow-packs in mountain areas, and was too large to be controlled by irrigation dams and reservoirs.

The single-peak El Arish flood of 1975 in the Sinai Peninsula occurred in an area with out dams and reservoirs. Produced by a 2-day regional storm system, the flow peaked at 1,650 m^{3}s^{-1} (59,000 ft^{3}s^{-1}) after the passage of an initial bore about 1 m high (Gilead 1975). The 1975 Wadi El Arish and the 1980 Salt River flows were typical of single peak events in that although the peak flow lasted less than a day in each case, relatively high stages persisted (in the Salt River case, for nearly a month).

Multiple peak floods result from multiple precipitation events typified by tropical storm or frontal systems that have stalled over a particular region. Multiple peaks may also originate as tributary contributions to the main stem, though it is common that flood flows in main streams in dryland areas are not supplemented by tributaries that did not receive precipitation. The ephemeral Finke River in the central drylands of Australia provides a useful example of multiple peak floods. In 1967 three incursions of monsoon precipitation generated rainfalls of 120–350 mm (4.7–13.8 in), ultimately producing three flood peaks at gauging sites with drainage areas ranging from several hundred to several thousand km^2 (Williams 1970). Multiple peak floods may be brief or last for several days or weeks even in dryland areas. The 1972 flood in Wadi Watir, southeastern Sinai, lasted only a few days but had numerous well-defined peaks with nearly instantaneous rises (Schick and Sharon 1974; Fig. 3.15). The 1969 Tunisian floods lasted several weeks as a result of steady incursions of moist unstable air into the usually arid and semiarid area (Stuckmann 1969).

Fig. 3.12

The multiple flood peaks on the Finke River might be considered seasonal in that monsoon circulation produced them. Unlike the ephemeral Finke, the Nile, Ganges, and Brahmaputra rivers are dryland streams that are throughflowing from sources elsewhere, but they too are influenced by monsoon circulation. The Nile River experiences low flows in May (mean minimum of 570 m³s⁻¹ or about 20,400 ft³ s⁻¹) and peaks in September (mean maximum of 8,440 m³s⁻¹, or about 300,000 ft³ s⁻¹ – nearly 15 times the mean minimum) at the Wadi Halfa measurement site. The single seasonal peak at Wadi Halfa is the product of a complex series of inputs from (1) the Blue Nile with a rapid autumn rise and rapid fall resulting from the rainy season, (2) similar contributions from the Atbara, which is ephemeral, and (3) more steady inputs from the White Nile, which has a more consistent flow due to the damping effects of lakes and swamps (Hurst and Phillips 1931). The Blue Nile contributes 69% of the total flow and most of the variability in the flow of the lower Nile, the Atbara contributes 17%, and the Blue Nile 14% (Fig. 3.16).

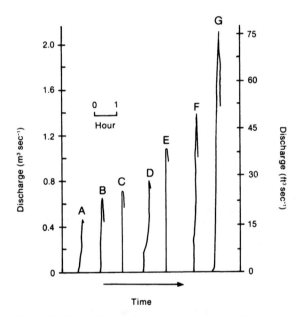

Fig. 3.13. Rising limbs of example hydrographs from flood events in the Negev, Israel, showing the nearly instantaneous nature of the rise in this extremely arid region. (After Schick 1970)

Fig. 3.12. Examples of flash floods from semiarid regions. *Upper*, the frothy front of a flash flood in Straight Wash Canyon, Emery County, Utah, 1930. The front is at the left of the frame following the bouldery channel. *Middle*, the same locality as above a few seconds later showing the frothy flow of the initial flood wave. *Lower*, view looking upstream into the front of a flash flood in Hunter or Little Salt Wash, Mesa County, Colorado, August 6, 1926, showing a mass of debris being pushed downstream by the flood wave. (Upper and middle photos by Aurthur A. Baker, lower photo by C. E. Erdman; U. S. Geological Survey Photographic Library, Denver)

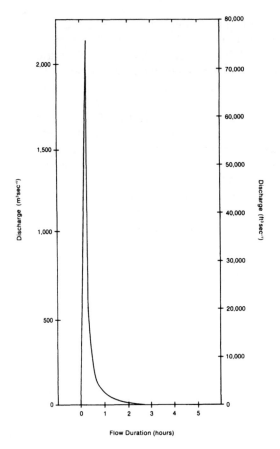

Fig. 3.14. The reconstructed hydrograph of the September 14, 1974, flood in Eldorado Canyon, Nevada, showing a nearly instantaneous rise from zero to 2,100 m^3s^{-1} (76,000 ft^3s^{-1}). (After Glancy and Harmsen 1975, p. 15)

3.3.4 Transmission Losses

Water that flows through dryland river channels is lost through evaporation and seepage into banks and beds. For flood flows on small and medium sized ephemeral streams the flows rarely last long enough for evaporation to become a major concern. For throughflowing large rivers, the evaporation is usually enough to cause a downstream reduction in discharge. Along the ephemeral streams which dominate the arid and semiarid landscape, transmission losses through seepage cause flood peaks and total discharge values to decline in the downstream direction (Babcock and Cushing 1941; Burkham 1970a, b). These losses are so great that eventually most flows decline to zero.

The presence of large amounts of alluvium beneath many dryland channels results in the loss of runoff volumes from the surface environment, but this water may be of economic importance as groundwater recharge. In 1965–1966 the Salt River in central Arizona discharged a peak flow of 1,900 m^3s^{-1} (67,000 ft^3s^{-1})

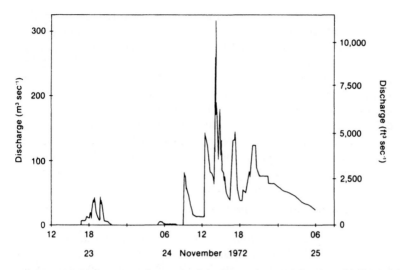

Fig. 3.15. Multiple peak hydrograph of the November 1972 flood on Wadi Watir, southeastern Sinai showing rapid rises and falls in discharge due in part to non-synchronous inputs from tributaries. (After Schick 1986)

into its previously dry channel for several weeks. The channel flowed over the surface of alluvial fills several hundred meters deep in a structure basin. Comparison of discharge records with measurements at the entry to the basin with those at the exit showed that 29% of the flow that entered the basin as surface flow disappeared into the groundwater reservoir as transmission losses (Aldridge 1970). In March 1978 another flood with a peak flow of 3,500 m³s⁻¹ (125,000 ft³s⁻¹) and lasting several days, transmission losses accounted for 17% of the flow (Aldridge and Eychaner 1984).

Several authors have proposed approaches for calculating the magnitude of transmission losses for small and medium channels (those with drainage areas

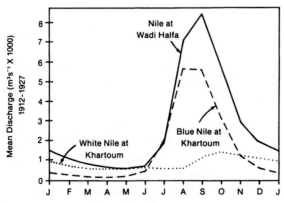

Fig. 3.16. Contributions from various sources to the annual flood peak of the lower Nile River, Egypt. (After Ward 1978, p. 26)

less than 1,000 km²). Burkham (1970a, b) and U. S. Soil Conservation Service (1972) suggested simplified loss rate equations that demand little input data. Jordan (1977) and Lane (1980) proposed deterministic perspectives with relatively simple differential equations, and Lane (1972), Wu (1972), and Peebles (1975) outlined cascading system models with built-in accounting methods and the assumption of leaking reservoirs. R. E. Smith (1972) offered a technique specifically for flood-wave attenuation employing kinematic wave theory.

In most dryland settings, the most successful method of estimating transmission losses is likely to be one that requires only modest amounts of data input. Lane (1980) provided a most complete summary of transmission loss equations and derived the following method. Lane et al. (1985) also report Lane's work on transmission losses. The fundamental statement is

$$V(x,w) = a(x,w) + b(x,w)V_{up} + F(x,w)V_{Lat/x}, \qquad (3.21)$$

where $V(x,w)$ = flow volume exiting from the downstream end of a channel reach of x length and w width, V_{up} is the flow volume entering the reach at the upstream end, $V_{Lat/x}$ is the lateral inflow from tributaries, and $a(x,w)$, $b(x,w)$, and $F(x,w)$ are parameters unique to each reach.

Definition of the reach-specific parameters relies on empirical evidence collected from semiarid streams in instrumented watersheds in the southwestern United States. The constant in (3.21) is

$$a(x,w) = [a(1-b)^{-1}][1-e^{-kxw}], \qquad (3.22)$$

where e = base of the natural logarithm system and

$$a = -0.00465 \ K \ D, \qquad (3.23)$$

where a = the parameter in (3.22), D = mean duration of flow (h), K = effective hydraulic conductivity (in h^{-1}) reviewed in Table 3.6, and k is a decay factor $(ft \ mi^{-1})^{-1}$ defined by

$$k = -1.09 \ln [1 - (0.00545 \ K \ D)V^{-1}], \qquad (3.24)$$

where V = mean flow volume (acre ft).

The first coefficient in (3.21) is

$$b(x,w) = e^{-kxw}, \qquad (3.25)$$

with symbols as above. The second coefficient, which is zero for short simple reaches or those with no tributary contributions is

$$F(x,w) = (1 - e^{-kxw})(kw). \qquad (3.26)$$

The estimates of K (Table 3.6), effective hydraulic conductivity, are at the heart of the method because the parameter is so pervasive in the calculations and because it is not likely to be directly measurable. The values suggested in Table 3.6 are from empirical results in natural channels by Lane (1980) and Wilson et al. (1980) and from unlined canals by (Kraatz 1977). The values indicate conditions with flashy, sediment-laden water that characterizes dryland channel floods and do not represent clear water, steady state conditons (Lane et al. 1985).

Table 3.6. Values for effective hydraulic conductivity for transmission losses in channel alluvium

Characteristics of bed material	Effective hydraulic Conductivity (in hr⁻¹)
1. Very clean gravel and coarse sand, mean particle size > 2 mm, very high loss rate	> 5.0
2. Clean sand and gravel under field conditions, mean particle size > 2 mm, high loss rate	2.0 – 5.0
3. Sand and gravel mixture with slight amounts of silt and clay, moderate to high loss rate	1.0 – 3.0
4. Sand and gravel mixture with large amounts of silt and clay, moderate loss rate	0.25 – 1.0
5. Consolidated bed material with high silt and clay content, low loss rate	0.001 – 0.1

Note: From Lane et al. (1985) with English units because empirical functions to be used with this parameter are calibrated in English units. If metric units are used, empirical constants in other equations must be altered.

Because of the large amounts of flow involved in transmission losses, accounting for seepage is a requirement in predicting flow volumes and in routing water and sediment through dryland river systems. The method outlined above is probably the best available approach for those streams with drainage basins of a few km² to a few hundred km². The family of equations has yet to be tested widely, though Lane (1980) derived the empirical constants from 14 channel reaches and 139 flow events in semiarid streams. For larger channels it may be possible to use soil survey data to estimate the initial rate of transmission loss per unit area of the channel floor. Modest reductions in the loss rate occur as the channel floor materials become saturated with water and the inter-granual spaces become clogged with fine sediment or organic debris.

3.3.5 Magnitude and Frequency

Magnitudes and frequencies of floods are useful variables in explaining the forms and processes of dryland rivers because floods appear to accomplish most of the morphology-changing work in such streams. Flood analysis deals with three primary types of data: annual flood series, partial duration series, and the probable maximum flood. The annual flood series consists of the maximum discharge values from a station for each year of record, with one peak recorded for each year. The partial duration series consists of a listing of all discharges from a station that were above a certain threshold value defined by the researcher. In a partial duration series, some years may have several values recorded while other years may have none. The probable maximum flood is an engineering and planning concept wherein an attempt is made to define the "worst case" when the maximum expectable rainfall occurs on a saturated watershed. The probable maxi-

mum flood is usually five or more times greater than the 100-year flood. The
following discussion is limited to the annual flood series.

Flood frequency analysis is useful for those streams with gauging data or reasonably long records constructed from physical evidence. This analysis provides
insights into the behavior of those streams for which data exist, but it is not useful for extrapolating discharge data to other streams in the region. Because of the variety of variables influencing flood discharges, and because those variables change
from one basin to another, regressions relating control factors to resulting peak
flows are more likely to be useful in predicting floods in ungauged sites (Sect.3.3.1).

Two common related methods provide geomorphologic researchers with
information about flood frequency: the flood frequency curve and probability
density functions. The flood frequency curve is a graph presenting annual flood
magnitudes plotted against recurrence interval. The probability density function
is a statistical statement of the probability that an annual flood of a given
magnitude will be equaled or exceeded. Other floods such as the partial duration
series could be analyzed in the same way. The flood frequency curve and the
probabilistic approach are related to each other by

$$P^{-1} = R.I.,\tag{3.27}$$

where P = the probability of a given flow magnitude being equaled or exceeded in
a single year and R.I. is the return interval of the flood of the same magnitude
(Costa and Baker 1981, p. 362). By this formula the flood with a 0.01 probability of
occurring in a year has a return interval of 100 years (Table 3.7 shows others).

Table 3.7. Formulae for plotting flood frequency curves

Formula	Reference
$R.I. = N\,m^{-1}$	California (1923)
First used only for California streams; produces probability of 100% and therefore cannot be used with probability paper.	
$R.I. = (2N)(2m - 1)^{-1}$	Hazen (1930)
Designed to plot at the centers of group intervals to accommo- date probability paper.	
$R.I. = (N + 1)(m)^{-1}$	Weibull (1939)
Found by Benson (1962) to be most satisfactory from the standpoint of statistical assumptions.	
$R.I. = [1 - (0.5^{-N})]^{-1}$	Beard (1943)
Formula applies only to the value $m = 1$; other values are inter- polated between this point and the value 0.5 for the median event.	
$R.I. = (N + 0.4)(m - 0.3)^{-1}$	Aleksayev (1955)
Empirical formula commonly used in the USSR.	
$R.I. = (N + 0.25)(m - 0.375)^{-1}$	Blom (1958)
Rarely used.	
$R.I. = (3N + 1)(3m - 1)^{-1}$	Tukey (1962)
Rarely used.	
$R.I. = (N + 0.12)(m - 0.44)^{-1}$	Gringorten (1963)
Designed for short records.	

Notes: Adapted from Chow (1964a, pp. 8/28 – 8/29).

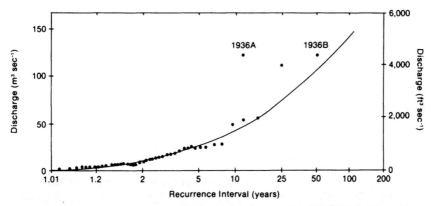

Fig. 3.17. Flood frequency curve for Bright Angel Creek, a stream with a drainage area of about 260 km^2 (100 mi^2) in the Grand Canyon, Arizona. The plotting formula for the points was the Weibull form; the line is a log-Pearson distribution. The largest flood of the record (which extends 1923–1969) appears to have a return period longer than those suggested by the data set which is constrained because of its length of record. To illustrate the impact of length of record on the method of analysis consider points *1936A* and *1936B*. Point 1936A is the plotting location based on the shorter 1923–1936 record, while point 1936B results from the longer 1923–1969 record. (After Cooley et al. 1977)

Return intervals must be calculated for the values in the annual flood series to create the flood frequency curve. Of several plotting formulae (Table 3.7), the one by Weibull (1939) has seen widest use:

$$\text{R.I.} = (N + 1)(m^{-1}),\tag{3.28}$$

where R.I. = return interval, N = total number of years of record, and m = rank of the discharge being plotted, with the largest discharge m = 1. Fig. 3.17 is a flood frequency curve for Bright Angel Creek, an example dryland stream near the Grand Canyon, Arizona, with a drainage area of about 260 km^2 (100 mi^2).

Each basin has its own peculiarities and thus its own flood frequency curve, making generalizations impossible. The data in Fig. 3.10 suggest that basins of less than 1000 km^2 in arid and semiarid regions tend to have larger floods than similar sized basins in humid regions. Therefore the flood frequency curves for dryland basins of this size are likely to rise more steeply than their humid-region counterparts.

The annual flood series for any gauge site is a record of discrete values, each for a year measurement, rarely exceeding 100 in number. If the data series is visualized as infinitely large and the class intervals infinitesimally small, then at the limit a smooth curve of the probability density function results (following material after Shaw 1983, pp. 302–307). The area under the probability density function is equal to 1.0 and is made up of two regions, P and F (Fig. 3.18). P is the inverse of return interval or the probability that an observed flood will be greater than some discharge X, while F (1.0 – P) is the probability that an observed discharge will be less than X. Given the nature of river hydrology, extremal functions are most likely to fit the observed data with a strong left skew and a long right tail to account for the rare but important large events.

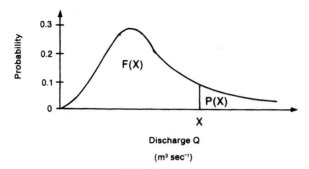

Fig. 3.18. An example probability density function showing the two regions of probability, area $F(X)$ depicting the probability of a flood being less than or equal to X, and area $P(X)$ depicting the probability of a flood exceeding the value of X

Four extremal functions see common use for the probability density function representing flood data: lognormal, Gumbel Type I, Gumbel Type III, and Pearson Type III. Because the density functions are based on stochastic assumptions and do not derive from first principles or physical considerations, the only criteria for choosing the most satisfactory function are goodness of fit and ease of application. In general application none of the options is clearly more successful than the others. The Pearson Type III distribution is more complicated than the others and is seeing increasing use in the United States (Benson 1971), but it does not appear to have significant advantages over the other options (Dunne and Leopold 1978, p. 306).

The Gumbel Type I distribution provides an example of the application of probability density functions to flood frequency analysis. Its general form is

$$F(X) = \exp\left[-e^{-b(X-a)}\right], \tag{3.29}$$

where $F(X)$ = probability that an annual flood will be equal to or less than discharge X, X is a discharge, and a, b are distribution parameters defined by the moments of the entire range of annual flood peaks in the record. Using standardized data, parameter b is

$$b = pi\left[v\,(6)^{0.5}\right], \tag{3.30}$$

where v = the variance, while parameter a is

$$a = m - 0.5772\,b^{-1}, \tag{3.31}$$

where m = the mean. Once established, the function provides an estimate of the return period of a particular discharge value X because the function can define $F(X)$, $P(X) = 1.0 - F(X)$, and (3.27) defines the return period from $P(X)$.

Flood frequencies vary from one basin to another because of spatial variation in the factors controlling floods, but the magnitudes and frequencies also vary along the same stream. In large throughflowing rivers, the magnitude of the flood of a given return period is smaller in upstream areas than the magnitude of the same return period in downstream areas because of the contributions of tributaries. There is a downstream limit to this increasing magnitude, however, because

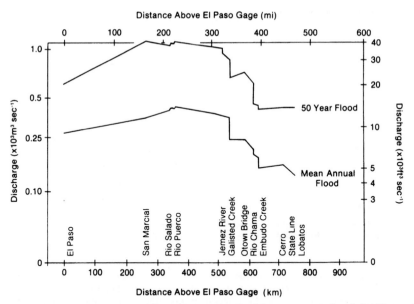

Fig. 3.19. Downstream variation in the magnitude of the mean annual and the 50-year floods for gauge sites along the Rio Grande, Colorado, New Mexico, and Texas, showing an initial increase in the downstream direction due to tributary additions followed by a decline from transmission losses. (After Wiard 1962)

of transmission losses. The magnitude of floods of a given return interval therefore decline in the lowest portions of through-flowing streams (Fig. 3.19). Similar variations probably occur on smaller ephemeral streams but are less well documented.

Flood magnitudes and frequencies in dryland systems are also affected by human manipulations. The installation and maintenance of reservoirs, even if they were designed primarily for irrigation purposes, substantially reduce flood peaks (e.g., Aldridge and Eychaner 1984). The installation of high dams causes a shift to the right of flood frequency curves for the downstream areas because floods of a given magnitude are smaller than before the construction of the dam (Fig. 3.20).

The impact of water resource development on magnitude and frequency is evident in the position and form of flow duration curves. These graphs depict magnitude and frequency by plotting discharge values against the percent of time for which those values are equaled or exceeded. The flow duration approach does not provide information about single flood events, nor does it provide sequential or time-based information (Hudson and Hazen 1964). The position of the curve in the resulting graph reflects the general magnitude of flow (Leopold et al. 1964), while the shape of the curve indicates the cumulative effects of streamflow regulation and water use (Eschner 1981). In an example from semiarid Nebraska, Kirchner and Karlinger (1981) found that the impact of water resource development was to radically alter the flow duration curves for the Platte River (Fig. 3.21). Irrigation and groundwater pumpage influence the magnitude and

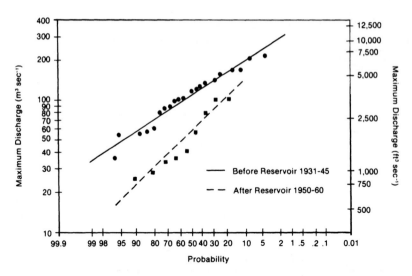

Fig. 3.20. Annual flood frequency curves for the Rio Duero at Penafiel, north central Spain. The 11,840 km² (4,570 mi²) basin produced reduced magnitudes after the installation of an upstream reservoir. Reductions were greatest for the smallest events. (Data from Michel 1979, after Thornes 1983)

frequency of low flows because of field drainage and return flows to the stream at times of the year when the channel is dry under entirely natural circumstances.

Finally, flood control works on major channels serve to speed up the conductance of water and substantially reduce lag time between the centroid of rainfall and peak discharge. Along the Jordan River in Israel and Jordan the installation of channelization works and the draining of lakes and swamps along the river's course have increased the annual mean maximum discharges from 57 m³s⁻¹ (2,035 ft³s⁻¹) to 96 m³s⁻¹ (3,430 ft³s⁻¹). The flood control efforts accelerated the rate of downstream translation of flood peaks through the middle reach of the river from 2–7 days under natural circumstances to 4–14 h (Inbar 1982).

Magnitude and frequency analysis provides useful information about the hydrologic behavior of rivers, but a major problem with the method is the question of the length of record. Because statistical approaches characterize the natural processes, sample size is critical, and if the subject of analysis is the annual flood series, there are few streams in the drylands of the world that have records lengthy enough to provide input for reliable inferences. The spotty nature of precipitation and the long periods between flows make the development of usable records in arid and extremely arid regions especially problematic as demonstrated in southern Israel by Schick (1971a, b). Jeppson et al. (1968) found in Utah that the percentage of error as measured by the 95% confidence interval in hydrologic predictions declined rapidly as the length of record lengthened from 2 to about 6 years, but thereafter the improvement was modest even to a length of 20 years. Between 2 and 6 years the error declined from about 57% to about 38%, but by the 20th year it had declined to only about 27%. Along with the problem of record length, major problems in flood analysis include regionalizing models,

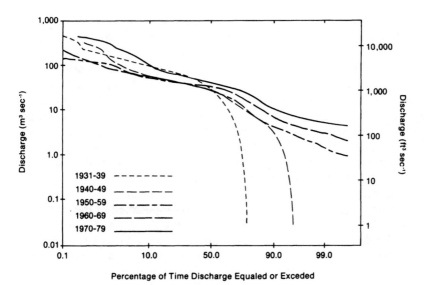

Fig. 3.21. Flow duration curves for five decades from the Platte River near Overton, Nebraska. The highest curves represent relatively moist periods, the lowest represent relatively dry periods. The change in shape from the 1931–1949 period to later decades represents the impact of streamflow regulation for irrigation. (After Kirchner and Karlinger 1981)

extending known records to ungauged sites, and investigating the process-based explanations for floods as opposed to statistical analyses of the events (Potter 1987).

3.3.6 Significance of Various Flow Magnitudes

The geomorphologic literature abounds with references to the special characteristics of discharges with particular frequencies. Although the most commonly mentioned special discharges, the mean annual flood, bankfull flow, and the 100-year flood, may have some usefulness in humid regions, this utility does not extend to dryland rivers. Extensions of generalizations based on these parameters to dryland rivers are not likely to be successful.

The mean annual flood is the arithmetic mean of all the values in the annual flood series. Because of the mathematical form of the Gumbel Type I distribution, the mean annual flood has the recurrence value of 2.33 years when this probability density function represents the annual flood series. In the Gumbel Type III probability density function the 2.33-year recurrence interval corresponds to the logarithmic mean of the distribution. Graphs designed with these functions can produce values for the mean annual flood by visual inspection of the intercept of the flood frequency curve and the recurrence interval value of 2.33 years, a standard U. S. Geological Survey practice (Dalrymple 1960). The mean annual flood has almost no practical or theoretical significance in dryland rivers, however easily it may be defined, because of the extreme variability of flow in such streams.

In arid and extremely arid areas channels may not experience flow for several years, so that the concept of an annual flow is nearly valueless for theory construction.

Wolman and Leopold (1957) suggested that the return interval for bankfull discharge for a wide range of streams is 1–2 years. Dury et al. (1963) maintained that bankfull frequency was 1.58 years, a proposition supported by data from a range of American streams (Dury 1973). Although a link between bankfull discharge, frequency, sediment transport, and flood-plain processes is a tempting invitation to develop the concept of a "dominant discharge", conditions in dryland rivers preclude the effort. Pickup and Reiger (1979) point out that simple relationships of the form where a morphologic variable is a simple function of a discharge are successful only for streams clearly in regime or some definable steady state. As outlined in Chapter 2, dryland rivers seem to defy this assumption.

.Further complications develop in the attempted application of the 1.58-year discharge as a theoretically useful bankfull discharge for dryland rivers. Field definition of bankfull is difficult in many environments (Woodyer 1968), but especially in dryland channels that are frequently incised, excessively broad and braided, or developed on bedrock. Bankfull discharge is not even the same throughout single drainage basins (Pickup and Warner 1976). Finally, extensive data collected by Williams (1978) showed that although the mean frequency of bankfull discharge is close to the traditional 1.5 years, the range of values is from 1.01–32.0 years, a breadth too great to inspire confidence in the reliability of the measure.

Discharge is related to certain geomorphologic characteristics of dryland rivers, but the development of mathematical or statistical relationships depends on the development of satisfactory underpinnings. The precise empirical relationships must first be defined, then the exact reason for the supposed connection between a discharge and the geomorphic response can be established. For example, in an investigation of bedload transport, Pickup and Warner (1976) calculated bedload transport associated with each discharge for several particle sizes using the Meyer-Peter and Muller equation. They found that the most effective discharge in terms of sediment transport was the 1.15–1.40-year flow, less than bankfull discharge. The bankfull discharge therefore is not likely to be the geomorphologically most significant flow in channel dynamics. The reconciliation of this situation with long-held views of the importance of bankfull discharges awaits further study. The 100-year flood has important implications for planning agencies, who need some reference discharge for the definition of flood hazard areas. The 100-year flood is difficult to define in may dryland rivers because of lack of long-term records. It does not seem to be correlated with any particular suite of geomorphic features and so is difficult to deduce from the physical evidence. It also lacks any particular significance from the standpoint of theory building except that it provides a satisfyingly round number.

The order of events of varying magnitudes may be more important in explaining the present observed geomorphic conditions than the exact nature of the flood frequency curve or the statistical properties of flood frequencies. Consider a large throughflowing river through two possible 10-year histories as an example. In the first case, the stream might experience a large flood with a 10-year recurrence interval during the first year of the 10-year history, followed by 9 years of relatively low flows. To an observer at the end of this sample decade, the channel is likely to be relatively narrow and to appear stable. In the second case, the stream might experience 9 years of relatively low flows, and in the 10th year a large flood. At the end of this second sample decade the observer would find a relatively wide, apparently unstable channel. In each case, the graphical or statistical description of the annual flood series would be the same, but because these approaches do not account for order of events they provide little useful explanation for the observed conditions. A catastrophe theory (Sect. 2.5.6) representation based on the interplay among force, resistance, and a morphologic measure would better represent the situation because the order of changes could then be taken into account.

Although it is difficult to specify the significance of discharges with particular magnitudes and frequencies, some measure of discharge magnitude is often useful in statistically explaining variation in geomorphologic features of channels. In this approach, measurements from many different sites include some sort of assessment of relative discharge magnitude, but the discharge measure is simply a means of comparing the amount of available energy for geomorphic work at various sites. Usually the mean annual flood is a reference point, but it is useful for comparative purposes only rather than explaining particular physical processes. Discharge characteristics statistically explain channel characteristics in several studies of semiarid streams in Nevada (Moore 1968), California (Hedman 1970), and New Mexico (Scott and Kunkler 1976). In an extensive study of 140 channel cross sections with high quality discharge data in the western United States, Hedman and Osterkamp (1982) found that they could predict discharges for floods of a variety of return intervals from channel dimensions with a standard error of estimate of about 50%. The relationships were least successful in the most arid regions. Because of the apparent lack of reliabilty of the characteristic discharge-channel relationships in extremely arid regions, Last (1974) substituted drainage area for discharge with statistically sound results in his work in the southern Negev and eastern Sinai.

3.3.7 Sheetfloods

Sheetwash is that portion of the runoff process wherein water moves downslope in a thin film (a few cm thick at most) of wide extent. Eventually the sheetwash collects and concentrates into rills, and the water behaves according to channel-related precepts. Sheetwash is virtually limited to dryland areas or artificially cleared zones in humid areas because the presence of dense vegetation breaks up the sheet of water as it moves downslope. Emmett (1970) found in semiarid areas of Wyoming that from the hydraulic perspective sheetflow behaved much like channel flows of infinite width. Leopold et al. (1966) found that sheetflows erode surficial material and transport it in drylands of the American West, contrary to the nonquantitative opinions of King (1953) based on his experiences in southern Africa. Kirkby and Morgan (1980) contended on the basis of widely collected evidence that the erosion capability of sheetflow is minor, though it influences sediment transport. Most of the transport takes place in relatively high velocity threads within the sheetflow.

McGee (1897) introduced the concept of sheetflood, a flow of less than a meter depth spread over a width of a few kilometers. McGee's vivid description of what he called sheetfloods in northern Mexico and southern Arizona fired the imagination of a generation of geomorphologists, who assumed that sheetflooding was a major factor in the formation of pediments (Rich 1935; Davis 1938). Unfortunately, observations of true sheetfloods are questionable, and photographs substantiating the existence of the phenomenon are rare. Cooke and Warren (1973, plate 3.12), for example, provide a photo of what appears to be standing water on the surface of the Atacama Desert of Chile, but it does not even approach what McGee described. Depths of flow in sheetfloods range from 4 cm reported by Joly (1965) in the northwest Sahara to 60 cm reported by Ives (1936) in Sonoyta Valley, Mexico.

There are at least four major considerations in determining the importance of sheetfloods in dryland fluvial processes (Cooke and Warren 1973, p. 203). First, sheetflow and unconfined flows of water occur across desert surfaces, but they are minor in depth and extent if pure sheets are required in the definition. Second, early descriptions of sheetfloods may be unintentionally exaggerated (Lustig 1969), a particularly important point given the lack of photographic evidence despite the recent expansion of dryland populations and the increased opportunities for observation. Third, even if sheetfloods exist, they are not likely to form surfaces, since the sheetflood depends for its own existance on a pre-defined planar surface. Sheetfloods could only modify surfaces (for an example from the Mojave Desert of California, see Lefevre 1952). Finally, because observations and measurements of sheetflows are generally lacking, no conclusions are possible about their physical impact.

Davis (1938) differentiated sheetfloods from streamfloods. Streamfloods occur on pediments and occasionally water may overflow the limited pediment channels to coalesce in a broad complex of flow. Rahn (1967) found sheetflood-like events occurring on the surfaces of basin fills in southern Arizona resulting from heavy frontal precipitation. Bull (1964) showed that these mingled waters from streamfloods deposit distinctive sedimentary units on alluvial fans of central California. It may be that sheetfloods do not commonly occur as defined, but that some streamfloods appear to be sheetfloods based on limited observations. Aldridge and Eychaner (1984) described a 1983 flood along the Santa Cruz River north of Tucson, Arizona, which disgorged from an entrenched portion of the channel onto an alluvial flat. The flow changed from a confined channel flood to a broad sheet about 4.8 km (2.9 mi) in width and less than 1.5 m (5.5 ft) deep. Further downstream the flow from this event, which appeared to observers to be a sheetflood, collected in a transverse canal which concentrated the flow once more into a single channel that experienced considerable scour (Haigh and Rydout 1985).

Graf (1987a) proposed that on the surfaces of valley fills in drylands, floods follow broad flow zones rather than easily defined channels. These flow zones conduct water in broad shallow flows that may be recorded by ground observers as sheetflood, but when viewed from orbital altitudes the flow appears as a broad ribbon with definable beginnings where water disgorges from mountain areas or spills from shallow collecting basins. The flow zone ends with the dissipation of the flow by transmission losses into the alluvial valley fill or when the flow is concentrated into defined channels as in the Santa Cruz case. Satellite imagery reveals large-scale evidence of such flow zones in the Sonoran and Mojave deserts of Arizona, California, and northern Mexico. If sheetfloods may be subsumed under this concept of a broad flow zone, theoretical generalizations about them may be accomplished as special cases of channel flow.

3.4 Paleohydrology

Paleohydrology is the study surface runoff under past climatic conditions. Leopold and Miller (1954) were among the first to use the term, though Schumm (1965) presented the first comprehensive statement of the presumed relationships among precipitation, temperature, runoff, and sediment yields based on previous work by Langbein et al. (1949) and Langbein and Schumm (1958). Gregory (1983) traces the historical development of paleohydrology in introducing a recent review of fundamentals. Ethridge and Schumm (1978), Church (1981), Gardner (1983), Maizels (1983), and especially Williams (1984) provide useful methodological summaries. Because climate is known to have been variable in the past, it is reasonable to expect that the surficial processes which respond to climatic inputs have also changed. Dryland fluvial processes, because of their discontinuous operation, are especially sensitive to climatic adjustments.

3.4.1 Paleohydrologic Evidence

The basic philosophy in paleohydrologic research is to define relationships between morphologic or sedimentologic evidence for present processes, and then to find ancient evidence which can inform on ancient processes using the same relationship. The extendable evidence in this method is either form-related (measurements of channel characteristics) or sedimentologic (measurements of sediment characteristics). The predictable variables are either discharge of a supposedly particular return interval or shear stress applied to the surface during ancient events.

As indicated in the previous section, it is difficult to establish reliable statistical relationships between channel form measures and discharges of particular return intervals under present conditons. Use of these statistical relationships to hindcast discharges in prehistoric circumstances is therefore open to some question, but at least broad estimates with a known standard error are possible. Evidence of paleochannels from which measurements may be taken to make estimates of prehistoric discharges is not unusual in drylands. As river courses adjust over time, abandoned channels are left on the landscape and survive for one to several centuries because of the relatively slow weathering and erosion processes in drylands. Flood-plain surfaces, surfaces of alluvial valley fills, and even some rock surfaces may retain channel forms that are easily measured. Hunt et al.

(1953, pp. 207–211) found extensive remnants of the pre-1896 channel of the Fremont River on alluvial terraces abandoned as the river incised itself. Schumm (1968) found similar evidence in central Australia as the Murrumbidgee River migrated laterally and abandoned its channels. Less common but equally useful are filled channels that are buried in subsurface alluvium but later exposed by entrenchment of active channels.

Alluvial materials themselves furnish evidence of prehistoric channel processes. As outlined in Chapter 4, particle size is directly related to the amount of stream power required for movement of the particle. Measurements of sedimentary particles in ancient deposits provide input for a variety of empirical and mathematical functions which can hindcast the amount of power required to move the particles. Further calculations provide estimates of the dimensions of the flows required to move the materials. Baker (1974) used this approach in analyzing the Quaternary history of Clear Creek in a semiarid portion of the Front Range of Colorado. He based calculations on the characteristics of alluvial materials stored in a sequence of Quaternary terraces.

3.4.2 Estimation of Paleoflow Parameters

Williams (1984) provided a most complete and useful review of functions used for paleohydrologic purposes. His multi-page table reviews the functions published by 1984 for the estimate of channel dimensions, flow dimensions, velocity of flow, meander lengths, sinuosity, and hydraulic parameters such as shear stress and stream power. Of greatest importance in the present discussion is the series of functions defined for estimates of discharges from morphologic evidence. Many of the published formulae derive from humid regions, so that their applicability to dryland systems is unproven. The formulae given below were derived from dryland rivers, but the complete review by Williams (1984) offers many functions developed in humid areas that could now be tested in drylands.

In an analysis of 33 sites in the semiarid United States and 3 similar sites in Australia, Schumm (1972; revision reported by Williams 1984, p. 354) found

$$Q_m = 0.029 \ W_b^{1.28} D_{max}^{1.10}, \tag{3.32}$$

where Q_m = average discharge or mean annual discharge (m³s⁻¹), W_b = channel bankfull width (m), and D_{max} = maximum depth (m). Osterkamp and Hedman (1982) derived a similar function for 252 channels in the arid and semiarid western United States:

$$Q_m = 0.027 \ W_b^{1.71}, \tag{3.33}$$

with symbols as before.

A function like (3.32) or (3.33) is likely to be useful for throughflowing rivers and for the few perennial streams in dryland systems, but its utility is limited by the fact that many smaller streams (with drainage areas less than 1,000 km²) have highly variable discharges and long periods of dry channel. Average discharge is virtually meaningless from the standpoint of geomorphological explanation in such circumstances. Hedman and Osterkamp (1982) provide alternative functions for various return intervals which may partially circumvent this issue, and Schumm (1972) offered a variation from his data for the mean annual flood ($Q_{2.33}$, in $m^3 s^{-1}$):

$$Q_{2.33} = 2.66 \ W_b^{0.90} D_{max}^{0.68}. \tag{3.34}$$

The major utility of these functions may be comparative in that they permit comparison of a representative discharge on the same stream at several points in time. Such a reconstruction provides an indispensable framework for the observations of present processes which may or may not be typical of the Quaternary.

Cheetham (1980) offered a relationship likely to be widely applicable in dryland rivers based on data originally published by Leopold and Wolman (1957) for braided streams:

$$Q_b = 0.000585 \ S^{-2.01}, \tag{3.35}$$

where Q_b = bankfull discharge ($m^3 s^{-1}$), and S = channel slope.

Interpretation of the variables in the function is easy, but the significance of the bankfull discharge is obscure. Again, the major value of the function may be in its comparative abilities.

Paleohydraulics refers to the analysis of sedimentary particles to deduce likely levels of shear stress or stream power in ancient rivers. Because these techniques are based on physical principles and use evidence only from the particles rather than landforms, the techniques are not region-specific. The first potentially useful function was by Shields (1936) who used the function to describe observed sediment transport processes in flumes:

$$\tau = 0.06 \ (\gamma_s - \gamma)d, \tag{3.36}$$

where τ = shear stress (Nm^{-2}), γ_s = specific weight of sediment particle (Nm^{-3}), γ = specific weight of water (Nm^{-3}), and d = intermediate diameter of the sediment particle (mm). The Shields forumla is widely used in engineering applications, but it is generally untested for gravels and larger particles. In geomorphic applications several authors have questioned the coefficient (Williams 1984, p. 356).

Church (1978) compiled data from a variety of laboratory and field settings in order to extend the Shields function. He found that in particle sizes greater than about 7 mm the relationship between shear stress and particle size was different than the relationship between the two variables for smaller particles. He recommended a simplified form of the function of the form

$$\tau_c = 1.78d, \tag{3.37}$$

where τ_c = critical tractive force or shear stress (Nm^{-2}) required to initiate the motion of the particle with diameter d (mm). The value of Church's function is that it is applicable to the relatively large particle sizes found in many dryland stream systems.

Another function that covers a wide range of data from a variety of environments and that includes the large particle size ranges (up to 3.29 m) is by Baker and Ritter (1975; as recalculated by Williams 1984):

$$\tau_c = 0.030 \ d^{1.49}, \tag{3.38}$$

with symbols as before.

An alternative approach to the problem of initial motion for particles from paleodischarges is to assess the velocity of flow instead of the shear stress. Williams (1984) reviews a number of formulae for the purpose, but the one by Costa (1983) provides an example likely to be applicable to dryland rivers:

$$V_c = 0.18 \, d^{0.49}, \tag{3.39}$$

where V_c = critical velocity (m s^{-1}) at which motion of the particle with diameter d (mm) begins.

3.4.3 Extension of the Flood Record

Although a major problem with flood frequency analysis for dryland streams is the usually short period of instrumented measurement, the flood record may be extended by paleohydrologic techniques. Early paleohydraulic analyses in dryland areas used evidence of prehistoric floods many times the size of present discharges, as suggested by the presence of large boulders in near-channel locations. Birkeland's (1968) investigation of Pleistocene flood velocities and sediment transport on the Truckee River in the semiarid eastern Sierra Nevada of California and Nevada used principles of force and resistance to deduce prehistoric flow conditions of floods never experienced in modern times. Baker's (1971, 1973) work on Pleistocene glacial floods in the now semiarid section of eastern Washington provided no information that might extend the modern flood record, but his approach and methods offer dryland researchers useful direction. Ballard (1976) used techniques similar to Birkeland's in assessing Pleistocene floods in the dryland mountains of Wyoming. These studies extended our understanding of the magnitude of floods that at least partially shaped the present landscape.

The analysis of slack-water flood deposits offers the possibility of extending the modern instrumented flood record a few thousand years. Kochel and Baker (1982) argued that in some river systems, slack waters that develop in tributary streams when the main stream is in flood deposit identifiable sediments that can be related to individual flood events. Floods frequently also deposit organic material which permits radiocarbon dating of various sedimentary layers. By extending the surface of the deposits for a particular event horizontally across the course of the main stream, the stage of the flood event can be reconstructed. Given assumptions about the roughness of the channel during flood, calculations for velocity of flow, and the equation of continuity (discharge = velocity x depth x width) it is possible to estimate the discharge of the event that deposited the sediments. After initial development of the technique along the Pecos River in semiarid southwest Texas, subsequent applications include the Escalante River in southern Utah, the Verde River in central Arizona, and streams in north-central Australia (Baker and Pickup 1987).

The case of the flood record for the Pecos River near Comstock, Texas, provides an indication of the usefulness of the method (Patton 1977). The instrumented history of 53 years provided reliable information on floods with discharges of less than about 4,200 m³s⁻¹ (150,000 ft³s⁻¹). A flood of nearly 28,000 m³s⁻¹ (1,000,000 ft³s⁻¹) resulted from a hurricane. The 1954 event was obviously a

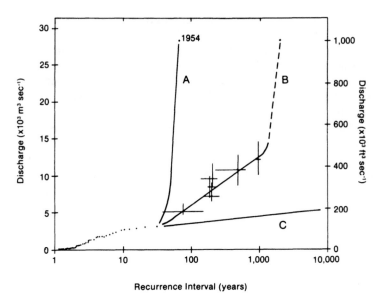

Fig. 3.22. The flood frequency curve for the Pecos River near Comstock, Texas, showing the instrumented record as solid dots and the data added by analysis of slack-water deposits as crosses with lines proportional to the maximum errors. (After Patton 1977)

rare one, with a return interval much longer than that suggested by the application of standard Weibull calculations using the period of record. Archeological material that was buried by deposits from the flood indicated that the recurrence interval was greater than 2,000 years. The actual recurrence interval of the event was a major question.

Evidence from slack-water flood deposits permitted an extension of the flood frequency curve derived from the instrumented record to about 1,000 years (Fig. 3.22). Extrapolation of the shorter instrumented flood frequency curve results in a predicted recurrence interval of the 1954 event of several million years (Fig. 3.22A). Extrapolation of the longer curve derived from the slack-water evidence results in a predicted recurrence interval of the 1954 event of about several thousands of years (Fig. 3.22B).

The experience with slack-water deposits on the Pecos demonstrates the advantages and the hazards of the method. Slack-water deposits offer significant extensions of the flood record for dryland rivers that can materially contribute to hydrologic and geomorphologic knowledge. With adequate deposits and datable materials, the method has acceptable errors. Discharge estimates in the Pecos case may have maximum errors of up to about 5,000 m³s⁻¹ (180,000 ft³s⁻¹) and date errors of up to several hundred years. The disadvantage of the method is that it is not applicable to all streams because many (perhaps the majority) rivers in drylands do not have usable slack-water deposits. Narrow, bedrock canyons with tributaries that do not flood at the same time as the main stream appear to be requirements that are not met in many areas.

There is also some question about the meaningfulness of a recurrence interval of several thousand years. Climatic change has occurred in most dryland areas over that time period to a considerable degree (for an example that includes the Pecos region, consider Knox 1983). The circumstances that produced the 1954 event on the Pecos were associated with a hurricane, but under synoptic patterns of global atmospheric circulation in some periods of the Late Pleistocene, similar discharges might result from totally different mechanisms. The ability to attach any recurrence interval to events like the 1954 Pecos flood is a major advance, however, even if it is only a reasonable estimate of the probability of occurrence regardless of cause. More recent research has made possible the extension of the record of flood power and permitted investigation of particles likely to have been moved in paleofloods (Baker and Pickup 1987).

4 Fluvial Sediment in Dryland Rivers

Sediment in dryland river systems is an active component of system processes and forms as well as serving as a passive recorder of past events. Sediment is shed from hillslopes into the channel networks in varying quantities depending on rock and soil types and climatic/vegetation variation within the dryland zone. When the sediment enters the channel system it becomes significant in analyses of river channel geomorphology, and it is at this point that the present chapter begins. The following pages provide a review of sediment characteristics with special attention to implications in dryland areas. Other subjects reviewed in this chapter include an areal approach to sediment by analysis of sediment yield from drainage basins and a more site-or river-reach-specific approach based on sediment transport in channels. The chapter concludes with a discussion of the relationship between sediment in dryland rivers and substances potentially hazardous to human health.

4.1 Sediment Characteristics

Individual sedimentary particles are best described by their lithology, size, and shape. Taken as a limited group, sedimentary particles exhibit sorting and fabric, a particular three-dimensional distribution of individual grains. Masses of sedimentary particles take on characteristics as groups that define them and separate them from other groups, leading to the separation of facies in depositional materials. At each of these stages of complexity, the characteristics of the sediment provide insights into the processes which transported and deposited them. In many cases in dryland rivers, the processes are so difficult to observe that the sedimentary evidence is the primary indicator of geomorphic activity.

4.1.1 Size, Sorting, and Shape

The size of any three-dimensional object can be characterized by measurements of its various axes. In the analysis of particles engaged in fluvial systems, three primary axes can be identified and measured, but the measure of the median axis most often serves as an indicator of particle size. The term particle diameter appears frequently in the geomorphologic literature, but it is not always clear which diameter has been used. In the following discussions, the assumption is that diameter equals the median axis measure.

There are no internationally agreed upon standards for the verbal description of particle sizes, though the most common scale is that of Wentworth (1922).

Table 4.1. Size classes used by geologists for sedimentary particles

Grade φ	limits mm		Grade names		
		Mammoth			
−12	4096	- - - - - - -			
		Very large			
−11	2048	- - - - - - -			
		Large	Boulders		
−10	1024	- - - - - - -			
		Medium			
−9	512	- - - - - - -			
		Small			
−8	256	- - - - - - - - - - - - - - -			
		Large			
−7	128	- - - - - - -	Cobbles	Gravel	
		Small			
−6	64	- - - - - - - - - - - - - -			
		Very coarse			
−5	32	- - - - - - -			
		Coarse			
−4	16	- - - - - - -			
		Medium	Pebbles		
−3	8	- - - - - - -			
		Fine			
−2	4	- - - - - - -			
		Very fine			
−1	2	- -			
		Very coarse			
0	1	- - - - - - -			
		Coarse			
+1	0.500	- - - - - -			
		Medium	Sand	Sand	
+2	0.250	- - - - - -			
		Fine			
+3	0.125	- - - - - -			
		Very fine			
+4	0.062	- -			
		Coarse			
+5	0.031	- - - - - -			
		Medium			
+6	0.016	- - - - - -	Silt		
		Fine			
+7	0.008	- - - - - -			
		Very fine			
+8	0.004	- - - - - - - - - - - - -		Mud	
		Coarse			
+9	0.002	- - - - - -			
		Medium			
+10	0.001	- - - - - -	Clay		
		Fine			
+11	0.0005	- - - - - -			
		Very fine			
+12	0.00025	- - - - - -			

Note: After Dietrich et al. (1982).

Table 4.2. Size classes used by engineers and soils scientists for sedimentary particles

Size (mm)	Engineering grade	Size (mm)	Soils grade
	Boulders		
305 – – – – – – – – – – – –			
	Cobbles		
76.2 – – – – – – – – – – – –		76.2 – – – – – – – – – – –	
	Gravel		Gravel
4.75 – – – – – – – – – – –		2.0 – – – – – – – – – – –	
	Coarse		Very coarse
2.00 – – – – – –		1.0 – – – – – – –	
	Medium Sand		Coarse
0.425 – – – – – –		0.50 – – – – – – –	
	Fine		Medium Sand
0.074 – – – – – – – – – –		0.250 – – – – – –	
	Silt		Fine
0.005 – – – – – – – – – –		0.100 – – – – – –	
	Clay		Very fine
		0.050 – – – – – –	
			Silt
		0.002 – – – – – –	
			Clay

Note: After Dietrich et al. (1982).

Recent modifications (Bouvee and Milhous 1978, p. 67) have extended the Wentworth scale into the large size ranges commonly encountered in dryland rivers (Tables 4.1 and 4.2; note the greater attention paid to sand-size particles as opposed to other sizes by engineers and soil scientists). Because particles from most sample environments have diameters that form log normal distributions, Krumbein (1934) proposed a logarithmic scale to normalize the arithmetic measurements of particle diameter. His φ scale is

$$\varphi = -\log_2 d, \tag{4.1}$$

where φ = the diameter expressed in phi units and d = particle diameter (mm). Description of particle diameters in terms of their phi measurement simplifies subsequent analysis because the measures are likely to form a normal distribution. Characterisitics of the method dictate that the largest sizes have negative phi values, when phi is zero the diameter of the particle is 1.0 mm, and in absolute values the difference of one phi unit in the larger sizes is greater than the difference of one phi unit in the smaller sizes (Briggs 1977, p. 58). This latter point is an accurate reflection of the circumstance that a small difference in particle size in the smaller ranges has radical process implications, while similar differences in the larger ranges are relatively unimportant (see Sect. 4.3 below).

Sedimentary particles may be sorted according to size, shape, density, or lithology, with each sorting process a possible key to understanding processes. The most common approach to sorting is through size measurements, with the standard deviation serving as measure of sorting in the population. Low standard

Mid-Channel Length (ft)

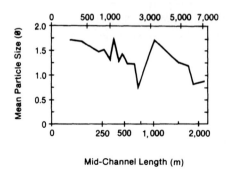

Mid-Channel Length (m)

Fig. 4.1. The relationship between downstream distance and particle size in a small stream near Lake Turkana, Kenya. Note the general decline to a distance of about 1000 m followed by a reversal due to local channel conditions that resulted in an increase in stream power. (After Frostick and Reid 1979)

deviations indicate high degrees of sorting, which might take place vertically through a sedimentary deposit or horizontally along a river channel. Maps of means and standard deviations of particle sizes provide valuable insight into the spatial variation of processes that deposited the particles. Generally, mean diameters decline and sorting improves (standard deviations decline) in the downstream direction, but local influences in the main channel and tributary contributions create many exceptions.

Particle shape may be an index of distance traveled, duration of transport, post-deposition weathering, or may simply reflect the original shape of the particle before it began its fluvial journey. Measures of shape include Zingg's (1935) analysis of the three major axes, Krumbein's (1941) sphericity index which compares particles with perfect spheres, and indices of flatness or roundness by Cailleux (1945), Powers (1953), and Kuenen (1955). Orford (1981) provides a useful entre to the far-flung literature. Dietrich et al. (1982) provide visual comparison charts of particle shape for field use.

Theoretical explanations for the spatial and temporal variation of particle parameters in dryland rivers are generally at an embryonic stage of development. Sternberg (1875) specified a downstream decline in particle size caused by reduction through abrasion:

$$w = w_0 e^{-bL}, \tag{4.2}$$

where w = particle weight, w_0 = initial particle weight, e = base of the natural logarithm system, b = an empirical constant, and L = length of transport. The constant b apparently reflects two processes, particle wear and selective transport of particles (Church and Kellerhals 1978). Evidence from ephemeral streams in the United States and Israel generally confirms Sternberg's generalization for mean and maximum grain sizes (Leopold and Miller 1956; Bluck 1964; Denny 1965; Beaumont 1972; Cherkauer 1972; Mayer et al. 1984). Evidence from sand-bed streams in New Mexico (Leopold et al. 1966) and in Kenya (Frostick and Reid 1979) indicates that particle size may increase in the downstream direction in response to increasing discharges and stream power occurring over nearly constant gradients (Fig. 4.1). Rhoads (1986a) found in small basins in the Mc-Dowell Mountains of Arizona that local variations in stream power and tributary contributions of materials introduced great variation in the downstream distribution of particle sizes.

Downstream trends in sorting have been elusive, though sorting might be expected to improve downstream as deposits are selectively reworked. In the drylands of the American Southwest Miller (1958) and Inderbitzen (1959) found no trends in sorting. In stochasitic simulations Rana et al. (1973) predicted improved sorting in the downstream direction in streams dominated by bedload transport, a conclusion partially supported by field evidence collected in arid-region basins by Rhoads (1986a).

Theoretical statements for spatial variation related to shape are generally lacking except for work by Knighton (1982) in humid Great Britain. He found that different portions of the stream system may be dominated by different processes, with the headwaters characterized by abrasion and sorting, middle reaches by sorting, and lower reaches by sorting and breakage. The extension of this regionalization of stream systems to dryland systems remains to be tested.

4.1.2 Fabric

Sedimentary particles, especially those with some flatness, may be deposited in ways that provide clues to the processes of emplacement. Fabric refers to the spatial arrangement and orientation of particles within a sedimentary deposit (Allen 1970, p. 89). Nearly spherical grains deposited intensively or with low velocity of fall produce deposits with many gaps and high porosity (Gray 1968). Once created, fabric controls permeability and porosity of channel deposits, and therefore partly controls transmission losses. Elongated particles deposited by gentle bedload transport produce imbricated deposits where the long axis of the particles are aligned normal to the direction of flow and with the particles inclined into the bed at an angle of 10–20° (Leeder 1982, p. 43). Highly turbulent flows cause deposition of elongated grains with their long axis parallel to the direction of flow.

4.1.3 Facies

Sedimentary deposits frequently exhibit definable subcomponents, or facies, that have characteristics directly related to the processes that produced them. In dryland regions sedimentary deposits provide evidence of past eolian, lacustrine, and fluvial activities, with each activity associated with particular facies. Eolian and lacustrine deposits are not of direct interest in this book, but brief descriptions are useful in order to distinguish them from fluvial deposits.

About 30% of the world's continental areas experiencing 15 cm (6 in) or less of annual rainfall have surfaces characterized by sand seas or ergs (Glennie 1970). In the past they had wider distribution and occurred in areas where they are not now active, especially in northern and western Africa as in Niger (Talbot and Williams 1979). Eolian deposits consist almost exclusively of sand because wind cannot entrain larger particles and disperses smaller ones. Eolian facies commonly have

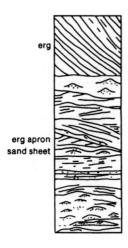

erg

erg apron
sand sheet

Fig. 4.2. Typical eolian facies from an erg, or sand sea, Great Sand Dunes National Monument, Colorado. The column, extending through several meters shows large-scale cross-stratification at the top with ridges, ripples, deflation hollows, lag deposits, and erosion surfaces below. (After Fryberger et al. 1979)

numerous inclined crossbeds and sheets with truncation surfaces sometimes crossing the beds. The migration of individual dunes produces much reworking of the materials that generates crossbeds, ripples, hollows, and erosion surfaces within the deposit (Fig. 4.2).

Lacustrine deposits in dryland areas are common in closed basin areas. These materials are laid down in relatively still waters, so that horizontal beds develop. Particle sizes smaller than coarse sand are usually found in lacustrine deposits of dryland regions because the coarser materials are deposited at the lake edges in deltas or on alluvial fans where flows are concentrated enough to transport the materials. Lake Turkana, Kenya, is typical in that varve-like sediments suggest cyclic contributions of material to the lake basin (Yuretich 1979). Closed basins indicate that lake waters are likely to be saline, as in the case of the Dead Sea (Neev and Emery 1967), so that if the lake evaporates completely crusts of evaporite materials become incorporated in the sedimentary record. The resulting lacustrine facies tends to have thin, fine-grained, horizontal beds, occasionally with evaporite deposits.

Buried soils sometimes provide datable materials and give clues to the environmental history of the region. Lacustrine facies of Lake Bonneville in the Great Basin of the western United States (precursor of the modern Great Salt Lake) exhibit soils and evaporite deposits that are linked with glacial sequences in the nearby mountains (studies initiated by Gilbert 1890). Newell (1946) made similar connections for Lake Ballivain (forerunner of Lake Titicaca) in Peru and Bolivia, as did Pumpelly (1905, p. 138) for two Kara Kul lakes and Prinz (1909) for Lake Issyk Kul in the Soviet Union. Lake Lahontan in Nevada experienced numerous overflows into a series of interconnected basins, producing a distinctive series of deposits (studies initiated by Russell 1885). Similar evidence appeared in deposits of Searles Lake, California (Smith and Haines 1964), and Lake Lisan (ancient Dead Sea) in southwest Asia (Farrand 1962). Flint (1971) reviews lacustrine studies on a global basis.

Fluvial facies in dryland environments have distinctive characteristics that separate them from eolian or lacustrine materials, though transitional forms

 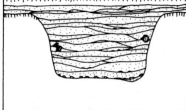

Fig. 4.3. Typical dryland fluvial facies from the alluvial fill of Chaco Canyon, New Mexico. The channel fills were derived from a meandering stream, *left*, and a braided stream, *right*. Each example includes some angular talus blocks from nearby canyon walls. (After Love 1983b, p. 200)

occur. Common fluvial forms include braided and meandering river channels, flood plains, and alluvial fans. Braided channels produce deposits of sand and gravel with cross sets created by migrating dunes, bars, and sand waves (Cant and Walker 1978). Massive beds of large particles mark the locations of mid-channel or channel-side bars. Particle sizes may be consistent through substantial portions of the vertical dimension of the deposit. Buried channels are sometimes evident as a product of lateral channel relocation. Lateral channel migration produces layers of similar particle sizes cross cutting previously established beds. The Brahmaputra River of India and Bangladesh, for example, develops these facies rapidly as its channel migrates horizontally as much as 1 km yr^{-1} (Coleman 1969). The forms are also developed in deposits of the Platte River, Nebraska, but with relatively little variation in particle size and a singular importance for mobile transverse bars (Smith 1971).

Meandering channels that transport sand or sand and gravel develop facies distinctive from those of braided channels (Fig. 4.3). Most analyses have been of facies from humid region rivers (e.g., Bluck 1971; Jackson 1976; Bridge and Jarvis 1976; and Levey 1978), but since the depositional processes are not climate-specific, the generalizations may extend to dryland rivers. Because meandering channels migrate during high flows, because those high flows also transport the larger-size fractions of sediment, and because point bar accumulations of coarse debris result from these processes, facies from meandering rivers commonly have a fining upward sequence of materials. Coarse materials occur in the lower portions of the deposit, with finer materials in the upper portions. The transition from coarse to fine may be gradual or abrupt, but the ensemble is easily identified in field exposures. Channel-fill materials and sand sheets deposited by mid-channel processes may be included in the deposits of meandering channels.

Flood-plain deposits are relatively fine-grained materials laid down in nearly horizon sheets. They may drape over other deposits and may be reworked into subsequent channel-related deposits. The high degree of variation in flood magnitudes in dryland river systems results in flood deposits at multiple levels above the channel floor in addition to the modern flood plain. In the Escalante River basin of southern Utah, Patton and Boison (1986) found a well-defined flood plain about 2 m above the channel, complemented by additional flood deposits covering prehistoric alluvial fill surfaces up to 10 m above the channel.

Alluvial fans formed by the emergence of streams from confined mountain drainages exhibit horizontal and vertical variation in materials. Fans created by ephemeral streams have facies with reworked materials, crosscut beds, filled channels, and complex inclined bedding. On any particular fan surface there may be fining upward sequences of materials produced by radical transmission losses, whereby large flows at the upper ends of the fans can transport large particles but the lower flows at distal points can transport only small particles. As the fan expands, a series of materials at any particular section shows a coarsening upward sequence because the section changes from a distal location when the fan is small to a more central one as the fan grows larger (Heward 1978). Coarsening upward sequences result from rejuvenated erosion in headwaters areas, the advance of subsidiary fans over preexisting ones, and the development of trenching in the upper portion of the existing fan which flushes large particles down the length of the form. Some fans may have jumbled deposits of large, partly oriented particles mixed with fine materials resulting from debris flows rather than stream flows. Cyclic development of alluvial fans in response to climatic changes in present dryland areas such as Niger result in complex forms (Talbot and Williams 1979).

Fans such as those created by perennial streams may be large and create deposits that take on the characteristics of the river channels that form them. The fan of the Kosi River, India, has been almost completely combed by channel migration over the past 250 years (Gole and Chitale 1966), and its deposits appear similar to those expected of a braided river.

Sedimentological evidence along dryland rivers provides indicators of processes in the very recent past. Love (1983a, b) used sedimentological evidence from the Chaco River in Chaco Canyon to trace changes in channel processes through a period of 6,000 years in association with archeological evidence. Deposits included those from braided and meandering streams because the river shifted back and forth between the two modes of operation. Hereford (1984, 1986) relied on over-bank deposits dated by the ages of riparian vegetation rooted at various depths in the sediments to deduce sedimentation processes related to environmental changes in the Little Colorado River of northern Arizona and the Paria River of southern Utah. Graf (1987b) used the total amounts of sediment from various units to explore 600 years of sediment yield and storage in canyons of the Colorado Plateau. These studies demonstrate the utility of sedimentological analysis in explaining changes in dryland rivers for which few direct observations of process are available.

4.2 Sediment Yield from Drainage Basins

Perspectives on the generation, transportation, storage, and ultimate deposition of sediment are as diverse as the researchers dealing with fluvial processes. Agricultural specialists are concerned with soil erosion from field-size areas of a few thousand m^2, watershed scientists investigate sediment yields from drainage basins up to several thousand km^2, while hydraulic researchers probe the issues of sediment mobility a grain at a time or in mass through a single channel cross section. Each of these perspectives develops its own techniques, research questions, models, paradigms, and body of literature. The geomorphologist might reasonably be expected to somehow link the entire group of perspectives together into a system-wide view of sediment-related processes from slope sources to ultimate sinks. Generalizations of this type are not yet in evidence for dryland river systems, though the components have begun to emerge from a variety of fields of research. Because this volume focuses on rivers, little attention is paid to slope processes, but the slopes cannot be completely ignored. The following two sub-sections review the explanations and predictive models for the production of sediment from slopes and field-size areas. The next two subsections deal with the drainage-basin perspective. Section 4.3 addresses the more restricted and specific in-channel processes.

4.2.1 Factors Controlling Erosion

From a theoretical perspective, explanation of soil erosion from small hillslope areas or basins less than 1 km^2 in extent is well established (reviewed for geomorphologists by Kirkby and Morgan 1980). The factors that control temporal and spatial variation of erosion are the characteristics of rainfall, vegetation, soils, and slopes (Evans 1980).

Characteristics of rainfall influence erosion because rain provides the basic energy for the erosion process. Raindrop size, velocity, and shape as well as storm duration and wind conditions, must be accounted for in an analysis of drop erosion (Gunn and Kinzer 1949). Rainfall intensity and drop size have direct influence on the amount of kinetic energy available for work in a given time period (Hudson 1963; Carter et al. 1974). Wischmeier and Smith (1958) and Wischmeier (1959) showed that kinetic energy of rainfall explained most of the erosion on the surface.

On a global scale, rainfall does not have a simple, direct relationship to the amount of erosion expected from slopes because of the confusing factor of vegetation. Vegetation intercepts rainfall and allows it to drip slowly to the surface, providing more gradual introduction of the moisture to soils than if vegetation were absent. Since soils can absorb moisture only at a limited rate, this influence combined with the anchoring effects of roots and stems is substantial in reducing runoff and erosion from slopes (Woodward 1943). When vegetation cover of the surface is greater than about 70%, variation in cover does not have substantial impact, but at coverages less than 70% runoff and erosion are sensitive to changes (Copeland 1965). In dryland areas that usually have less than 30% cover, runoff and erosion (to a lesser degree) are nearly directly related to the amount of bare ground (Branson and Owen 1970).

Biomass, the amount of biologic material per unit area, is an effective combined measure of vegetation cover and density (Table 4.3). Ecologists have used biomass estimates for several decades in attempts to estimate biological productivity of portions of the earth's surface (Whittaker 1975; Lieth 1975). Despite its relative ease of definition, biomass has yet to be statistically related to erosion on a general basis. Biomass can be an important component of geomorphologic explanations and models because as biomass increases, resistance to erosive forces increases, while runoff and erosion decrease. Graf (1979a) used

Table 4.3. Biomass for common dryland vegetation communities

Bioclimatic/soils formation	Biomass (kg m^2)
Subboreal, semiarid regions	
Steppe, leached chernozems	2.5
Steppe, ordinary chernozems	2.0
Steppe, solonets chernozems	2.0
Steppe, solonets	1.6
Halophytic formations, solonchak	1.2
Psammophytic formations, sand	1.8
Dry steppe, dark chestnut soils	2.0
Desert steppe, light chestnut soils	1.3
Dry/desert steppe, chestnut/solonets complexes	1.4
Dry/desert steppe, solonets	1.4
Subboreal, arid regions	
Steppe desert, brown semidesert soils	1.2
Steppe desert, brown/solonets complexes	1.0
Steppe desert, solonets	0.9
Desert, gray-brown desert soils	0.45
Psammaphytic formations, sand	3.0
Desert, takyr soils	0.3
Halophytic formations, solonchak	0.15
Riparian formations	8.0
Mountain desert, brown semidesert soils	0.9
Mountain desert, highland soils	0.7
Subtropical, semiarid regions	
Xerophytic forest, brown soils	17.0
Shrub-steppe formations, gray-brown soils	3.5
Shrub-steppe formations, gray-brown solonets	2.0
Shrub-steppe formations, subtropical chernozems	2.5
Psammophytic formations, sand	2.0
Halophytic formations, solonchalk	0.15
Riparian formations	25.0
Mountain xerophytic forest, brown mountain soils	12.0
Mountain shrub-steppe formations, gray-brown soils	3.0
Subtropical arid regions	
Desert steppe, serozems	1.2
Desert, subtropical desert soils	0.2
Psammophytic formations, sand	0.3
Desert, takyr soils	0.1
Halophytic formations on solonchak	0.1
Riparian formations	20.0
Mountain desert, mountain serozems	1.5
Mountain desert, subtropical mountain desert soils	0.3
Tropical semiarid regions	
Xerophytic forest, brownish-red soils	25.0
Grass/shrub savanna, red-brown soils	4.0
Grass/shrub savanna, tropical black soils	3.0
Grass/shrub savanna, tropical solonets	2.0

Table 4.3 (continued)

Bioclimatic/soils formation	Biomass (kg m^2)
Meadow and swamp savanna, red and meadow soils	6.0
Riparian formations	20.0
Xerophytic mountain forest, brownish-red soils	20.0
Mountain savanna, red-brown mountain soils	4.0
Tropical arid regions	
Desert savanna, reddish-brown soils	1.5
Desert, tropical desert soils	0.15
Psammophytic formations, sand	0.1
Desert, tropical coalesced soils	0.1
Halophytic formations, solonchak	0.1
Riparian formations	15.0
Mountain desert, tropical mountain desert soils	0.1

Notes: Data from Rodin et al. (1975); also published in Lieth (1978).

biomass as a measure of resistance to be compared with the shear stress developed on alluvial surfaces, but the measure could also be used in general slope and basin studies where resistance by vegetation is a significant consideration.

Because of the influence of vegetation, erosion rates on slopes do not have a simple, direct relationship to rainfall on a global scale. Langbein and Schumm (1958) and Schumm (1965) suggested a relationship between sediment yield and effective precipitation defined as the annual precipitation required to generate the given annual runoff at a standardized annual mean temperature of 50°F. Sediment yield in the relationship showed a maximum where effective precipitation was about 400–500 mm. In many parts of the world, areas with annual precipitation close to this amount have semiarid vegetation communities. Langbein and Schumm argued that low sediment yield occurred in the lowest precipitation ranges because there was not enough water (and energy) to drive erosion. In the higher precipitation ranges, they argued that grasslands and forests provided enough cover to retard erosion processes. Their oft-repeated diagram showing the peak sediment yield in the 400–500 mm effective precipitation zone was plotted with only six points representing group averages from subdivisions of 94 sample stations (Fig. 4.4).

Walling and Webb (1983) have pointed out two major criticisms of the Langbein-Schumm curve showing a decline in sediment yield with an increase in precipitation. First, Walling and Webb indicate that the group-averaging method causes the plot to be an abstraction of the data subject to alteration given small amounts of new data. The data used were all from the United States, and thus represented a limited range of climatologic conditions. Second, although relationships between sediment yield and mean annual runoff from American data developed by Judson and Ritter (1964) and Dendy and Bolton (1976) reveal curves somewhat similar to the Langbein-Schumm model in the lower precipitation ranges (Fig. 4.4), additional data from a variety of climates and conditions do not substantiate the original formulation.

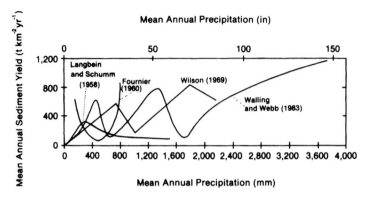

Fig. 4.4. The relationships between precipitation and sediment yield plotted with points from grouped data means. Raw data show no relationship. (After separate figures by Walling and Webb 1983)

Walling and Kleo (1979) assembled precipitation and suspended sediment yield data for 1,246 drainage basins from all the major continents and found that there was no relationship between the two variables. The grouping of data according to classes of precipitation reduced the variability and produced a definable relationship with maxima of sediment yield in semiarid conditions, seasonal mediterranian conditions, and tropical monsoon areas (Fig. 4.4). The second and third maxima were not anticipated by Langbein and Schumm, though Schumm (1977, pp. 29-30) indicated that the expected decline in sediment yield with increasing precipitation observed in the American data might be less applicable to seasonal areas.

Scale of analysis plays an important role in generalizations about the relationship between sediment yield and precipitation. Most of the aggregated data sets used in the global analyses suggest that in semiarid regions at some point a decline in sediment yield is to be expected with increasing precipitation, but this expectation is a product of the grouping processes. Apparently there are no data sets collected on a restricted spatial basis that show the decline – in every case yield increases with precipitation. Heusch and Millies-Lacroix (1971) found exponentially increasing sediment yield with mean annual precipitation in Morocco. Even accounting for the influences of geologic variation and land use, Sharma and Chatterji (1982) in India, and Dunne (1979) in Kenya found steadily increasing yields with increasing precipitation across the range of precipitation values typifying semiarid conditions.

The studies of the sediment yield and precipitation relationship provide three lessons for researchers dealing with dryland rivers. First, data from one region, even one large region, is not likely to provide valid generalizations for global use. The Langbein-Schumm curve based on American data is not necessarily valid for other continents, especially those with high seasonal variability in precipitation. The explanation offered for the American data also does not appear to be completely transferable. Second, the construction of theory from grouped data appears to be risky. The conclusions drawn from grouped data strongly reflect the

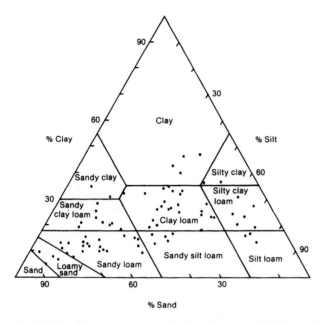

Fig. 4.5. Particle size characteristics for soils deemed highly erodible showing the importance of clay content. (After Evans 1980)

grouping process which obscures the natural variability in the data. Successful theory must grow from the complexity of the data and account for the complexity. Finally, the nearly 30 years of research into the question typifies the way in which science in general and geomorphologic science in particular make progress. Langbein and Schumm (1958) proposed a tenuous relationship, Schumm (1965 and later works) expanded the relationship and developed supporting theory, and finally a host of subsequent researchers provided wider-ranging data, criticisms, and further refinements of explanation. It is in the collective process of research rather than in the individual efforts that success ultimately resides.

In addition to vegetation and precipitation, soil characteristics directly influence erosion and contribution of sediment to channels. Generally, soils with low clay content tend to be most erodible. In a review of erodible soils from England, Canada (Ripley et al. 1961), India (Mehta et al. 1963), and the United States (e.g., Megahan 1975), Evans (1980) found that soils with more than 30–35 per cent clay are unlikely to be classified as highly erodible (Fig. 4.5). The stability of soils in dryland areas is dramatically affected by monovalent cations such as sodium, a relatively weak binding agent that is associated with rapid erosion and piping. In other dryland areas the presence of a stoney surface layer retards erosion because water infiltrates more readily around the individual stones (McIntyre 1958) and the stones provide stabilizers for the surface (Lamb et al. 1950).

Finally, slope characteristics control erosion in part. Slope angle controls the process of splash erosion (Ellison 1944). Erosion increases with increasing slope angles, especially at angles of less than 10°. Erosion also increases as slope length increases, especially in field-size areas because more runoff is available from upslope areas for the work of erosion (Wischmeier et al. 1958). Storage of materials on footslopes limits the generalization.

4.2.2 Estimations of Erosion

Functional estimation of erosion and contribution of sediment to channel systems is difficult because many factors control the process and because accurate measurements are scarce. The functions most commonly available are designed from data from agricultural areas and are questionable in natural terrain often of interest to the geomorphologist dealing with dryland situations. The following pages review as examples the functions by Zingg, Musgrave, and Hudson, the Universal Soil Loss Equation, the Pacific Southwest Inter-Agency Committee Equation, and the Soil Loss Estimator for Southern Africa.

Based on experimental plots and simulated conditions, Zingg (1940) proposed

$$A = C \, S^{1.4} \, L^{0.6}, \tag{4.3}$$

where A = average soil loss per unit area from a plot of unit width, C = an empirical constant that accounts for rainfall and vegetation conditons, S = slope, and L = horizontal length of the slope (ft). Zingg's equation might be transferable to dryland areas but only for small areas of a few thousand m^2 or less.

Musgrave (1947) used data from field sites to generate a predictive function applicable to the central and eastern United States. Mitchell and Bubenzer (1980) formalized the function in metric units:

$$E = 0.00527 \, I \, R \, S^{1.35} \, L^{0.35} \, P_{30}^{1.75}, \tag{4.4}$$

where E = soil loss (mm yr^{-1}), I = inherent erodibility of a soil at 10 per cent slope and 22 m slope length (mm yr^{-1}), R = a vegetation cover factor, S = slope, L = length of slope (m), and P_{30} = maximum 30-minute rainfall (mm). The Musgrave function has seen wide use in estimating gross erosion from small watersheds, but it is probably not well suited to use in drylands without modification.

Hudson's (1961) erosion equation is similar to the Universal Soil Loss Equation in the use of "factors" that must be defined by local conditions compared with established standards. The function derived from sub-tropical southern Africa is

$$E = T \, S \, L \, P \, M \, R, \tag{4.5}$$

where E = erosion, and the remaining variables are factors describing controls on erosion: T = soil type factor, S = slope gradient, L = slope length, P = factor for land management practice, M = mechanical protection factor, and R = rainfall factor. Hudson (1961) defined the nature and problems associated with the various factors specifically for southern Africa.

The Universal Soil Loss Equation has a form similar to Hudson's:

$$A = 0.224 \, R \, K \, L \, S \, C \, P, \tag{4.6}$$

where A = soil loss (kg $m^{-2}s^{-1}$), R = a rainfall erosivity factor, K = soil erodibility factor, L = slope length factor, S = slope gradient factor, C = cropping management factor, and P = erosion control practice factor (for extended information on the equation, see Wischmeier and Smith 1965).

The rainfall erosivity factor, R, of the Universal Soil Loss Equation is an indicator of the erosivity of rainfall events and reflects two precipitation characteristics: kinetic energy and the maximum 30-minute intensity (Wischmeier 1959, provided the mathematical derivation). The factor ranges from an average annual value of more than 2000 in tropical rainy climates to about 20–50 in dryland areas. Wischmeier and Smith (1978) mapped the approximate values of R for the entire United States. Similar maps are available for a number of dryland regions: by Roose (1977) for western Africa, Masson (1972) for Tunisia, Delwaulle (1973) for Niger. In areas where spatial variability of precipitation events is high such as in most dryland regions, calculated values of R are not reliable (Renard and Simanton 1975).

The soil erodibility factor, K, indicates the varying degree of erodibility of soils given different particle size characteristics and organic content. The value of K varies from about 0.05 for sandy soils with almost no organic material to more than 0.60 for some silty soils. Direct measurement of K from soils is possible experimentally, but usually not cost-effective. In the United States, standard government soil surveys define K for each soil. For other areas, Wischmeier and Mannering (1969) developed a regression to predict K based on 24 soil properties, and Wischmeier et al. (1971) offered nomograms for predictive purposes. The erodibility factor is primarily defined by experience with agricultural soils, and the extension of its use to nonagricultural soils is not well supported experimentally.

The slope length factor (L) and the slope gradient factor (S) may be combined into a topographic factor (LS). For a slope 100 m long, the value of the topographic factor varies from about 0.18 for a 1 per cent gradient to about 15.0 for a 30 per cent slope. Complex slopes require complex methods for the estimation of the LS factor (Onstad et al. 1967; Foster and Wischmeier 1974).

The cropping management factor, C, is the ratio between the amount of soil lost under a particular cropping scheme and the amount lost if the subject plot were to remain fallow. Wischmeier and Smith (1978) provide tables as guides for the values of C likely to include conditions encountered in most agricultural areas of the world. Wischmeier (1974) provided a guide for application to nonagricultural lands based on percentage of vegetation cover of various types, with values likely to be observed in dryland areas ranging from 45 for areas of no vegetation to values of about 4 for areas about 80 per cent covered with shrubs. Elwell and Stocking (1976) developed a classification guide for Zimbabwe, as did Roose (1977) for western Africa, though in each case the emphasis is not on drylands.

The erosion control practice factor, P, is the ratio between the soil loss experienced with a given cropping scheme and an up-and-down-hill culture. Values defined by Wischmeier and Smith (1978) range from about 0.90 for contouring on 21–25 per cent slopes to about 0.1 for terracing on 1–2 per cent slopes. The factor seems to have little application to dryland natural slopes, though it probably would be in the 0.1–0.3 range.

The Universal Soil Loss Equation is the most widely used estimator of slope erosion and slope sediment production because it is well defined, has published guidelines, and offers compatibility with numerous studies in many parts of the

world. Wischmeier (1976) enumerated the proper uses of the equation, including
the prediction of long-term average soil losses under specific conditions, prediction of possible effects of changes in land management, and estimation of soil
losses in nonagricultural areas. For geomorphologic research, the equation has
limited utility. It is designed for long-term averages rather than analysis of single
storm events or even specific years. In many dryland areas, empirical research has
not verified the values of the various factors, so that application of the function
sometimes involves little more than educated guesses. In some cases the equation
fails to make accurate predictions (e.g., Elwell 1977, 1984). The expense of
developing accurate evaluations for the various factors is sometimes prohibitive
for developing countries (Wendelaar 1978). The rainfall erosivity index is unknown for most dryland areas, and its concept is especially difficult to apply in
extremely arid areas. The basis of the equation is from standard measurement
plots that are relatively small (about 22 m long), so that it is not likely to be an

Table 4.4. Rating values for erosion factors in the Pacific Southwest Inter-Agency
Committee Method

Factor	Minimum	Maximum	Characteristics
Surface geology	0	+10	Rock type, hardness, weathering, fracturing
Soils	0	+10	Texture, salinity, organic matter, caliche, shrink-swell, particle size
Climate	0	+10	Storm frequency, intensity, duration, snow, freeze-thaw
Runoff	0	+10	Volume per unit area, peak flow per unit area
Topography	0	+20	Steepness of upland slope, relief, fan and flood-plain development
Ground cover	−10	+10	Vegetation, litter, understory
Land use	−10	+10	Percentage cultivated, grazing intensity, logging, roads
Upland erosion	0	+25	Rills, gullies, landslides, eolian deposits in channels
Channel erosion	0	+25	Bank and bed erosion, flow depths, active headcuts, channel vegetation

Note: From Pacific Southwest Inter-Agency Committee (1968).

accurate model of complex processes on larger scales. For example, the equation cannot account for storage of sediment on footslopes.

The most fundamental issue with use of the Universal Soil Loss Equation by geomorphologists is that the equation is for agricultural applications. From its inception, the equation was viewed by its designers as primarily a predictor of sediment production from fields, and the hundreds of decisions that went into its development were made with a view toward agricultural applications. To use the equation over large drainage basins in nonagricultural dryland settings is an intellectual extrapolation that may not be valid and that may violate many hidden underlying assumptions in the method.

The Pacific Southwest Inter-Agency Committee, an American governmental group, developed an alternative to the Universal Soil Loss Equation specifically to accommodate the nonagricultural dryland conditions in the southwestern United States. Recognizing the futility of accurate prediction of erosion from small plots in such areas, the agency settled for the definition of five classes of erosion to be applied over large basins (Pacific Southwest Inter-Agency Committee 1968; 1974):

1. Greater than $14.3 \text{ m}^3\text{km}^{-2}$ (less than 3.0 acre-ft mi^{-2})
2. $4.8 - 14.3 \text{ m}^3\text{km}^{-2}$ (1.0 – 3.0 acre-ft mi^{-2})
3. $2.4 - 4.8 \text{ m}^3\text{km}^{-2}$ (0.5 – 1.0 acre-ft mi^2)
4. $1.0 - 2.4 \text{ m}^3\text{km}^{-2}$ (0.2 – 0.5 acre-ft mi^2
5. Less than $1.0 \text{ m}^3\text{km}^{-2}$ (less than 0.2 acre-ft mi^{-2}).

These broad classes provided a base from which nine factors controlling erosion could be evaluated. The agency provided guidelines for each of the factors (reviewed in Table 4.4). For a particular area the factors are summed and used to directly estimate sediment yield from the function

$$Y = 0.0816 \ e^{0.0353 \ X}, \tag{4.7}$$

where Y = annual sediment yield (acre ft mi^{-2}), e = base of the natural logarithm system, and X = the sum of the assigned values for the nine rating factors.

The value of the Pacific Southwest Inter-Agency Committee Method is that its design is specifically for nonagricultural areas in a dryland zone. In a test of several sediment yield prediction equations in semiarid conditions in southern Arizona, Renard and Stone (1982) found that the predictions of the Pacific Southwest Inter-Agency Method most closely agreed with measured data. Its disadvantages are that estimations must be compared with known or measured sediment yield as a way of checking the success of the user in assigning meaningful values to the rating factors. Unlike the Universal Soil Loss Equation, the Pacific Southwest Inter-Agency Committee Method is suited to large areas (greater than 26 km^2 or 10 mi^2) and is not applicable to small basins or individual slopes. The reliability of the Pacific Southwest Inter-Agency Committee Method is not widely established, and it has not generally seen use in areas other than the American Southwest. Its most successful employment is as a broad-gauged planning tool to identify and outline hazardous areas.

Another alternative to the Universal Soil Loss Equation is the Soil Loss Estimator for Southern Africa. This estimator is the product of model development, some measurements, and accumulated professional experience in the highveld of Zimbabwe. The estimator is based on the relationship between sediment yield from established cultivation methods and adjustments that occur in that yield as a result of changes in control factors. The basic model defined by Elwell (1978) is

$$Z = K \times C, \tag{4.8}$$

where Z = predicted mean annual soil loss (t ha^{-1} yr^{-1}), K = mean annual soil loss from a conventionally tilled plot 30 x 10 m at a 4.5° slope for a soil of known erodibility (t ha^{-1} yr^{-1}), X = the ratio of soil loss from a field slope of length and slope different from the standard length, and C = the ratio of soil loss from a cropped plot to that from one which is fallow. Elwell (1978, 1984) provided a series of functions to predict the values of X and C depending on vegetation and management conditions. The Soil Loss Estimator for Southern Africa may not be especially useful at present for the geomorphologist, but it might be adapted through modest amounts of experimentation and verification with measured data. The method is probably best suited for small basins or slope-sized areas of analysis and is probably not applicable to extremely arid conditions.

A sediment yield function specifically designed to accommodate large drainage basins in the dryland areas of India was based on a combination of the areal coverage of vegetation communities with physical characteristics of the basins. Miraki (1983) used measurements of sedimentation from 32 reservoirs to define a series of regressions using control factors to predict sediment yield. The most complex regression was

$$V_{SA} = (1.182 \times 10^{-6})\, A^{1.026} P^{1.289} Q^{0.287} S^{0.075} D_d^{0.398} F_c^{2.422}, \tag{4.9}$$

where V_{SA} = sediment yield (m^{3}yr^{-1}), A = drainage area (km^2), P = annual rainfall (cm), Q = mean annual runoff (m^3 x 10^6), S = basin slope, D_d = drainage density (km^{-1}), and F_c = an areally weighted erodibility factor defined by the vegetation communities. Individual community values for F_c were protected forest = 0.2, unclassed forest = 0.4, arable areas = 0.6, scrub and grass = 0.8, and waste area = 1.0. Miraki's method provided predictions of the sediment yields for 28 of the 32 reservoirs to an accuracy of $\pm 30\%$. Problems with its use elsewhere include the likelihood that the coefficients are regionally specific and the lack of precision in assigning the erodibility values to various communities. The function's relative success in India indicates that given enough measured sediment yields for calibration of the model, it can predict sediment yield in ungauged basins to a level of accuracy useful for the geomorphologist.

For planning purposes in India, Khosla (1953) used a more simple regression:

$$V_s = 0.00323\, A^{0.72}, \tag{4.10}$$

where V_s = volume of sediment yield (10^{6}m^3) and A = drainage area (km^2). Functions of this simplistic type are useful for general planning but are not likely to be sensitive enough to variations in topography, soils, geology, and vegetation to support most geomorphologic investigations.

All the estimators of slope erosion have limitations in application to dryland conditions. This proviso does not mean that they should be automatically eliminated from consideration, because the alternative is to ignore the issue of slope sediment production altogether. The reservations about each method indicate that results obtained from them in drylands must be used with caution and that the limitations must be explicitly identified in any conclusions drawn from their use.

4.2.3 Sediment Yield From Basins

The prediction of sediment yield from drainage basins with their complex arrangements of slopes and channels could be accomplished by modeling each slope segment and each channel reach, but an approach of this detail would be too costly for most dryland river studies. Statistical relationships between control factors and sediment yield are cheaper and easier to apply but sacrifice mathematical explanation (Weber et al. 1976). Anderson (1975) developed a complex regression model to predict the sediment yield from a drainage basin emptying into a southern California reservoir. Many of the 34 predictive factors contributed little to the statistical explanation of sediment yield, but the most important variables included slope and vegetation characteristics. There are no wide-ranging tests of the function.

Flaxman (1972), Dendy and Bolton (1976), and Renard (1972) provided relatively simple models for the prediction of sediment yield from small dryland basins. Flaxman (1972) developed a more general relationship that described sediment yield from drainage basins in the dryland western United States. The predictive equation is useful for small drainage basins (less than 26 km², 10 mi²) and has the form

$$\log_{10}(Y + 100) = 524.3 - 270.6 \log_{10}(X_1 + 100)$$
$$+ 6.4 \log_{10}(X_2 + 100)$$
$$- 1.7 \log_{10}(X_3 + 100)$$
$$+ 4.0 \log_{10}(X_4 + 100)$$
$$+ 1.0 \log_{10}(X_5 + 100), \tag{4.11}$$

where Y = average annual sediment yield (t mi^{-2}), X_1 = ratio of average annual precipitation (in) to average annual temperature (°F), X_2 = watershed slope, X_3 = per cent soil particles greater than 1.0 mm, X_4 = per cent soil aggregation or dispersion less than 0.002 mm, and X_5 = 50% of the mean annual flood (m^{3}s^{-1}). Despite its odd mixture of units the Flaxman model has enjoyed reasonably wide usage in the United States, and has been successful in intensive tests in semiarid conditions (Renard and Simanton 1973). Extension of the model to other parts of the world demands local verification against instrumented records and adjustment of the coefficients.

Dendy and Bolton (1976) used sedimentation data from over 800 small reservoirs in the United States to develop the function

$$S = 1280 \, Q^{0.46}(1.43 - 0.26 \log_{10} A), \tag{4.12}$$

where S = sediment yield (t mi^{-2}yr^{-1}), Q = annual runoff (in), and A = drainage area (mi²). The problem with application of the Dendy-Bolton function to dryland areas is that it is highly general and reflects relationships on a continent-wide

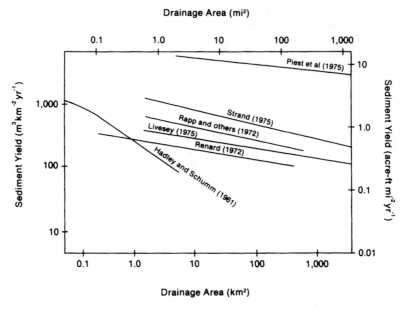

Fig. 4.6. Relationships between sediment yield per unit area and total drainage basin area. (Expanded from a diagram by Branson et al. 1981, p. 146, who converted some original data from weight to volume)

basis. It is not likely to be useful for predicton of individual basin yields. Renard (1977) found that modification of the function to meet local dryland relationships between runoff and drainage area improved its applicability.

Renard (1972) proposed a combined model to predict sediment yield from small dryland drainage basins. The model combined a stochastically defined runoff component with a deterministic sediment transport relationship. Several experimental watersheds in southern Arizona provided data for calibration, which resulted in a simple statistical relationship

$$Y = 0.00186 \, A_a^{-0.1187}, \tag{4.13}$$

where Y = sediment yield (acre ft acre^{-1}yr^{-1}) and A_a = drainage area (acres). The relationship reflects by the negative exponent the impact of transmission losses and related losses in sediment transport capacity in small stream networks.

When instrumented records are available they may be used to interpolate (or more hazardously, extrapolate) relationships between sediment yield and basin area into unmeasured basins. Hadley and Schumm (1961) established that sediment yield per unit area declined as total basin area increases in their analysis of the Cheyenne River drainage in semiarid eastern Wyoming. Their analysis extended to subbasins of up to about 5.2 km^2 (2 mi^2).

A relationship that predicts sediment yield only on the basis of drainage area is necessarily region-specific because the simple function cannot account for the effects of climate, vegetation, geology, and soils. Analyses of the yield/area relationship for a variety of environments reveals that yield per unit area always declines

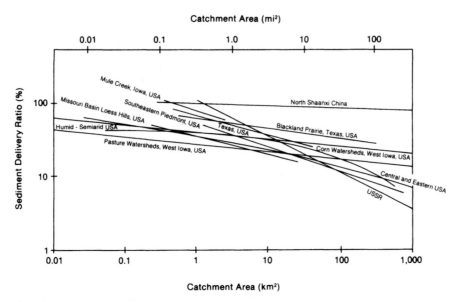

Fig. 4.7. Relationships between sediment delivery ratios and total drainage basin area. (After Branson et al. 1981, p. 143)

with increasing drainage basin size, but the rate of decline varies from one region to another (Fig. 4.6). The relationships plotted in Figure 4.6 derive from two general climatic regions. The curves by Hadley and Schumm (1961), Renard (1972), and Strand (1975) depict conditions in semiarid conditions of the American Southwest. Data from Livesey (1975) represent the upper Mississippi River basin (partly semiarid) while data by Piest et al. (1975) are from a humid-semiarid transition zone in southwestern Iowa. The data by Boyce (1975) is particularly wide-ranging, but does not include arid or extremely arid regions. The relatively high sediment yield for the Iowa data reflects the input of soils and geology because the basins are developed in highly erodible loess materials.

Although Branson et al. (1981, p. 143) characterized a subgroup of the curves shown in Fig. 4.6 as "extremely varied", the slopes of the curves in Fig. 4.6 are remarkably similar, discounting the Hadley and Schumm data, which covers a different scale range than the others. Data from Tanzania (Rapp et al. 1972) suggests wide applicability of this general rate of decline in unit yield with increasing area. The reason for the scale-related decline in yield is that sediment eroded from slopes does not immediately make its way to the basin outlet. Temporary storage on lower slopes of sediment eroded from upper slopes is common to the drylands of the American West (Leopold 1966) and Tanzania (Rapp et al. 1972). Even that material which escapes the slope system is not likely to reach the exits of large channel systems without some storage along channels (Trimble 1975). More opportunities for storage present themselves as we consider increasingly large basins.

The sediment delivery ratio is the ratio of the amount of material eroded from slopes of a drainage basin to the amount of material that exits the basin. If the ratio is equal to 1.0, the amount of material lost from slopes (soil loss) equals the amount leaving the basin as fluvial sediment yield. Sediment delivery ratios, like raw sediment yield values, decline with increasing basin area. Because the ratio is dimensionless and does not account for absolute magnitudes, there is greater likelihood that similar ratio/area relationships will obtain for different areas (Fig. 4.7). The shapes of the curves probably reflect the geomorphologic characteristics of the slopes and channels of the basins which determine the nature and distribution of storage sites. There are probably limiting curves determined by the combined possible basin and network geometries. As yet these curves are undefined, and research from dryland areas is noticeably lacking.

Of the established relationships between sediment delivery ratio and drainage basin area (Fig. 4.7), the most widely based is that by Roehl (1961, central and southwestern U.S.A.). The curve by Renfro (1975) is from data from Texas, while the ones by Piest et al. (1975) and Beer et al. (1966) are from a humid-semiarid transition area. As a general statement, the available data (which do not strongly represent dryland areas) suggest that

$$R_s = C A_d^{-0.2}, \tag{4.14}$$

where R_s = sediment delivery ratio, A_d = drainage area, and C = an empirical constant.

Most presently available data reveal sediment delivery ratios of less than 1.0, indicating varying amounts of internal storage. The maintenance of this arrangement on a geologic time scale would not be possible because sediment in storage would indefinitely increase. Those materials stored along channels as valley alluvium eventually leave the basin during intensive erosion episodes. This episodic erosion is well known in dryland settings. Schumm and Hadley (1957) established the process for small channels, and Graf (1985b) demonstrated that the cumulative effects of accelerated erosion episodes in the upper Colorado River Basin during the period 1926–1962 produced sediment delivery ratios as high as 1.5 for some dryland streams.

4.2.4 Spatial Variation of Sediment Yield

The processes of sediment yield from slopes and drainage basins have distinctive spatial characteristics. The functional relationship between sediment yield per unit area and total drainage basin area (depicted in Fig. 4.6) is

$$Y_s = a A_d^b, \tag{4.15}$$

where Y_s = sediment yield per unit area (e.g., $m^3km^{-2}yr^{-1}$), A_d = drainage basin area (e.g., km^2), and a, b = empirical constants. The absolute total yield from a basin of a particular size then is

$$Q_s = (a A_d^b)(A_d), \tag{4.16}$$

or

$$Q_s = a A_d^{(b + 1)}, \tag{4.17}$$

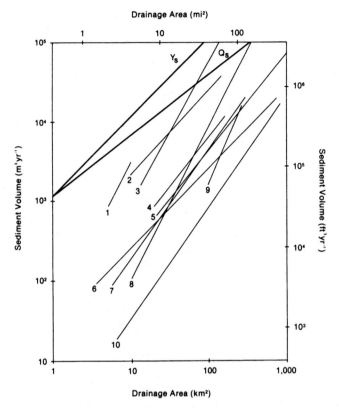

Fig. 4.8. The relationship between drainage basin area and total amount of sediment stored in a particular stratigraphic unit for ten sample basins in the Colorado Plateau, Utah and Arizona. Basins *1–3* drain high sediment-yield areas with erodible sandstones. Basin *10* includes large areas of crystalline rock. Basins *4–9* have a wide range of rock types. All the curves are similar in form to the curves representing slope sediment production (Y_s) and fluvial sediment yield from basins (Q_s). (From data associated with Graf 1987b)

where Q_s = absolute sediment yield (e.g., m³yr⁻¹). Because the units of drainage basin area are length² and the units of sediment discharge are length³, an isometric relationship would have an exponent of 1.5. Most data suggest that (4.17) is allometric, however, because the exponent is much less than 1.5. The data of Strand (1975) are representative of dryland areas in the United States and produced a value of b = −0.229, so that the exponent in (4.17) is (b + 1) = 0.771. Sediment yield increases with drainage basin area, but at a fractional rate because of internal storage.

In the simplest case, slope sediment production would increase directly with drainage basin area in an isometric form with f = 1.5 in the relationship

$$S_s = c\, A_d^f, \tag{4.18}$$

where S_s = slope sediment production (e.g., t yr⁻¹), A_d = drainage area (e.g., km²), and c, f = empirical constants. The exponent probably is less than 1.5 because sedimer̅ ̅ ̅ ̅ footslopes and because, as larger basins enter the analysis, the nature of the slopes involved changes from upland slopes only to a more mixed population that includes valley side-slopes, valley floors, and internal interfluves.

The relationships between absolute sediment yield or slope sediment production on one hand and drainage basin size on the other hand have implications for the distribution of sediment storage. A relationship between total absolute amounts of sediment stored internally in drainage basins is likely to be related to basin size by functions similar to (4.17) and (4.18) because they control the amount of materials available for storage. The resulting function is

$$D_s = i\, A_d^j, \tag{4.19}$$

where D_s = total volume of sediment stored internally in the basin (m^3yr^{-1}), A_d = drainage basin area (e.g., km^2), and i, j = empirical constants with j broadly similar to $(b + 1)$ and f above. In an analysis of ten drainage basins $10-1790\ km^2$ ($3.9-690$ mi^2) in extent in the Colorado Plateau, Graf (1987b) found that j in (4.19) had a mean value of 1.42 for a depositional unit that began accumulating about 650 BP and a value of 1.74 for another depositional unit that began accumulating about 37 BP (Fig. 4.8).

Detailed sediment budgets that account for sediment sources, storage, and yields are rare because of the difficulty in collecting complete data sets that span the entire system. In an intensive study of the sediment budget for a humid basin complex covering less than $50\ km^2$ (about $20\ mi^2$) Trimble (1983) found that the amount of sediment leaving the basin was less than 10% of the amount of sediment reaching the channels from the slopes during the 1853–1938 period. Trimble and Lund (1982) contend that after 1938 the movement of sediment in the basin was altered by conservation measures, but the amount leaving the basin was still less than 10% of the amount entering channels from slopes. In the Soviet Union, Zaslavsky (1979) found similar results for similar environments. Studies of forested basins show similar small amounts of sediment on a percentage basis exiting from the drainage basins (Dunne 1979).

The extension of these results to dryland areas is problematic. In a partly forested nearly semiarid mountain basin in California, Lehre (1982) found that more than 50% of the slope sediment production left the basin. In ten arid and semiarid Colorado River tributary basins, Graf (1987b) found that about 52% of the probable slope sediment production was being stored along channels, implying that as much as 48% was leaving the basins as fluvial sediment yield. The study was less precise than those in other areas, but it suggests that conclusions from humid basins are not necessarily transferable to dryland conditions.

On the continental scale, variation in sediment yield from drainage basins depends partly on geology and mostly on climate. Maximum sediment production appears to come from semiarid zones in concert with the first maximum on diagrams relating sediment yield to mean annual precipitation (Fig. 4.4). In a map compiled from gauging station data and a variety of documentary sources, Meade and Parker (1984) found that in the United States maximum yield areas were almost exclusively in semiarid portions of the country. Exceptions were areas with high degrees of tectonic activity, erodible loess, or with poor agricultural practices. Estimates from similar data by the U. S. Water Resources Council (1968) showed

Table 4.5. Estimated sediment yield from basins less than 260 km^2 (100 mi^2) in various parts of the United States

Area	Approximate annual rainfall (cm)	Estimated sediment yield (t km^{-2}yr^{-1})
Humid		
Columbia-North Pacific	150–200	140
South Atlantic	100–150	280
Tennessee	100–150	250
Lower Mississippi	100–150	1,820
Delta	100–150	20
North Atlantic	50–100	90
Great Lakes	50–100	40
Ohio	50–100	300
Upper Mississippi	50–100	280
Texas-Gulf	50–100	630
Drylands		
Missouri	25–50	530
Arkansas	25–50	770
Rio Grande	25–50	460
Upper Colorado	25–50	630
Lower Colorado	<25	210
Great Basin	<25	140
California	25–150	460

Note: Modified from Dunne and Leopold (1978).

semiarid areas producing the greatest amount of sediment. Arid regions produced low to moderate quantities (Table 4.5). This geographic pattern of sediment yield and climate is repeated in central Africa (Rapp 1972) and most parts of the world, especially taking into account geologic variation.

The fundamental operation of drainage basins in the generation, storage, and yielding of sediment at scales less than continental represents a major challenge to fluvial specialists working in drylands. Simple budget-related data are lacking for most dryland regions of the world, and the development of suitable theoretical explanations for the amounts and locations of storage are in embryonic stages of development (Sect. 5.5). Preliminary results cited above suggest that relatively less sediment is stored internally in dryland basins than in humid basins, but the firm establishment of even this broad generalization and full explanation of it awaits further research.

From the spatial perspective, the notion of the variable source area concept appears to be applicable to conceptualizing sediment processes as well as water runoff processes. During any given runoff event, sediment enters the channel system from only a limited portion of the total drainage basin. The contributing area for sediment is probably larger for events of longer return intervals, so that the sediment entering channels and ultimately stored or transported from the basin has a complex set of origins.

4.3 Sediment Transport in Dryland Rivers

The movement of sediment through dryland fluvial systems might be viewed from an areal perspective as in the case of sediment yield from slope units or from drainage basins, or alternatively it might be considered as a mechanical process taking place at a series of cross sections in a channel network. The following section explores the geomorphologic implications of sediment transport in dryland river channels by reviewing general principles, suspended sediment processes, bedload processes, total load considerations, and some spatial aspects of sediment transport. The section does not represent an attempt to duplicate widely available engineering treatments of sediment transport, but rather seeks to critically analyze the available methods in view of possible applications to dryland research problems. For detailed expositions of the fundamental concepts and mathematical models, see Graf (1971); Shen (1971, 1979); Garde and Raju (1977); Simons and Senturk (1977); Bogardi (1978); Stelczer (1981). Of these sources the one by Walter Hans Graf (1971), a researcher at Ecole Polytechnique Federale in Lausanne, Switzerland, is probably the most readable for geomorphologists.

4.3.1 Sediment in Dryland Rivers

Because of the high rates of erosion and sediment yield in basins with semiarid climates (in addition to high rates in some other climatic circumstances – see Sect. 4.2), rivers draining semiarid regions have high sediment loads. In a review of representative world rivers, Holeman (1968) found that the greatest sediment load

Table 4.6. Sediment yield from major world rivers

River	Area (km^2)	Total yield (10^6 tonnes)	Production (t km^{-2})
Yellow, China	673	1,890	2,640
Ganges, India	956	1,450	1,400
Bramaputra, Bangledesh	666	730	1,300
Yangtze, China	1,943	500	490
Indus, Pakistan	969	440	455
Ching, China	57	410	7,180
Amazon, Brazil	5,776	360	60
Mississippi, USA	3,222	312	100
Irrawaddy, Burma	430	300	820
Missouri, USA	1,370	220	160
Lo, China	26	190	7,070
Kosi, India	62	170	2,800
Mekong, Thailand	795	170	430
Colorado, USA	637	140	380
Red, Viet Nam	119	130	1,080
Nile, Egypt	2,979	110	40

Note: Data from Holeman (1968, p. 738).

Table 4.7. Relationship between bedload and total load

Concentration Suspended load (ppm)	Channel Bed material	Nature of Suspended Material	Portion of Total load as bedload
< 1,000	Sand	Sand	0.60
< 1,000	Gravel	Small amount of sand	0.11
1,000 – 7,500	Sand	Sand	0.26
1,000 – 7,500	Gravel	25% sand or clay	0.11
> 7,500	Sand	Sand	0.13
> 7,500	Gravel	25% or less sand	0.07

Note: Adapted from Lane and Borland (1951), who based their work on efforts by Maddock. The final column lists only maximum values and assumes that the unmeasured load in suspended sampling efforts equals bedload.

was transported by the Yellow River of China which drains semiarid loess-based terrain (Table 4.6). Other major sediment transporters derive water from mountain source areas and flow through dryland areas which contribute large quantities of sediment. Because of its size, extent of dryland and seasonal precipitation areas, extent of erodible materials, and highland water sources, Asia contributes up to 80% of the sediment reaching the world's oceans (Holeman 1968, p. 745).

Four general processes characterize sediment transport in river channels: flotation, suspension, saltation, and traction, with the latter two usually considered together as bedload. Relatively little is known about flotation, especially in dryland areas, but its significance is minor in comparison with the other processes. Air-borne dust undoubtedly descends onto flowing water surfaces and moves through fluvial systems while temporarily floating, but the magnitude involved is small compared to suspended and bedload transport.

Table 4.8 a. Percentages of total load as suspended load for dryland rivers in the USA

River	Mean particle size (mm)	% Total as suspended
Rio Grande at San Marcial, New Mexico	–	0.86
Five Mile Creek near Riverton, Wyoming	0.24	0.81
Colorado River at Yuma, Arizona	0.10	0.80
Moore Creek above Granite Creek, Idaho	0.25	0.75
Snake River near Burge, Nebraska	0.29	0.67
Boise River near Boise, Idaho	0.10	0.65
Middle Loup River at Dunning, Nebraska	0.33	0.53
Cour d'Alene at Rose Lake, Idaho	–	0.49
Niobrara River near Cody, Nebraska	0.30	0.49
Niobrara River near Valentine, Nebraska	0.27	0.47

Note: Table adapted from Garde and Raju (1977, p. 262), data from Stevens (1936), Benedict and Matejka (1952), and Love and Benedict (1948).

Table 4.8 b. Ratios of maximum to minimum discharges

River	Maximum/Minimum
Ujh at Chak Basti, India	2,400.0
Ravi at Madjopur, India	411.0
Sutlej at Rupar, India	133.0
Mahanadi at Naraj, India	108.0
Irrawadi at Saiktha, Burma	48.8
Indus at Kalabah, Pakistan	53.0
Nile at Cairo, Egypt	48.5
Rhone below the Durance, France	38.0
Mekong at Kratie, Cambodia	35.3
Ganga at Farakka, India	34.6
Volga at Rybinsk, USSR	32.0
Columbia at Portland, USA	28.7
Mississippi at St. Paul, USA	23.5
Danube at the delta, Rumania	14.3
Elbe at Hamburg, Federal Republic of Germany	7.8

Note: Adapted from Garde and Ranga Raju (1977, p. 341) based on data from
UNESCO (1950).

Although suspended load and bedload account for almost all the sediment transported in most fluvial systems (solute loads are usually minimal), the relative importance of each process is not the same from one system to another. The ratio of suspended load to total load varies from nearly 1.0 to less than 0.33 (Table 4.7) depending on the types of material in suspension and on the bed (Lane and Borland 1951). Generally, large through-flowing perennial streams carry large percentages of their loads in suspension while the smaller ephemeral tributaries carry relatively small percentages in suspension and bedload is the dominant process (Table 4.8). In extremely arid regions exemplified by the Nahal Yael watershed in southern Israel, bedload may account for as much as 87% of the total load (Lekach and Schick 1980, p. 39).

4.3.2 Suspended Sediment Transport

Suspended sediment in river channels varies over space and through time in response to varying characteristics of the sedimentary particles and of the flow which transports them. Dryland rivers (and those subject to highly seasonal runoff) present additional predictive problems because of their high variability (Table 4.8). Understanding of the suspended load is further complicated by its two origins: (1) wash load derived from direct contributions from slopes, and (2) material derived from the bed and banks. This distinction is especially important in dryland rivers because the input from the first source is likely to be highly discontinuous over time and space, while the second is not. In the Rio Grande, New Mexico, Nordin and Beverage (1965) found that the suspended load had three maxima for particle sizes. Two were in the clay-silt range and derived from

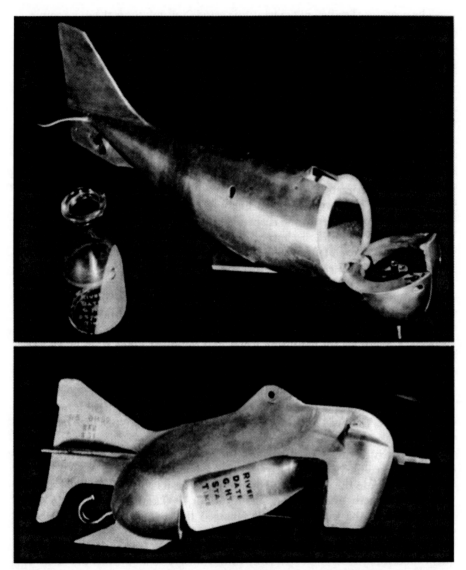

Fig. 4.9. Examples of field samplers for suspended sediment. *Upper*, US D-49 depth integrating device that is lowered into the flow from a reel and cable. *Lower*, US DH-59 sampler designed for hand-held use. (Photos by US Geological Survey)

wash load while the third corresponded to the sand bed and banks of the channel. Similar results appeared in a study of small (less than 5 km²) watersheds in extremely arid southern Israel. Lekach and Schick (1982) found that in 82 flood flows, when suspended concentrations were below an apparent threshold of 30,000 ppm, about two-thirds of the sediment was washload-derived clay and silt. In flows with concentrations above the 30,000 ppm threshold, almost all the added material was sand derived from bed and bank erosion.

The product of sediment concentration and flow volume is the most common measure of suspended load. Concentrations measured in mg l^{-1} or ppm (a weight measure) are interchangeable up to about 7,000 mg l^{-1} (Richards 1982, p. 90). The concentration of suspended sediment is directly related to discharge by the function

$$c_s = a \, Q^b, \tag{4.20}$$

where c_s = sediment concentration (ppm or mg l^{-1}), Q = flow discharge ($m^3 s^{-1}$), and a, b = empirical constants, with b usually between 1 and 2, implying that increasing discharges result in increasing concentrations. In dryland streams this is a likely arrangement, since high discharges also have higher stages with an increased wetted perimeter and greater opportunity to entrain materials. In dryland areas increased discharges are also likely to be associated directly with slope runoff which contributes to the suspended load by wash load additions. Unlike streams in humid regions, discharge does not continuously increase with distance downstream because of transmission losses, so discharge and distance are not simple direct substitutes for each other in dryland applications.

Measurement of suspended load consists of collection of water and sediment samples from a variety of points in the cross section of the flow in the field, determining the mean concentration of sediment using laboratory techniques, and multiplying the concentration times the simultaneously measured flow volumes. Containers for capturing water and sediment range from small hand-held devices for waded streams to heavily weighted, streamlined containers for suspension from cables into large rivers (Fig. 4.9). Guy and Norman (1982) provide complete descriptions and instructions for use. Some sampling systems use pumps (Walling and Teed 1971), turbidity meters based on light transmission (Fleming 1969; Truhlar 1978), or x-ray penetration of the dynamic flow (McHenry et al. 1970).

Instrument changes pose hazards for researchers relying on instrumented records for sediment load in rivers. The sediment transport record for the Colorado River in the Grand Canyon, for example, has three prominent sections with divisions between sections occurring about 1930 and 1943. In 1930 a minor modification in the sampling bottle changed the manner in which the bottle opened to admit water and sediment (Howard 1947). In 1943 a streamlined collecting device replaced bottles in relatively unwieldy frames. Although subsequent laboratory tests indicated only minor differences between the latter two samplers (Benedict and Nelson 1944), implications of the changes for field use are unknown, and the resulting sediment transport record is at least questionable and at most in serious error.

The vertical distribution of suspended sediment in the flowing water depends on the particle size (which influences settling velocities) and turbulence. When the particles are small and/or the turbulence great, the distribution of suspended sediment is relatively uniform throughout the depth of flow. If the particles are large or the flow not relatively turbulent, a relatively steep gradient of sediment concentrations results. When this gradient is steep, suspended sediment measurements are difficult because devices cannot adequately sample close to the bed where the high concentrations are located.

As an example, the location of the sampler in the Colorado River case may in part explain the change in the sediment load record about 1930. After 1930 the annual measured sediment load was only about half the pre-1930 values. Before 1930 operators lowered the sample bottles for collection of suspended sediment to bed level, whereas after 1930 operators lowered the bottles only part of the vertical distance from water surface to bed (Howard 1947). It is possible that the early samples showed higher values simply because that portion of the suspended load near the bed that may have been of high concentration was included. In post-1930 samples the high-concentration parts of the vertical profile were excluded, perhaps leading to lower measurements.

Three models of instantaneous suspended sediment transport in channels are likely to be useful to the geomorphologist: a simple power function, Lane's model, and Einstein's model. The simple power function is

$$G_s = a \, Q^j, \tag{4.21}$$

where G_s = suspended sediment discharge, Q = water discharge, and a, j = empirical constants [j = b from (4.20) plus 1.0]. Leopold and Maddock (1953) found from a wide range of sample sites (including dryland sites) that generally $2 < j < 3$. For dryland rivers the value of j is probably close to 2 since reported j values include 2.16 for the Missouri River in Kansas (Campbell and Bauder 1940), 2.07 for the Red River in the central USA (Stanley 1948), about 2.00 for the Colorado River in the southwestern USA (Leopold and Maddock 1953), and 2.00 for the Chenab River at Khanki, Pakistan (Mao and Rice 1963). The coefficient a must derive from local measurements, but the function is probably accurate enough for regional geomorphologic investigations.

Lane et al. (1941) provided a much more highly detailed model for suspended sediment transport than the general power function model. They proposed that the diffusion of sediment is constant throughout the vertical profile of water flow and that the suspended load rate per unit time and width is

$$g_{ss} = q \, C_a \, P_L \, e^{[(15v_{ss}a)][(Du*)^{-1}]}, \tag{4.22}$$

where g_{ss} = suspended-load rate per unit time and width, q = water discharge, C_a = sediment concentration at vertical elevation a off the bed, P_L = a factor related to particle size, bed roughness, particle settling velocity, and velocity of flow (Lane 1941, provide a graphical solution), e = base of the natural logarithm system, v_{ss} = grain size of suspended particles, a = reference depth where the sample concentration was measured, D = diameter of bed particles, and u* = shear velocity. Equation (4.22) describes the rate of sediment transport for only one particle size, and must be recalculated for the full range of observed particle sizes. From the geomorphological perspective with its demands for sampling in remote locations, the method has at least one major advantage in that it has a built-in assumption concerning the vertical distribution of sediment and therefore requires sampling of the sediment concentration at one elevation, a, off the bed. Application to dryland rivers may be appropriate in view of empirical tests of the formula by its originators on many wide, shallow streams of the type frequently found in drylands.

The most detailed model of suspended load discharge is that by Einstein (1950). It includes relatively few simplifying assumptions and involves numerical solutions of complicated integrals. The basis of the method is the function

$$g_{ss} = 11.6 \, C_a \, u_*' \, a \, [2.303 \log (30.2 \, D)(n)^{-1} \, I_1 + I_2], \qquad (4.23)$$

where g_{ss} = suspended-load rate per unit time and width for a given size fraction v_{ss}, C_a = measured concentration, u_*' = shear velocity due to grains only, a = reference depth from which the concentration sample was drawn, D = depth of flow, n = a roughness parameter provided by Einstein (1950), and I_1, I_2 = factors calculated using nomograms supplied by Einstein (1950) and reproduced by Graf (1971, pp. 192–193). The Einstein function provides a physically based model with some stochastic aspects to describe suspended sediment loads, but because of its detailed nature its data demands are substantial. Computationally, it is easily adapted to computerized operations, so the geomorphologist may find the function useful if the data is available.

At this point it may be prudent to address the question of accuracy versus precision in using predictive models in dryland fluvial geomorphologic research. The simple power function model for suspended load is not likely to produce predicted values that precisely match actual values. The power function model may therefore not be precise, but because it makes few assumptions, requires little questionable measurements, and is statistical in nature, it may provide accurate mean values when applied to large populations. The Lane and Einstein models produce supposedly highly accurate values that may be close to observed values in controlled situations. When these latter functions apply to natural rivers in dryland environments, however, their highly precise predictions may be far off the true mean values for a population because of (1) the excessive data quality demands that cannot be met in field conditions, and (2) the range of field values for some variables that exceed the design range of the original applications of the models.

4.3.3 Bedload Sediment Transport

Analysis of suspended load for streams provides an incomplete view of the total sediment transport for those streams that have their origins in drylands. Unlike the large, throughflowing streams, these regional rivers are almost always ephemeral and burdened with large quantities of sediment of highly variable sizes. Because bedload forms a small percentage of the total load of most humid-region rivers where stream gauging techniques have generally developed, understanding of bedload processes and methods of measurement in dryland rivers have lagged behind similar efforts directed toward the suspended load. After briefly reviewing bedload measurement, the following section addresses several conceptual models of bedload transport that may be of use to the geomorphologist. The models include those based on shear stress, flow parameters, fluid mechanics, and basic physical principles.

Collection of bedload samples is complicated by the unstable nature of many natural streambeds and by the disturbance of the transport process by the sampling instrument. Instrument designs to minimize these problems include boxes, trays, pressure-difference approaches, slots or pits, and tracing. Early box samplers, known in Europe as Nesper and Ehrenberger samplers (Hubbell 1964), had open sides facing upstream and open tops, with solid bottoms, backs, and sides. Considerable material appears to be transported over the backs of the boxes, so wire basket traps are more effective. The trapping efficiency of wire baskets which include an open mouth and cage-like top, back, and sides is still only about 45% for particles up to about 250 mm (10 in) (Hubbell 1964). The disruption of flow processes by simple boxes and baskets makes them difficult to use, but if their efficiency can be accurately calibrated they may be useful in exploring in a preliminary way bedload transport in relatively steady, low flow conditions.

Pan or tray samplers have gently sloping upper surfaces which act like ramps. The bedload material moves up the ramp and falls through a slot-like opening into the interior of the sampler. Losiebsky or Polyakv samplers (Garde and Ranga Raju 1977, p. 272) are pan or tray designs in common use in the Soviet Union. Their efficiency is below 50% but they may also be useful for low bed discharges and low velocities (Einstein 1948).

In order to increase the efficiency of bedload samplers, design improvements have included a narrowing of the entrance chamber in order to accelerate the inflow by means of the Venturi effect. The Goncharov and Arnhem (or Dutch) samplers consist of a rigid rectangular entrance that widens slightly before it connects to a rubber neck. The neck conducts sediment to a mesh collecting basket, with the entire apparatus stabilized by a rectangular framework which rests on the bed and a vertical fin to maintain orientation toward the flow. Trap efficiency is about 70%.

The Helley-Smith sampler is probably most widely used in the United States and represents a modification of the Arnhem device. Helley and Smith (1971) adapted the rectangular collector, widening conductor, and mesh sample bag to a streamlined framework resembling a metal fish. Vertical and horizontal fins provide stablity. The efficiency of the device is greater than 100% for particles in the 0.20–0.50 mm range, about 100% for the 0.50–16.0 mm range, and about 70% for particles greater than 16 mm (Emmett 1979). The utility of the sampler in field applications depends on the particle sizes in transport in the study stream.

More ambitious (and costly) efforts at measurement of bedload include the excavation of slots or trenches across the channel. Slot width must be at least 40 and preferably 100 to 200 times the diameter of the particles to be trapped (Einstein 1948). Leopold and Emmett (1976), sponsored by the U. S. Geological Survey, have developed an extensive slot sampling system on the East Fork River in semiarid Wyoming. The slot is 0.25 m wide and extends 14.6 m across the channel. Bed material falls through the slot onto an endless belt which conducts the collected material to a sump in the bank. A system of belt-mounted buckets collects the sediment, lifts it above the bank, and dumps it into a hopper for weighing. Although the system collects all the bedload in transport (Klingeman and Emmett 1982), its level of sophistication and expense limit its use to experimental work at optimal sites rather than general geomorphologic sampling.

Tracing of individual particles in the bedload is relatively simple in ephemeral streams because each particle can be marked with paint and then relocated after a flow (Leopold et al. 1964). A major problem with using visual markers is that many of the transported particles become buried after the flow and cannot be recovered. Of potentially greater utility is a magnetic marking scheme whereby ceramic magnets inserted into drilled holes identify individual particles. Ergenzinger and Conraday (1982), Ergenzinger (1982), and Ergenzinger and Custer (1983) used a fixed device to measure the movement of tagged particles past a given location during transport in semiarid Montana. In an alternative approach, Hassan et al. (1984) and Schick et al. (1985) tagged particles in an extremely arid watershed near Beer Sheba, Israel, during a period when the channel had no flow. They relocated the particles with a magnetic locator device after a flow. Their method was a valuable improvement because it showed that it was possible to relocate 93% of the original particles, and that 53% were buried below the surface. Use of visual methods would have resulted in loss of more than half of the sample particles.

Most mathematical and statistical models of bedload transport rely on laboratory flume experiments rather than field investigations. One of the earliest attempts at generalization of bedload processes was by DuBoys (1879), who began with the premise that

$$\tau_0 = \gamma \, D \, S, \qquad\qquad\qquad (4.24)$$

where τ_0 = shear stress or tractive force as DuBoys termed it per unit area of the bed $(N \, m^{-2})$, γ = unit weight of water $(9,807 \, N \, m^{-3})$, D = depth of flow (m), and S = gradient. DuBoys reasoned that the amount of bedload sediment transported depended on velocity shear at the base of the flow, the difference between the velocity of flow at the bed and the maximum velocity of flow in the water. He assumed that the sediment in the deep bed was stationary and that the overlying sediment moved in sheets at increasing velocities until in the maximum velocity section of the flow moved the particles at the same velocity as the water. He suggested that particle motion began at that point at which shear stress was great enough to overcome some critical value. The resulting fundamental statement by DuBoys on bedload transport rates became

$$q_s = \chi \, \tau_0(\tau_0 - \tau_{cr}), \qquad\qquad\qquad (4.25)$$

where q_s = bedload sediment discharge $(N \, m^{-2})$, χ = a characteristic sediment coefficient related to the thickness of the layers of sediment and the critical shear stress or tractive force required for particle movement (m), τ_0 = shear stress or tractive force generated by the flowing water $(N \, m^{-2})$, and τ_{cr} = critical shear stress or tractive force, the amount needed to initiate motion of the particles $(N \, m^{-2})$. Donat (1929) used a different line of reasoning but derived the same equation. Schoklitsch (1914), Donat (1929), Straub (1935), Chang (1939), and U. S. Waterways Experiment Station (1935) derived empirical functions for the definition of chi.

The DuBoys function (4.24) appears to be theoretically incomplete. Many subsequent researchers found that sediment does not move in the form of sliding layers as he suggested (Schoklitsch 1914). Further, an implicit assumption in the model is that all the available potential energy in the flow is expended in overcoming the resistance of friction at the bed, but it is now known that energy is also expended in internal friction, vortices, air resistance, and bank friction (Bogardi

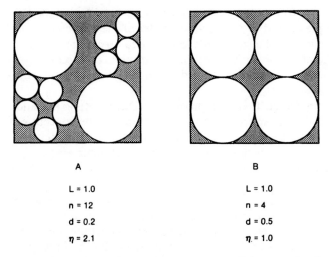

A

L = 1.0

n = 12

d = 0.2

η = 2.1

B

L = 1.0

n = 4

d = 0.5

η = 1.0

Fig. 4.10 A, B. Examples of White's packing coefficient showing the influence of particle size and arrangement on the value of the coefficient. Text contains definitions of symbols and their functions. Note that loosely packed large particles have a lower packing coefficient than numerous small particles nestled among a few large ones

1978, p. 83). The simplicity of the approach and the sometimes close agreement with flume and field data apparently explain the popularity of the DuBoys function among researchers: it has been the most widely used model.

In some cases the geomorphologist is concerned with identification of those places or those times at which the fluvial system crosses a threshold of instability. The basic DuBoys equation [Eq. (4.24)] provides a convenient measure of the amount of shear stress or tractive force (τ_0) exerted by the water flow on bed materials. The calculation of τ_{cr} provides an assessment of the resistance offered by the bed particles, when the available stress exceeds the available resistance, particle motion and geomorphic instability result. Critical shear stress for a wide range of particle sizes loosely arranged is

$$\tau_{cr} = 1.78 \ d, \tag{4.26}$$

where τ_{cr} = critical shear stress or tractive force (N m^{-2}) and d = particle diameter (mm) (Church 1978).

A more complex definition of critical shear stress or tractive force is that by White (1940):

$$\tau_{cr} = (\alpha \ \eta \pi \rho_1 \ d \tan \varphi)(6)^{-1}, \tag{4.27}$$

where τ_{cr} = critical shear stress, α = a constant (ranging from 0.3–0.4), η = a packing coefficient, π = the constant, ρ_1 = density of the sedimentary particles, d = mean diameter of the particles, and φ = angle of repose of the bed particles, usually about 45°. The packing coefficient is related to particle diameter and arrangement on the bed:

$$\eta = (n \ d^2)(L^{-1}), \tag{4.28}$$

where (in reference to Fig. 4.10) η = the packing coefficient, n = the number of particles in a square sample area, d = the mean diameter of the particles in the sample area, and L = the length of a side of the sample area. From the geomorphological perspective the utility of White's approach is its ability to account for this packing or arrangement of particles on the bed which may display significant spatial variation.

Schoklitsch (1926) and Donat (1929) improved the DuBoys function and generated an entire family of transport functions that relied on the concept of excess power in one form or another. Many of the functions rely on the velocity of flow as the primary predictor of bedload transport rates. A later version in nondimensional form by Barekyan (1962) is

$$q_s = 0.187 \, \gamma \, [(\gamma_s)(\gamma_s - \gamma)^{-1}]qS[(u - u_{cr})(u_{cr})^{-1}], \tag{4.29}$$

where q_s = bedload discharge rate, γ = unit weight of water, γ_s = unit weight of sediment, q = water discharge, S = energy gradient, u = velocity of flow, and u_{cr} = critical velocity of flow at which particle motion begins. According to Barekyan (1962) predictions with (4.29) compared favorably with data collected in the Soviet Union and with data by Gilbert (1914), but Simons et al. (1965) reported more limited success.

Meyer-Peter and Muller (1948) proposed further modifications of the Schoklitsch approach after more than a decade and a half of experimentation. Their complex function includes accountings for grain resistance and bedform resistance and relies in part on the difference between the unit weights of water and sediment. The predictive function is

$$[\gamma \, R_h(k/k')^{3/2}S][d(\gamma_s - \gamma)]^{-1} - 0.047 = [0.25(\rho)^{0.33}(g_s')^{2/3}][d(\gamma_s - \gamma)], \tag{4.30}$$

where γ = unit weight of water, γ_s = unit weight of the sediment, R_h = hydraulic radius, k/k' = a measure of the size of bedforms (ranges between 0.5 for strongly developed bedforms and 1.0 for a planar bed), S = energy gradient, d = mean grain size, ρ = density of the fluid, and g_s' = the rate of sediment transport. The sediment transport in terms of weight per unit width of the channel per unit time (g_s) is

$$g_s = g_s'[(\gamma_s)(\gamma_s - \gamma)]. \tag{4.31}$$

The originators of the function did not test it in field conditions in dryland rivers, but Meyer-Peter (1949, 1951) addressed the special aspects of alpine and subalpine rivers which may have some bearing on potential dryland applications. The function has the added advantage that it has been tested for large grains and is dimensionally homogeneous (Graf 1971, p. 157).

Einstein (1950) approached the problem of bedload transport from a statistical perspective. His guiding principles were the following experimental results regarding bedload (Graf 1971, p. 140):

(1) A steady and intensive exchange of particles exists between bedload and the channel bed.
(2) Bedload moves slowly downstream, while the movement of individual particles is by quick steps with long intermediate stops.
(3) The average step made by a particle is always the same and appears to be independent of flow condition, transport rate, and bed composition.

(4) Different transport rates result from different average time between steps and different thicknesses of the moving layer.

From these precepts Einstein (1950) developed complex probabilistic functions to describe the motion of average particles. These mean transport conditions then combine in a forcast of total transport rates. Graf (1971, p. 139–153) provided a summary of the development of the functions and critical comments; Simons and Senturk (1977, p. 610–615) provided an extensive worked example. The statistical nature of the Einstein approach makes it attractive for geomorphological use, but its complexity and numerous assumptions concerning processes in the field situation seem to preclude its wide application.

4.3.4 Total Load Estimations

Total load estimations result by summing calculations for suspended load and bedload using formulas similar to those sampled in the previous sections. Total load formulae are also available: the simplest is probably Bagnold's (1966):

$$i = \omega \left[(e_b)/(\tan \alpha) + 0.01(u/v_{ss}) \right], \tag{4.32}$$

where i = total load transport rate, ω = unit stream power, e_b = a bedload efficiency factor (defined by Bagnold 1966, who also provided a nomogram for specification based on flow velocity and particle size; values usually range from 0.11 to 0.15), $\tan \alpha$ = coefficient of solid friction, u = mean velocity of flow, and v_{ss} = settling velocity of the particles. Bagnold also provided a nomogram for solution of the value of $\tan \alpha$ based on particle size, particle density, and shear stress.

Bagnold's total load formulation has much to recommend its use by geomorphologists. Its derivation is from first physical principles, so that the formula contains some explanation in the form of energy being transferred into the work of sediment transport. It is relatively simple to apply and has readily identifiable bedload and suspended load components. Application to river data has been successful. Reservations on the use of the Bagnold method include the underlying assumption that the velocity of flow and velocity of particle movement are the same. Researchers must also contend with the limitation that the formula appears not to be applicable for particles less than 0.015 mm in diameter.

4.3.5 Definitions and Dimensions

One of the more difficult aspects of sediment transport calculations, especially those involving stream power relationships, is the question of definitions and dimensions. An example is DuBoys' use of the term tractive force to describe his basic equation. The equation does not actually describe force, and the units of the equation are the same as those for shear stress. The term tractive force is embedded in the literature, however, and is likely to see continued use. A similar difficulty arises with regard to Bagnold's work, in which dynamic transport rates

Table 4.9. Conversions to SI units common in fluvial geomorphology

Conventional units	System international units
Length	
SI: meter	
inch (in)	0.0254 m
foot (ft)	0.3048 m
yard (yd)	0.9144 m
rod (rd)	5.0292 m
chain (ch)	20.1168 m
engineer's chain	30.48 m
mile (mi)	1,609.344 m
nautical mile (n.mi)	1,852.00 m
Area	
SI: square centimeter, square meter, square kilometer	
square inches (in^2)	6.45 cm^2
square feet (ft^2)	0.093 m^2
square yards (yd^2)	0.836 m^2
acres	4047 m^2
square miles (mi^2)	2.59 km^2
Volume	
SI: cubic meter	
cubic inch (in^3)	16.3871 x 10^{-6} m^3
cubic foot (ft^3)	28.3168 x 10^{-3} m^3
cubic yard (yd^3)	0.7646 m^3
US gallon (gal/US)	3.7854 x 10^{-3} m^3
UK gallon (gal/UK)	4.5406 x 10^{-3} m^3
acre foot (ac ft)	1,236 m^3
Mass	
SI: kilogram	
ounce (oz)	28.350 x 10^{-3} kg
pound (lb)	0.4536 kg
ton (t)	907.185 kg
slug	14.594 kg
Velocity	
SI: meter per second	
mile per hour (mi hr^{-1})	0.4470 m s^{-1}

have dimensions and quality of work rates. In fact they are not work rates, because the principal equations combine stress and velocity even though they are not exercised in the same direction (Graf 1971, p. 209).

Terminology aside, dimensions and units of measure bedevil the fluvial geomorphologist because the literature includes three different systems: metric engineering, System International (SI), and British (or American) engineering. There is a general shift among the sciences to SI units, hence their dominant use

Table 4.9 (continued)

Conventional units	System international units

Density

SI: kilogram per cubic meter

pound per cubic foot (lb ft^{-3})	16.0185	kg m^{-3}
slug per cubic foot	515.4	kg m^{-3}
pound per cubic inch (lb in^{-3})	27,680.0	kg m^{-3}
ounce per cubic foot (oz ft^{-3})	1.0011	kg m^{-3}

Force

SI: newton

tonne (t)	9,806.65	N
dyne	10^{-5}	N
foot pound (ft lb)	4.448	N

Specific Gravity

SI: newton per cubic meter

kilogram force per cubic meter (kg m^{-3})	9.807	N m^{-3}
dyne per cubic centimeter (dyn cm^{-3})	10	N m^{-3}
pound force per cubic foot (lbf ft^{-3})	157.087	N m^{-3}

Volume Discharge

SI: cubic meter per second

cubic foot per second (ft^3 s^{-1})	0.0283 m^3 s^{-1}
US gallon per minute (gal US min^{-1})	63.09 x 10-6 m^3 s^{-1}

Mass Discharge

SI: kilogram per second

ton per minute (t m^{-1})	16.667 kg s^{-1}
pound per second (lb s^{-1})	0.4334 kg s^{-1}

Shear Stress

SI: newton per square meter

gram per square centimeter (gm cm^{-2})	98.0665 N m^{-2}
kilogram force per square meter (kg m^{-2})	9.8067 N m^{-2}
dyne per square centimeter (dyn cm^{-2})	0.1 N m^{-2}
pound force per square inch (lbf in^{-2})	6,894.75 N m^{-2}
pound force per square foot (lbf ft^{-2})	47.8803 N m^{-2}

Note: Adapted from Stelczer (1981) which has a more complete listing.

in this volume, but much classical literature is in one of the other systems, requiring laborious conversions. Table 4.9 contains some of the more common conversions likely to be required by the geomorphologist dealing with the engineering or agricultural literature.

The most troublesome units of measure in fluvial processes are those associated with force. Carson (1971), Sumner (1978), and Stelczer (1981) provide the basis for the following points. The confusion is probably a result of fundamental

Table 4.10. Units of measure related to force

Item	System international	Metric engineering		British engineering	
Mass	1 kg	0.0685	slugs		
Density	1 gm cm^{-3}	1.940	slugs ft^{-3}		
Unit Weight	9807 N m^{-3}	1000	kg m^{-3}	62.42	lb ft^{-3}
Force	10 N	1.02	kg	2.248	lb
Pressure	1 bar	1.02	kg cm^{-2}	14.5	lb in^{-2}

Particular equivalencies (SI units)
Acceleration of gravity = 9.80665 m s^{-2}
Specific gravity of water = 9810 N m^{-3} or kg m^{-2}s^{-2}
Specific gravity of bed load = 25996.5 N m^{-3} or kg m^{-2}s^{-2}

Note: Adapted from Carson (1971) and Stelczer (1981).

differences between the measurement systems: in engineering systems pounds and kilograms are units of weight, but in the SI system kilograms refer to mass. There are also problems with editorial policies regarding the use of the term newton, which is a basic SI unit defined as the force that accelerates a stationary body of one kilogram mass in one second to the velocity of 1 meter per second. For this reason, the SI unit of stress is one newton per square meter (N m^{-2}). It is not written "per second" or "s^{-1}" because the concept of time in the sense of per second is already included in the definition of newton. For an example of correct usage, see Church (1978). Table 4.10 reviews some common associations among measures of force.

4.3.6 Movement of Sediment in Pulses

Engineering approaches to the modeling of suspended load, bedload, and total load depend on the assumption of a continuously operating process with a representative mean condition. An additional common assumption is also that the channel system is in a given regime or some sort of equilibrium so that the maximum rate of transport (transporting capacity) is a dependable representation of the instantaneous rate. A major problem with the transfer of the engineering approaches to the geomorphologic applications is that sediment in natural rivers does not appear to move at a general maximum rate over even brief periods. Although the water flow moves through representative cross sections in a continuous stream, sediment moves through the system as a series of slugs, pulses, or waves.

At the individual grain level, Einstein (1950) recognized that particles in the bedload tend to travel in a series of movements with intervening rest periods. He used this concept to construct a statistical description of the length of rest periods and the distance of travel during periods of motion. When the countless number of individual grains move by a start-stop process, the movement of the general population might result in a consistent stream of sediment, but the little research

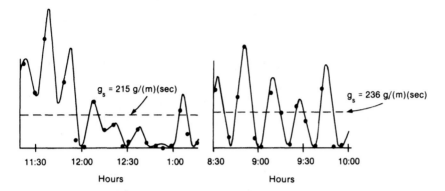

Fig. 4.11. Hourly unsteady bedload transport in the Danube showing successive pulses of approximately equal period but variable amplitude. (After Ehrenberger 1931)

reported to date indicates that sediment in mass also moves in an unsteady manner. Ehrenberger (1931) found that bedload at a measurement station in the Danube River moved as a series of highly regular waves with an almost constant period of about 20 minutes but variable amplitude (Fig. 4.11). Ehrenberger explained his observations as the products of the migration of bedforms along the channel. An additional explanation for similar observations on the Inn River by Muhlhofer (1933) and the Rhine by Nesper (1937) is the passage of successive dense clouds of particles separated by moving zones of lesser concentrations.

Even in small dryland rivers sediment moves in pulses or waves during time periods of hours. In the extremely arid southern Negev, Lekach and Schick (1983) found that during infrequent flow events the total load moved past measurement points in waves that were unrelated to variations in water discharge (Fig. 4.12). Most of the wave-like behavior involved coarse particles, but this may be a simple reflection of the domination of the total load by coarse materials. The significance of the work is that it demonstrates that the pulsating characteristics of sediment transport extend to small streams (drainage area less than 0.5 km^2).

On longer time scales in a dryland river, Meade (1985) found that sediment moved through the East Fork River, Wyoming, in three pulses over a 1-year period. The movement of each pulse was related to a pulse of water discharge resulting from melting snow packs in the mountain source area for the stream. Sediment in East Fork River is stored in pool areas separated by riffles. When discharge increases, scour excavates each pool and the material transfers across the riffle into the next pool downstream. At particular cross sections the movement of the sediment on an annual basis therefore appears as a series of waves (Fig. 4.13). Meade's work demonstrated the importance of scale of analysis in description and explanation of fluvial processes. His interpretation of sediment transport relied on annual data and his explanation addressed the issue on an annual scale. At the same measurement sites, Emmett et al. (1983) rely on traditional hydraulic engineering principles for explanation of bedload transport without reference to wave-like patterns because they deal with problems of instantaneous rates.

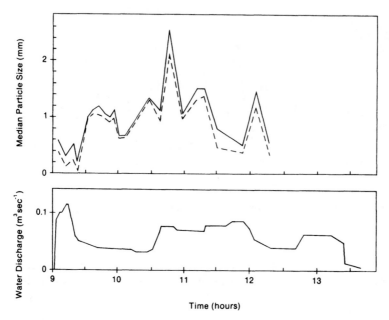

Fig. 4.12. Above, hourly unsteady total load (*solid line*) and coarse fraction load (*dashed line*) transported in a small stream in the extremely arid Negev Desert. Below, water discharge. (After Lekach and Schick 1983)

The East Fork River drains an area of about 500 km² (193 mi²), but the scour and fill processes and wave-like movement of sediment related to the annual flood also occur on larger streams. In the canyon of the San Juan River (drainage area of 23,000 km² – 8,900 mi² in Colorado, New Mexico, and Utah), Miser (1924, p. 64) found that during annual low flows sediments covered the canyon floor but that during high floods almost all the material became mobile. Leopold and Maddock (1953) documented the rise and fall of the bed and coincidental evacuation and refilling of sediments at a stream gauge site on the San Juan at Bluff, Utah. Viewed from the perspective of one or several years of observation, movement of sediment in pulses is the result.

Lowering of the bed elevation concurrently with increases in the water surface elevation is characteristic of many dryland throughflowing streams. The sediment along substantial lengths of such streams may be mobilized during the annual flood, so that nearly the entire system is occupied by a moving slug of material which is deposited on the falling stage of the flood (Lane and Borland 1953). In addition to the San Juan River example, detailed surveys of river-bed and water-surface elevations confirm this behavior on the Colorado River at Lee's Ferry with about 3 m (10 ft) of scour (Leopold et al. 1964, p. 228), the Rio Grande at San Marcial with about 18 m (60 ft) of scour (Lane and Borland 1953, p. 1072), and the Yellow River at Chiang-Kou, China, with about 3 m (10 ft) of scour (Fig. 4.14; Freeman 1922). Possibly the maximum depth of reported scour is that at the site of Hoover Dam on the Lower Colorado where excavation revealed a sawed plank 12 m (40 ft) below the bed elevation (Lane and Borland

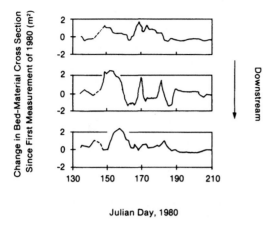

Julian Day, 1980

Fig. 4.13. Annual unsteady bedload transport in the East Fork River, Wyoming, showing successive pulses of sediment driven by pulses of water discharge. (After Meade 1985)

1953, p. 1075), but where total scour may extend to 38 m (126 ft) (Leopold et al. 1964, p. 229).

These great depths of scour and associated movement of large quantities of bed sediments during seasonal scour periods are characteristic of dryland rivers but not of humid-region streams. Leopold et al. (1964, pp. 229–230) explained the behavioral difference by speculating that ephemeral streams in drylands do not have perennial flows that winnow away fine materials. Leopold et al. suggested that dryland streams are not armored with large particles and are therefore subject to radical scour during high flows. This explanation is unlikely for at least two reasons. First, humid-region rivers frequently have beds of fine particles and yet they do not generally exhibit scour as radical as that in dryland streams. Second, dryland streams with large particles do exhibit radical scour. Graf (1983a) found in the Salt River, Arizona, that floods of 5,000 m³s⁻¹ (180,000 ft³s⁻¹) resulted in scour to a depth of 7 m (23 ft) despite the boulder-sized particles involved. A more likely explanation for the radical scour along dryland channels is the high degree of variability of the discharge. Because dryland streams must accommodate seasonal changes in discharge of one or more orders of magnitude, the only possible method of adjustment is radical scour of bed materials. Humid-region streams experience lesser degrees of discharge change, and require less radical changes in bed configuration due to scour.

Sediment also moves through channels as pulses on a decadal to century-long basis. In what is probably the best-known study of its type, Gilbert (1917) found systematic changes in bed elevations in the Sacramento River system, California, after the introduction of huge quantities of hydraulic mining debris to the channels (Fig. 4.15). He concluded that "the flood of mining debris is analogous to a flood of water in its mode of progression through a river channel. It travels in a wave..." (Gilbert 1917, p. 31). He found that in the main river there were a series of waves of sediment moving downstream, with each wave having its origin in a different tributary. Subsequent research showed that the waves were not as

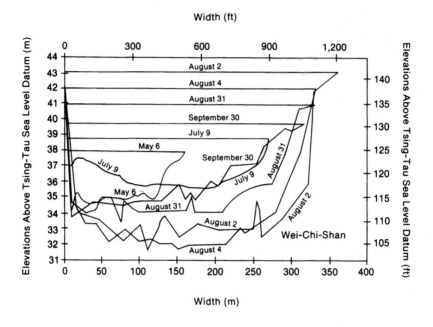

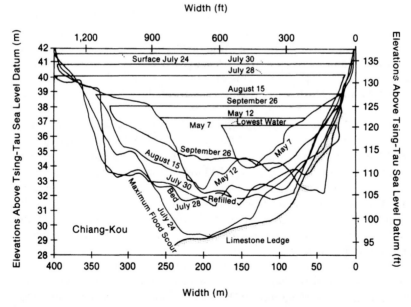

Fig. 4.14. Changes in the cross section of the Yellow River at Wei-Chia-Shan and Chaing-Kou, China, 1919, showing deep scour and subsequent redeposition and illustrating the consequences of the movement of sediment in pulses related to flood discharges. (After Freeman 1922, p. 1436, and Lane and Borland 1954, p. 1077)

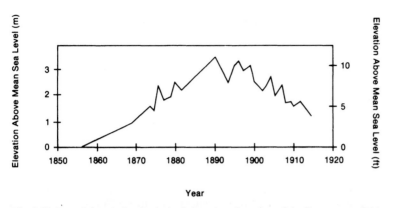

Fig. 4.15. Decadal variation in the bed elevation elevation of the Sacramento River at Sacramento showing the passage of a wave of sediments initiated by hydraulic mining in upstream areas. (After Gilbert 1917, p. 30)

symmetrical as Gilbert suspected because mining areas continued to discharge sediment into the system (James 1986). In the case of the California mining debris, the pulses are products of the time of input of materials, whereas in the previous cases the pulses were products of the inherent characteristics of the transport processes.

Pulses or waves of sediment may also appear on a decadal or century-long basis in small dryland streams (those draining less than a few hundred km²) as a result of complex responses as outlined by Schumm (1973, 1977). As the system alternatively erodes and deposits materials internally along the channels of the system, pulses of sediment output are likely.

On the time scale of millions of years, sediment discharge may again be in the form of pulses. As drainage networks expand by headward erosion, newly developed tributaries may discharge increased quantities of material into the main stem of the channel network. Field measurements of such a process are not possible, but depositional sequences might reveal it. In a laboratory simulation of drainage network development using a large sediment tank with a developing miniature network, Parker (1976) found suggestions of waves or pulses of sediment leaving the outlet of the system (Fig. 4.16).

The foregoing observations indicate that from a geomorphologic perspective, theoretical explanations of dryland river behavior regarding sediment transport that rely on steady rates of transport or yield are likely to be incomplete. At some point, models and theories must provide for pulsing behavior, though there has been relatively little research effort in this direction. There are at least three approaches that might handle pulsing behavior: kinematic wave, normal mass conservation, and dispersion models.

Langbein and Leopold (1968) proposed that kinematic wave theory provides a useful explanation for the construction and a maintenance of pool and riffle sequences. Based on transportation theory developed for vehicular traffic (Lighthill and Whitham 1955), the kinematic wave theory specifies that particles in motion in a one-dimensional system develop varying concentrations per unit

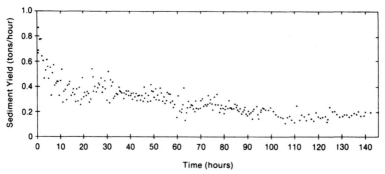

Fig. 4.16. Sediment yield in pulses from an experimental drainage network developing in an artificial landscape as measured at an outlet flume. (After R. S. Parker 1976)

distance because of mutual interference. These pulses of particles (or vehicles) define waves with characteristics that depend on the total transport rate and total linear concentration. As the linear concentration increases from zero, the transport rate increases as greater numbers of particles pass the measurement location. At some point the transport rate reaches a maximum, beyond which the linear concentration is so great that the particles interfere with each other in their motion. Further increases in linear concentration result in increasing interparticle interference and concomitant jamming that produces lower transport rates. At extremely high linear concentrations the jamming interference is so great that the transport rate declines to zero or the process completely changes (Bagnold 1954).

The graphic relationship between transport rate and linear concentration has two useful properties defined by the chord from the origin to the point on the curve and the tangent to the curve at the same point (Fig. 4.17). Any point on the curve represents the system state at a particular time. The gradient of the chord (OC) represents the average velocity of an individual particle, while the gradient of the tangent (AB) represents the velocity of the wave itself. If the gradient of the tangent is negative, the wave moves upstream in the direction opposite from the direction of motion for the individual particles.

Transformation of the theory into mathematical terms applicable to fluvial processes could be accomplished in a number of ways, but in each case the resulting curve has the general shape as depicted in Fig. 4.17. Of several functions reviewed by Haight (1963), one that is applicable for sand transport in river channels is

$$T = v_o W [1 - (W/W')], \tag{4.33}$$

where T = transport rate per unit of width, v_o = individual particle velocity (directly related to velocity of flow), W = areal concentration of material in mass per unit area, and W' = areal concentration at which transport ceases. Transport probably ceases when dunes on the channel floor become so high that they block the flow, so that the value of W' varies with depth of water flow. Experiments by Hubbell and Sayre (1964) suggest that the transport processes break down when W

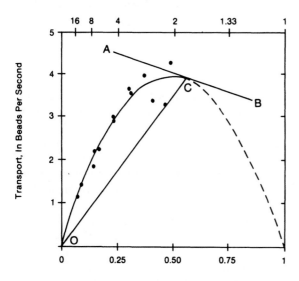

Fig. 4.17. A kinematic wave representing the measured movement of glass beads in a flume experiment with the dashed portion of the curve representing the theoretical extension. The gradient of chord OC represents the velocity of individual beads while the gradient of tangent AB represents the velocity and direction of the movement of the wave. (After Langbein and Leopold 1968, p. 5)

is about 243 kg m^{-2} (50 lb ft^2), so that value may be taken as an approximation for W'.

Kinematic wave theory is attractive for use in modeling and explaining the transport of materials in streams in pulses or waves because of its flexibility and because the analogy between sediment grains and vehicular units is intuitively attractive. Experiments with glass beads in flumes confirm the interference effects among grains in flowing water (Langbein and Leopold 1968). However, application of the theory to moving waves or pulses of sediment in the channel are rare. Most applications of kinematic wave theory to rivers focus on the distribution of pool and riffle sequences or on the morphology of dunes on the channel floor. In the case of pool and riffle sequences, the kinematic waves are virtually stationary, because in most streams the riffles are relatively immobile. According the theory, then, the transport rate must be at a maximum so that there is no wave migration (that is, the tangent in Fig. 4.17 would have a slope of zero and would be at the top of the curve). It seems unlikely that transport rates would be at a maximum in many streams because of the inefficiencies of sediment transport due to energy losses and irregularities imposed by tributary inputs. The combination of kinematic wave theory and analyses of stream power may offer explanatory possibilities.

A fundamental alternative to kinematic wave theory in explaining successive pulses of sediment in the system is reliance on the concepts of conservation of energy and mass (outlined by Bennett 1974). Rather than use detailed statements of conservation of mass and momentum, a simplified version provides the basis for analysis of sediment movement through a series of reaches (Hydraulic Engineering Center 1977):

$$w (1 - p)(dy/dt) + (dQ_s/dX) = 0, \tag{4.34}$$

where w = channel width, p = porosity of deposited sediment, dy/dt = change in depth per unit time, and dQ_s/dX = change in sediment transport (expressed as a volume) per unit distance X. Computer operations based on (4.34) tend to "blow up" when deposition or erosion increases dramatically, but Higgins (1979) used moving average techniques and limiting decision rules to control the problem. The conservation of mass approach may be useful in a predictive sense to estimate the end product of pulsating sediment movement, but it does not accommodate individual moving waves in the system that have wave lengths less than X in (4.34).

Pickup et al. (1983) developed a second alternative to kinematic wave approaches by relying on a series of dispersion functions to track individual waves as they pass through the system. The model is more descriptive than explanatory, but it proved useful in depicting the movement of mine tailings through a steep river system in Papau, New Guinea. Theoretical background for definition of some parameters of their model is lacking, so empirical calibrations provided critical definitions. Pickup et al. (1983) tested their dispersion model against observed data in the Kawerong River and found that it was 50 per cent more accurate than the continuity approach.

The continuity and dispersion approaches offer descriptions of sediment transport if the sediment injection into the stream is in pulses, but they do not offer explanations of the development of wave or pulse patterns in systems that have relatively steady supplies of sediment. The kinematic wave approach offers promise in this regard, but is largely untested. Geomorphologists dealing with dryland rivers dominated by pusating sediment transport require models and theories that are as yet undeveloped.

4.4 Contaminants

Analyses of environmental quality increasingly involve investigation of geomorphic processes as distributors of hazardous materials. The study of water quality and the dispersion of substances by flowing water lie outside the area of interest of this volume, but a brief exploration of contaminants in sediments is a logical extension of the foregoing review of sediment transport and deposition. Salinity (dissolved solids) and dissolved chemicals such as pesticides and herbicides have mobilities related to hydrologic characteristics of the surface and near-surface environment, but the mobility of heavy metals and radionuclides is more directly related to sediment transport processes. Despite this obvious connection and the direct importance of geomorphology to understanding heavy metal cycles in the environment (Fortescue 1980, p. 249), recent reviews of the subject by Whicker and Schultz (1982) and Salomons and Forstner (1984) show that geomorphologic research in the subject is generally uncommon and especially rare in drylands.

4.4.1 Heavy Metals and Placers

Heavy metals (sometimes referred to as trace elements) are important to life processes in most organisms, but only in small quantities. When the metals reach unusual concentrations they become toxic for organisms, with the toxic threshold for some metals being at low concentrations. For example, humans typically ingest about 10 to 20 mg of zinc per day and apparently require at least a few mg per day for good health. Intake of 1,000 mg, however, results in acute poisoning (Sittig 1981, p. 366). The geomorphologic connection is that fluvial processes distribute and sometimes concentrate zinc which enters the food chain through uptake by plants. Bioamplification can result in concentrations in living tissue that are 20,000 times the concentration in sediments. Zinc-laden sediments in contact with water may also contribute to high zinc levels in drinking water through solution. Heavy metals likely to occur in sediment transport processes include arsenic, boron, cadmium, chromium, copper, gold, lead, manganese, mercury, molybdenum, nickel, selenium, silver, and zinc. Adriano (1986) provides a most complete review of most of these metals in the earth-surface environment.

Heavy metals at and near the earth's surface experience two types of mobility: chemical and physical. Chemical mobility refers to the ability of metals to change compounds given varying temperature, water, and pH conditions. The chemical mobility of cadmium in soil, for example, is most completely explained by the pH of the soil (Page et al. 1979). At pH values of 6.5 or greater, plant uptake of cadmium is minimal, but as pH decreases below 6.5 increasing amounts of cadmium make their way into plant tissues. Solubility of mercury compounds changes the mobility of the element. Mercury sulfide (cinnabar) commonly occurs at the earth's surface and is highly soluble so that it readily dissolves in runoff (Hem 1970). In dryland areas this dissolved mercury then recombines with chlorine ions to form mercury chloride, which precipitates out from solution to form relatively insoluble crystals. The mercury then moves through the system physically as salts associated with sediment.

In dryland settings, physical sediment transport is more closely associated with heavy metal transport than in humid environments. First, heavy metals most frequently concentrate to the greatest degree in fine sediments (Levinson 1980). Although fine sediments occur in the loads of throughflowing dryland streams, coarse sediments dominate ephemeral streams and play a relatively larger role in explaining sediment transport in dryland streams. For this reason, investigation of only fine-grained sediments that might suffice for characterizing heavymetal transport in humid-region rivers provides only an incomplete picture for the dryland streams.

Second, heavy metals frequently concentrate in organic materials (e.g., Wershaw 1970). In dryland streams organic materials are present in throughflowing streams but are nearly absent in ephemeral streams. This arrangement implies that understanding sediment dynamics without regard to organics is possible in dryland streams.

The concentration of heavy metals in fluvial sedimentary structures produces placer deposits. The placers are of economic value if the concentrated materials are precious metals, but they represent hazardous areas if toxic heavy metals are involved. In either case, generalizations about the vertical and horizontal distributions of placers in fluvial sediments are desirable. There are three common vertical arrangements for placers likely to appear in dryland rivers: channel bottom deposits, lag deposits within alluvial accumulations, and in alluvial fans.

In an extensive investigation of the origin of placers, Cheney and Patton (1967) concluded that the most common emplacements for placers are as deposits at the contact between alluvium and underlying bedrock. For example, streams in southern Arizona concentrate heavy metals into placers that form in bedrock channels beneath the alluvium, an apparently common situation in dryland streams (Schraeder 1915). Gunn (1968) suggested that these "pay streaks" occurred because streamflow agitated the alluvial sediments causing the heavy metals to sift to the bottom of the deposits. In view of the mechanisms of bedload transport, deep scour during flooding, and the common occurrence of horizontal undisturbed bedding structures in stream alluvium, this explanation seems unlikely. Tuck (1968) argued that heavy metals concentrate on channel floors during downcutting periods, and that when the stream later begins to aggrade, burial of the heavy metals occurs under the alluvium. Shepherd and Schumm (1974) confirmed the process in a laboratory using a physical model. Given that dryland rivers scour their fills to bedrock occasionally, redeposition of the heavier materials first seems likely.

Concentrations of heavy metals also occur within the alluvial deposits beneath river channels. Scour and subsequent redeposition may not extend to bedrock and therefore may leave a lag deposit within the alluvium. These "false bottoms" are probably common in dryland streams. Lindgren (1911) observed economically significant concentrations of gold and silver in lag units 8 and 18 m (25 and 55 ft) above the alluvium-bedrock contact in streams of the Sierra Nevada, California. Wilson (1961) observed similar concentrations of copper 5 and 13 m (16 and 40 ft) above the contact near La Cholla, Arizona. The in-alluvium deposits also occur in terraces if major climatic or tectonic activities have produced adjustments in the river regime.

Alluvial fans may also be the site of heavymetal concentrations (Macke 1977), probably as a product of scour and redeposition as well as of general reworking of sediments during the development of fan-head trenches and segmented fans (Sect. 5.6). Whereever they occur, placers result from repeated reworking of sediments to sort the heavier from the lighter particles (Adams et al. 1978). Bateman (1950) suggested that because reworking is required for placer development, streams subjected to radical changes in climate or tectonic setting are unlikely to create placers. He postulated that relatively low gradient streams were likely candidates for placers, with gradients of about 0.006 as optimal. The optimal gradient probably varies depending on the hydrologic characteristics of individual streams and the particle sizes that must be sorted.

Deposition of heavy metals in overbank deposits results in concentrations of metals but usually not to the degree found in placers. In humid environments, investigators have found that heavy metals provide datable marker horizons in flood-plain sediments (Davies and Lewin 1974; Lewin et al. 1977). Similar studies of dryland rivers are generally lacking, in part because dryland rivers have much more restricted flood plains than their humid-region counterparts. In drylands valley floors have braided channels between terrace edges or have channels entrenched in alluvium, so that overbank processes are relatively less important.

At the local scale horizontal distribution of heavy metals along stream channels is highly variable. In active wide and shallow channels, heavy metals may be concentrated at the upstream ends of point-, mid-, or side-channel bars (Schumm 1977, p. 222), at the confluences of tributaries (Adams et al. 1978), or along the channel at bends, wide sections, and in scour pools (Crampton 1937). At more lengthy scales, mathematical models that include distance downstream from the source provide possible descriptions of metal distribution and concentration. Each model involves a particular theoretical perspective on the process of metal transport.

Glover (1964) proposed a series of distance decay or diffusion functions to describe the dispersion of materials suspended in flowing water. Wolfenden and Lewin (1978) found that a simple distance decay function adequately described the along-stream distribution of several metals in Welsh streams:

$$c = e^{[a - b(d_{ij})]}, \tag{4.35}$$

where c = metal concentration in sediments, e = base of the natural logarithm system, d_{ij} = the downstream distance between the source at i and the measurement point at j, and a, b = empirical constants. The underlying assumption for the distance decay function is that there is a continuous decrease in concentration as the heavy metals drop out of the transport process, and the remainder are diluted with uncontaminated sediments.

In a slight variation, Wertz (1949) proposed that metal concentrations declined at a more gradually decreasing rate described by a logarithmic function:

$$c = \ln a - b \ln (d_{ij}), \tag{4.36}$$

with symbols as before. Another alternative uses a power function to describe more precipitous declines in concentration with distance (Glover 1964; Marcus 1983, for copper in an arid-region stream):

$$c = a (d_{ij})^{-b}, \tag{4.37}$$

with symbols as before.

Two problems exist regarding the distance-related descriptions of heavy metals along streams. First, almost all the investigations are for humid-region systems. The work of Lewin et al. (1977) and Wolfenden and Lewin (1977, 1978) represents pioneer efforts that suggest possible generalizations, but the extention of results to dryland rivers remains untested. The second related problem is that sediment, especially bedload, in dryland rivers moves in pulses or waves, and none of the generalizations outlined above is particularly well suited to accommodate pulsating behavior. Present on-going research at Arizona State University is exploring the utility of polynomial and Fourier functions linked to dynamic process models as distance-related explanations of metal concentrations.

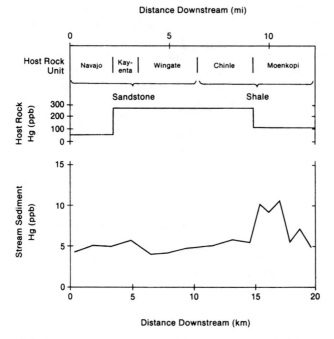

Fig. 4.18. The downstream distribution of mercury from the channel of North Wash, Utah, showing mercury concentration in bed sediments and concentrations in bedrock exposed in and along the channel. The maximum concentration in the bed sediments coincides spatially with the Moenkopi Formation, a unit that contains a modest amount of mercury but that yields fine-grained materials. (After Graf 1985b)

Broadly generalizable models for horizontal distribution may not be possible for those areas lacking a single or a few identifiable sources of metals. Graf (1985b) found that the distribution of bedrock units explained the along-stream distribution of mercury in a sample river in the Colorado Plateau. High concentrations occurred where the source rocks had moderately high concentrations and were of moderately fine-grained materials (Fig. 4.18).

In a broader sense, basin-wide perspectives might characterize heavy metal distributions. For example, Hawkes (1976) showed that a useful geochemical exploration concept was the fundamental relationship

$$Me_m A_m = A_a (Me_a - Me_b) + A_m Me_b, \tag{4.38}$$

where Me_m = concentration of the metal source, A_m = surface area of mineralization, A_a = total surface area of the drainage basin, Me_a = concentration of metal in a sample drawn from the stream draining the basin area, and Me_b = the background concentration for the basin. The function served to identify concentrations in mineralized zones in the semiarid southwestern United States as well as in humid regions in Central and South America. Basin approaches may be useful in describing regional metal cycles by identifying certain basins as primary sources of heavy metals. Graf (1985b) found that in the upper Colorado River Basin, some

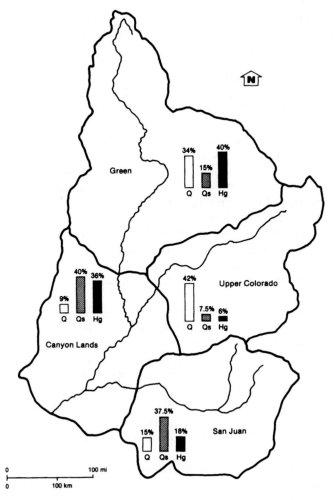

Fig. 4.19. Mass budgets for the upper Colorado River Basin draining into Lake Powell, the reservoir behind Glen Canyon Dam, Arizona. The Green River and Canyon Lands (an interbasin area) contribute most of the mercury, the Green and Colorado rivers most of the water, and Canyon Lands and the San Juan River most of the sediment. (After Graf 1985b)

sub-basins were primarily mercury suppliers, while others supplied mostly water or sediment (Fig. 4.19). Similar studies in other dryland systems are lacking.

4.4.2 Radionuclides

The study of the transport and storage of radionuclides in dryland sediment systems is at its formative stages. The advent of nuclear weapons in 1945, with subsequent weapon-testing, loss of fissionable materials in transport, releases from mining and milling operations, and accidents at nuclear power stations, has resulted in world-wide distribution of radionuclides in the surface environment. Atmospheric fallout distributes radioactive particles in a haphazard fashion over the landscape, but erosion and sedimentation processes concentrate these materials in depositional units in specific locations. The development of connective studies between geomorphology and radiochemistry takes on added importance in dryland areas where surface materials are especially mobile.

The transuranic elements most likely to be of significance from a hazard perspective are neptunium, plutonium, americium, and curium (Watters et al. 1983). Neptunium-237 is a decay product of nuclear detonations, and although it occurs only in small quantities, it is the most mobile of the transuranic elements in ecosystems and organisms. Plutonium is an element that does not occur naturally in the environment, but it is increasingly common because it is a component of weapon-testing. The operation of uranium-fueled nuclear power stations results in spent fuel that includes plutonium-239 which may be reused as fuel after reprocessing (see Table 4.11 for isotopes and properties). Nuclear weapons testing has also widely distributed americium-241, the most common transuranic element in the environment. Weapon-testing has produced only small amounts of curium-244, but nuclear power stations are generating increasing amounts as waste material.

According to estimates by Watters et al. (1983, p. 91), atmospheric processes have distributed up to about 2 mCi of plutonium-238, -239, and -240 per sqare km in the latitudinal bands of 20–40° N and S, the predominant dryland portions of the earth. No studies have been published to measure the concentration of these materials in stream sediments of dryland rivers, though Foster and Hakonson (1983, 1984) estimated that in the arid and semiarid Rio Grande basin in the American Southwest slope erosion would deliver 39 per cent of the plutonium fallout on the watershed to streams in 20 years. They also estimated that only 1 per cent would exit from the fluvial system. Internal storage at unknown locations accounts for the difference.

As with other heavy metals, the actinides (those elements with atomic numbers of 89 – 103) have an affinity for fine particles in fluvial sediments

Table 4.11. Radionuclides from atmospheric fallout likely to be entrained by fluvial processes

Element	Isotope	Emission	Half-life (yrs)	Amount (kCi)
Neptunium	237	α	2,100,000	–
Plutonium	238	α	86.4	24
	239	α	24,400	154
	240	α	6,580	209
	241	β	13.2	9,721
Americium	241	α	458	336
Curium	244	α	17.6	–

Notes: Amounts refer to global totals. Data from Watters et al. (1983); does not include releases from the Chernobyl, USSR, power station accident, 1986.

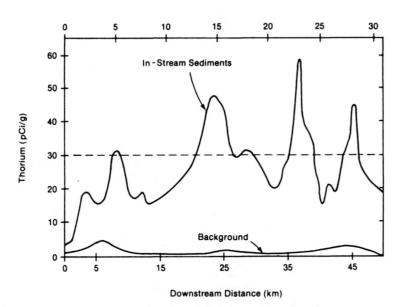

Fig. 4.20. The downstream distribution of thorium-230 in the Puerco River, New Mexico after a release of materials from a tailings mill pond and the passage of several floods. (Data from Weimer 1981, after Graf 1985a)

(Purtymun et al. 1966). Because transport and depositional processes tend to sort particles according to size, the metal-bearing fines may become increasingly separated from other materials, resulting in a contaminant enrichment process (Lane et al. 1985). Massey and Jackson (1952) proposed as a simple description of this enrichment process with the function

$$Q_c = ER\, C_s\, Q_s, \tag{4.39}$$

where Q_c = rate of contaminant discharge (mass per unit time), ER = the enrichment ratio, C_s = mean concentration of soil contaminant, and Q_s = sediment discharge rate (mass per unit time). The enrichment ratio is

$$ER = [\Sigma\, C_s(d_i)\, Q_s(d_i)][Q_s\, \Sigma\, f_i\, C_s(d_i)]^{-1}, \tag{4.40}$$

where $C_s(d_i)$ = mean concentration of soil contaminant associated with particle size class i, $Q_s(d_i)$ = sediment transport rate for particle size class i, Q_s = total sediment discharge rate, and f_i = proportion of the channel bed material in size class i. (Lane et al. 1985) used (4.40) and (4.41) in conjunction with other sediment yield and transport relationships to analyze the storage and transport of plutonium in stream sediments in a semiarid basins in northern New Mexico. In basins ranging from 18 to 150 km² (7 – 58 mi²) they found enrichment ratios averaged 5.5 with a range of 1.4 – 13.3.

None of the presently available analyses of radionuclides in stream systems accounts for the pulsating nature of sediment transport. An on-going project at

Arizona State University is addressing the issue using data on the downstream distribution of thorium-230 in the Puerco River, New Mexico and Arizona. The data show a wave-like pattern of deposition of radionuclides that might be expected as declining flood flows deposit their transported pulses of material (Fig. 4.20). Chemical mobility of radionuclides remains a confusing factor in understanding the role of physical transport in the movement of the materials through dryland fluvial systems.

4.5 Dryland Perspectives

Much of the discussion in this chapter has focused on the differences between dryland and humid-region river systems. Many of the techniques available to the geomorphologist working in drylands are derived from humid-region research, and the transfer of the technology from one region to another without testing is questionable. There are limits to this issue, however, because at some point the processes under consideration are so basic and fundamental that they operate the same everywhere. The gravitational forces and lifting forces operating on individual grains in fluid flow, for example, are the same no matter what the location of the process. At some point research begins to consider increasingly complex systems that include the climatic inputs to the explanation. At that point, dryland systems must be considered separately and new explanations are required for observed behavior. Definition of this point of complexity when the combination of fundamental physical principles is no longer universal is critical to the successful construction of theory for dryland rivers.

5 Process-Form Relationships

Process and form are not distinct and separate in geomorphological systems but are complex in their interactions. Sometimes one influences the other, sometimes there is mutual adjustment. The purpose of this chapter is to explore the relationship between process and form for dryland fluvial systems and to identify the causal factors that control process and form changes. The chapter has a geographical organization, beginning with badlands in headwater zones, proceeding to pediments and alluvial fans, channels and flood plains, and valley fills. Each of these subsystems has identifiable forms and poses special challenges for human users of the environment.

Table 5.1. Coverage of dryland surfaces by per cent

Surface type	Sahara	Libyan Desert	Arabia	SW USA
Mountains	43	39	47	38.1
Volcanic features	3	1	2	0.2
Badlands	2	8	1	2.6
Wadis	1	1	1	3.6
Fans	1	1	4	31.4
Bedrock pavements	10	6	1	0,7
Riparian zones	1	3	1	1.2
Desert flats	10	18	16	20.5
Playas	1	1	1	1.1
Sand dunes	28	22	26	0.6
	100	100	100	100

Note: Adapted from Cooke et al. (1982, p. 194), data by P. G. Fookes.

Our knowledge about various components of the dry landscape is uneven, and published data and explanations largely ignore the mountains which are the most prevalent landforms in deserts (Table 5.1). Much more information is available concerning badlands, for example, yet badlands make up only a small fraction of the total dryland areas. Fluvial lowlands occupy relatively insignificant percentages of the total dry landscape, but they represent areas probably most important to human activities, so that the literature pertaining to them is more voluminous than that for other landscape types.

5.1 Badlands and Piping Processes

Badlands are intricately dissected landscapes formed by dense drainage networks eroding poorly consolidated sediments that lack protective vegetation cover. In Australia the term badlands applies to relatively flat unvegetated lands that have low density rill networks (Bryan and Yair 1982, p. 1), but these terrains are not the subject of the present discussion. Naturally occurring badlands are located in areas with limited precipitation and long dry periods. Subtropical areas of dryland conditions with strongly seasonal rainfall and mid- to high-latitude drylands in continental locations with dry winter seasons encompass most natural badlands. Artificially induced badlands occur in almost all climates where human activities have removed vegetation cover.

5.1.1 Badlands

The perpetuation of natural badlands depends on the specific combination of seasonal precipitation including lengthy dry periods with highly erodible surficial materials. Schumm (1956b) demonstrated the importance of materials in controlling processes and forms in badlands by comparing two geologic units in the White River Badlands of South Dakota. The Brule Formation consists of poorly consolidated shales that generate a clay-rich regolith upon weathering. The surfaces of the Brule Formation do not readily absorb water and are scored by numerous rills and desiccation cracks. The Chadron Formation also contains clays, but weathers in a different fashion by generating a surface accumulation of loose clay aggregates. These aggregates readily absorb water.

Schumm (1956a) showed that the material differences between the two formations influenced badland processes. On the Brule Formation, rainsplash and associated runoff and erosion create relatively steep slopes with maximum slope angles of about 44° and sharp interfluves. The slopes erode rapidly and consume hill masses by a process of parallel slope retreat similar to that postulated by Walter Penck. On the Chadron Formation, the surface aggregates absorb water and move downslope by sliding over unweathered subsurface layers in a creep process. The result is more slowly eroding slopes with a mean maximum angle of about 33° and gently rounded interfluves. The slopes on the Chadron Formation consume hill masses by a process of downwasting similar to that postulated by William Morris Davis (Fig. 5.1).

In addition to weathering properties, chemical characteristics influence processes and forms. Soils with high amounts of exchangeable sodium erode rapidly because their particles disperse in solutions that have low electrolyte concentrations (Quirk and Schofield 1955). Chemical characteristics also influence the tendency of soil masses to swell and shrink during periods of wetting and drying (e.g., Roswell et al. 1969). The influence exerted on final forms is illustrated by three badland areas in the Rif Mountains area of Morocco (Imeson et al. 1982). The first area has v-shaped gullies and moderate to steep slopes that are developed on colluvial soils and flysch weathering products. The materials are nonsaline and nonsodic. Overland flow is the dominant process. The second area has u-shaped

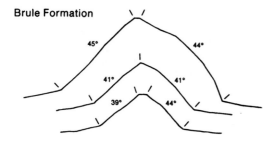

Brule Formation

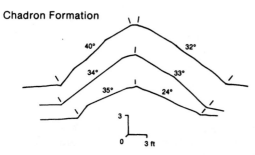

Chadron Formation

Fig. 5.1. Slope profiles from the White River Badlands, South Dakota, showing the morphologic differences between slopes developed on the Brule Formation (*above*) and the Chadron Formation (*below*). (After Schumm 1956b, pp. 702–703)

gullies and moderate slopes developed on marine sediments rich in saline and sodic materials. The dominant process is overland flow, but some piping is also present. The third type of badland occurs in valley fills along wadis in sediments without saline components but with large amounts of sodium. Piping is almost the only process. The Moroccan data show that soil chemistry plays an important role in badland development (Imeson and Kwaad 1980).

Montmorillonite and kaolinite clays appear frequently in badland terrains, and probably promote erosion where their shrinking and swelling properties are pronounced. Shrink-swell characteristics are not prerequisites for badland development, however, as illustrated by the Zin Valley Badlands of the northern Negev in Israel. Yair et al. (1980b) found that although montmorillonite and kaolinite were common in the area, swelling was less than 4 per cent and apparently had little to do with the erodibility of the materials.

In an investigation of the general importance of chemistry in determining badland development, Alexander (1982) investigated Italian badlands developed in semiarid environments. His analysis revealed significant differences between calanchi (chaotic badland areas developed on cohesive sediments exposed with large mass movements), and biancane (badlands which develop through the extension and incision of channel networks into masses of erodible sediments). The calanchi had closely spaced rills and v-shaped channels, while the biancane had cones and hummocks with intervening spaces of relatively gradual slopes and had extensive piping (Bombicci 1881).

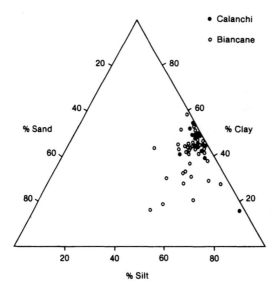

Fig. 5.2. Grain-size characteristics of materials underlying calanchi and biancane badlands, Italy, showing that although the two types have different forms and processes, their particle-sizes are similar. (After Alexander 1982, p. 80)

Although Cori and Vittorini (1974) and Vittorini (1977) reported significantly more clay and less sand in the material in calanchi, subsequent data failed to substantiate the difference (Fig. 5.2). Alexander (1982) found that statistically significant differences in exchangeable sodium percentages separate the materials of the two badland types. For calanchi the mean percentage was 30, while for the biancane it was 51. High erosion rates associated with the dispersive qualities of sodium result in the biancane areas where surface wash removes weathering products almost as rapidly as they are formed.

Spatial variation in materials influences the contributions of sediment and water to downstream fluvial systems. Because the chemical and physical properties of the materials control the amount of runoff, the variable source concepts of runoff generation are sometimes not applicable to badlands (Hodges and Bryan 1982). Runoff and sediment production have maps that coincide with subtle variations in lithology, and may reflect seemingly minor factors such as the frequency and depth of desiccation cracks (Haigh 1978). For example, although the Red Deer River Badlands in Alberta contribute most of the sediment loading to the main river (Campbell 1977), their contribution is highly irregular within the badland area (Bryan and Campbell 1980).

Processes on badland slopes that contribute sediment to the channel systems below include fluvial and mass movement activities. Kirkby and Chorley (1967) have questioned Hortonian overland flow as an idealized view of the series of events involved in runoff processes (Horton 1945). In badland areas it appears that runoff processes are closely controlled by variations in surface particles (Yair and Lavee 1974), structure of surface materials (Bryan et al. 1978), and the distribution of grain sizes on and near the surface (Bork and Rohdenburg 1979; Scoging 1982).

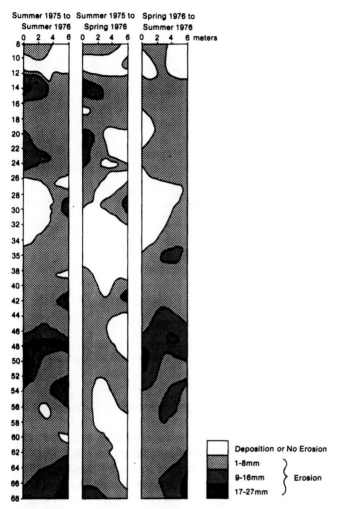

Fig. 5.3. The distribution of erosion on a badland slope, southeast Spain through three time periods showing the degree of spatial and temporal variability in the erosion process even in a limited area. (After Scoging 1982, p. 97)

In those areas where overland flow develops, fluvial erosion eventually delivers sediments to the channels.

Mass movement processes including debris flows, dry mass movements, and freeze-thaw-related activities supplement fluvial transport. Saturated materials that flow in the form of miniature debris flows carry sediment-laden waters over relatively impermeable surfaces after even moderate rainfall events (Hodges and Bryan 1982, p. 40). Between rainfall-runoff events when desiccation occurs during long dry periods, drying material may lose cohesion and slide or roll downslope. Because many badlands have steep slopes terminating in channels, this dry material comes to rest in locations where it will be transported downstream during the next runoff event. Freeze-thaw activities loosen surficial material which

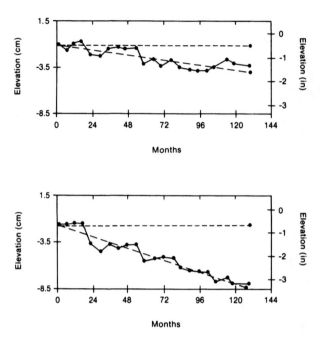

Fig. 5.4. Downslope passage of pulses or waves of sediment represented by elevation changes on badlands plots in the Red Deer River Badlands, Alberta. (After Campbell 1982, p. 230)

reaches channels by mass movement. The first spring runoff event continues the transport process in channels (Schumm 1956a).

Over any brief period of time such as a single erosion event or even through a season of events, badland erosion is unequally distributed on the surface. Detailed mapping of erosion on badland slopes in southeast Spain by Scoging (1982) showed that through a single year the location of intense erosion changed from one part of the slope to another (Fig. 5.3). Wise et al. (1982) came to similar conclusions in nearby areas in Spain. In detailed measurements of several years duration on the Red Deer River Badlands of Alberta, Campbell and Honsaker (1982) analyzed 1 m² plots to observe the variation of erosion and sedimentation on badland slopes. A series of computer-generated animated diagrams summarizing the measurements and shown by Campbell at the Twelfth Annual Geomorphology Symposium, Urban, Illinois, in 1982 revealed the passage of successive waves of sediment down the slope. The process is similar to the passage of pulses of materials through river channels as discussed in Chapter 4 above. Because of the passage of these waves or pulses, any given point on the slope alternatively experiences erosion and sedimentation, but over time periods longer than about 2 years, each point is likely to experience net erosion (Fig. 5.4).

Campbell and Honsaker (1982) indicate that the explanation for this pulsating behavior is that badlands represent nonequilibrium landforms. The resistance offered by their surfaces is very close to the threshold beyond which erosion occurs. Even during stable periods the state of resistance is never far from a failure mode, and the landform is "transport limited" (Thornes 1976). This designation signifies that erosion is not limited by the availability of erodible material, but only by the availability of forces to accomplish the work. The rate of badland erosion therefore depends on the magnitude and frequency of transporting events.

5.1.2 Rates of Badland Erosion

Badlands experience high rates of erosion relative to surrounding vegetated land surfaces, but badlands in various parts of the world have highly variable rates of erosion. Because they are transport-limited landforms and the transport mechanisms are controlled by climatic factors, variation in world-wide rates of badland erosion correspond to variations in climatic regimes (Table 5.2). Removal of material either does not occur or is very slow in winter in the Canadian badlands, while in the Hong Kong badlands readily available precipitation year-round provides continuing energy for erosion. In dryland areas, the availability of precipitation is also important, as indicated by the difference between rates for semiarid cases in South Dakota and New Mexico contrasted with extremely arid Israel.

Table 5.2. Example rates of badland erosion

Badlands area	Rate (mm yr^{-1})	Reference
Southeast Spain	0.1 – 10	Scoging (1982)
Zin Valley, Israel	0.5	Yair et al. (1980a, b)
Red Deer River Valley, Alberta	2	Campbell (1982)
Chaco Canyon, New Mexico	3 – 20	Wells and Gutierrez (1982)
Southern Italy	5 – 12	Alexander (1982)
Hong Kong	17	Lam (1977)
White River, South Dakota	18	Schumm (1956b)

Note: The example from Hong Kong is from an artificial landform.

Rates of badland erosion are not likely to be consistent over long periods of time because climatic conditions are likely to be variable. In the Chaco Canyon area of New Mexico, for example, present rates of denudation in badland areas appear to be nearly two orders of magnitude greater than they were 5,000 years ago (Wells et al. 1983, p. 183). In the same area, fluvial geomorphic rates appear to have experienced a similar acceleration. A warm, dry period 5,000 years ago may have slowed surface process rates and forced them into circumstances similar to those now observed in extremely arid Israel.

5.1.3 Piping

A distinctive process associated with badlands is piping (also known as tunnel erosion in Australia: Downes 1946), the development of subsurface drainage ways

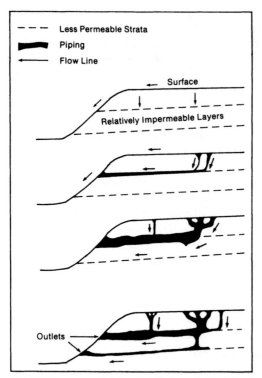

Fig. 5.5. Sequence of piping development in the Big Muddy Badlands, Saskatchewan, showing the influence of relatively impermeable layers. (After Drew 1982)

with intakes on upper surfaces and outlets in gully, arroyo, and channel walls (Fig. 5.5). Rubey (1928) in one of the earliest detailed papers on piping outlined their genesis in the semiarid plains of the United States. Inlets develop as water percolates into the subsurface. The water flows at a shallow angle nearly parallel to the surface and downslope, developing ever-enlarging tunnels. Eventually the tunnels become so large that their roofs collapse, creating a gully or entrenched channel on the surface. Other early observations of piping include those by Fuller (1922) in the loess terrain of China and by Leonard (1929) in southern Arizona. Except for a slight elaboration of Rubey's explanation by Buckingham and Cock-field (1950), subsequent research into the characteristics of pipes and their formative processes did not develop in appreciable quantity until the 1960s.

Pipes form subsurface conduits of varying sizes and lengths, but generally in cross section their height is two to six times their width (Heedee 1971). Well-developed pipes may have interior heights of up to 3 m (10 ft). In long profile, their floors describe exponential curves similar to typical long profiles of "graded" streams adjusted to a base level (Jones 1968). The long profile connects the inlet, usually located on a hillslope or on the surface of a terrace deposit, with the floor of a nearby channel. If the outlet is located some vertical distance above the channel floor, a cone or fan of debris from the pipe is usually located at the outlet.

Collapse of the roof of the pipe usually takes place in segments, resulting in a line of holes and miniature blind valleys on the upper surface that may include narrow natural bridges across the sunken pipe. At the surface, the phenomena produces a terrain occasionally described as karst-like (Mears 1963; Zaborski 1963). As these segmented valleys connect with each other, a new gully develops tributary to the original one (Anthony 1971). Berry (1970) found that piping is instrumental in the extension of stream courses in arid and semiarid portions of the Sudan and Tanzania, and Bishop (1962) observed it in gully erosion in Uganda.

Piping is a common feature of dryland terrains with fine sediments and intensive erosion of stream channels, and it is frequently (though not always) found in association with badlands. Piping occurs outside dryland areas where-ever mixed silt and sand sediments occur over a relatively impermeable layer and the surface configuration is conducive to the arrangement of relatively high inlets and low outlets. Gibbs (1945) reported the phenomena in grasslands of New Zealand, Slaymaker (1982) in glacio-lacustrine silts of Okanagan Valley, British Columbia, and Anderson and Burt (1982) in Great Britain.

Wherever it is located, piping dissects soils rich in clay-or silt-sized particles with some sand content; often montmorillonite, illite, or bentonite are present (Parker 1963). These materials may be part of a geologic unit undergoing erosion and forming badland topography, giving rise to coincidental development of pipes and badlands (Bell 1968). Volcanic materials in dryland settings also support piping, as illustrated by piped volcanic materials in the semiarid portion of Molokai, Hawaii (Kingsbury 1952) and the tuffs and ash in semiarid eastern Oregon (Parker et al. 1964). High sodium content enhances the probability that piping will develop if other conditions are favorable because of the lack of cohesion in sodic soils (Brown 1962). Local variation of piping is related to variation in compaction, permeability, and cohesiveness of the materials (Atchison and Wood 1965; Fletcher and Carroll 1948; and Zoslavsky and Kassiff 1965).

The presence of a subsurface layer that is relatively impermeable beneath the permeable materials may promote piping as surface water percolates downward to a perched water table (Parker 1963). If a hydraulic head develops and lateral flow terminates by exiting from the deposit at a wall, piping is likely to develop. The incision of stream channels into materials likely to support pipes may initiate the process by creating the hydraulic head between the former level of the channel and the newly established lower level.

As a common process in badlands in arid and semiarid environments, piping is often intimately connected with mass movement and channel processes. Re-search in the White River Badlands of South Dakota indicates that fluvial erosion undercuts and oversteepens slopes, producing slumping in near-channel areas (Bell 1968; Salisbury and Parson 1971). The portion of the slumped material that reaches the channel may not be soon removed by large channel flows, but small channel flows can initiate piping in the debris resting on the channel floor. Eventually piping disrupts or at least weakens the material; flows break down the mass and transport it downstream as part of the load of the stream. The portion of the slumped material that does not reach the channel also becomes riddled with pipes which weaken it and leave it susceptible to further transport down the slope by mass movement, surface wash, or transport through the pipes themselves.

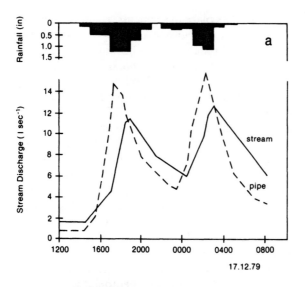

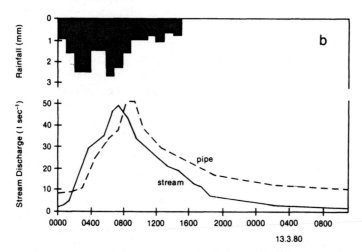

Fig. 5.6 a, b. Pipe and stream response to a precipitation event in the Shiny Brook catchment, Southern Pennine Hills, west Yorkshire. Horizontal axes represent hours. (After Anderson and Burt 1982, p. 351)

Piping results in a second drainage network connected to the first network on the surface. Channel studies in drylands have yet to account for this dual arrangement, but it undoubtedly has implications for the conductance of water and sediment through the system. In a humid region, Anderson and Burt (1982) found that hydrographs representing channel and pipe flow from a precipitation event were similar, but that the lag times were different (Fig. 5.6).

Piping poses significant hazards to human construction activities in drylands by weakening foundations of structures. Highways, bridge abutments, and artificial grades are especially susceptible to the development and collapse of pipes

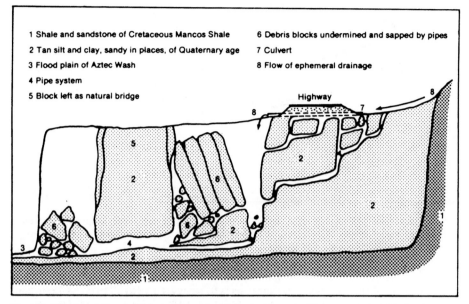

1 Shale and sandstone of Cretaceous Mancos Shale   6 Debris blocks undermined and sapped by pipes

2 Tan silt and clay, sandy in places, of Quaternary age   7 Culvert

3 Flood plain of Aztec Wash   8 Flow of ephemeral drainage

4 Pipe system

5 Block left as natural bridge

Fig. 5.7. Development of piping and associated destruction of a highway grade and base fill, US Route 140 near Aztec Wash, southwestern Colorado. (After Parker and Jenne 1967)

(Bell 1968). Highway builders frequently use nearby materials to develop fills and grades, but if this material contains clay and silt mixed with some sand, piping is likely to develop because runoff from the hard-surfaced road provides the input and the built-up arrangement of materials provides the necessary hydraulic head (Fig. 5.7). A relatively impermeable base level may be provided by the pre-construction surface. Prevention of destructive piping in artificial landforms depends on adequate drainage structures to prevent uncontrolled percolation into pipe-forming sediments. Dams constructed of earth fill are candidates for destruction by piping unless proper surface seals are included. In the catastrophe of Teton Dam, Idaho, where 14 people died, water from improperly sealed bedrock joints piped the loess fill material in the dam and destroyed it (Boffey 1977).

5.2 Pediments

Pediments, relatively gently sloping surfaces of low relief and partly covered by a thin veneer of alluvium (Hadley 1967, p. 83), are closely associated with dryland fluvial processes, but the nature of that association is the subject of seemingly endless debate. Gilbert (1877) first described the form in detail based on his observations of the Henry Mountains, Utah, but McGee (1897) coined the term to describe surfaces in southern Arizona. From the standpoint of the geomorphologist investigating modern processes, the pediment is a controlling landform for water flow. For the geologic geomorphologist concerned with long-term evolution of landscapes, water may be the agent that controls the pediment landform.

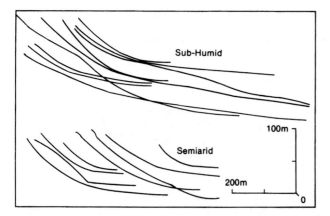

Fig. 5.8. Example pediment profiles from South Africa. The upper family of profiles represent sub-humid region slopes from Natal, the lower family of profiles represent semiarid Karroo. (After Fair 1948)

The purpose of the following paragraphs is to review the fluvial process-pediment landform connection. Dresch (1957), Hadley (1967), and Weise (1980) offered comprehensive analyses of the overwhelming body of literature on the general subject of pediments.

5.2.1 Definition, Description, and Distribution

The pediment is a connective landform between mountain mass and alluvial lowland near the local base level. The pediment almost always has a long profile of bedrock that can be described by logarithmic (to the base 10) curves as shown by Dury's (1966b) work in Australia. The mean slopes of pediments in drylands may be somewhat greater than mean slopes for similar forms in humid regions (Fig. 5.8). Slopes high on the pediment may be as steep as 10° but at the foot of the form slope may have declined to as little as 0°15' (Mabbutt 1977, p. 84). Irregularities in the bedrock profile may be filled by alluvial material to create a relatively smooth surface (Langford-Smith and Dury 1964). The pediment connects with the mountain mass at its upper end at a piedmont junction often formed by a radical change from gradual pediment slopes to steep mountain slopes. At its lower end, the pediment often grades imperceptibly into the surface of alluvial or lacustrine deposits. The primary process function of the pediment from a general systems perspective is to act as a transfer zone for erosion debris from mountains to basin storage sites.

Pediments can cut across and bevel bedrock structure, but they most often reach their best definition on granitic rocks that weather to gruss (e.g., Warnke 1969). In semiarid Colorado and Wyoming Eggler et al. (1969) showed that pediment surfaces were almost exclusively located on granite rock types. Pediments are not limited to granite rock types, as indicated by their development on limestone in Australia (Jennings and Sweeting 1963), on alternating resistant and non-resistant rocks in the Sahara Desert (Dresch 1957; Birot and Dresch 1966), and on alluvium in southern Arizona (Cooke and Warren 1973, p. 191).

The veneer of alluvium over the bedrock surface of the typical pediment is a few m thick and represents the material in transit from mountain source area to basin sediment sink. The alluvial cover typically contains buried paleochannels 1–3 m (3-10 ft) deep. These channels represent former water courses that were probably abandoned and filled in response to climatic changes or to changes in channel location during floods. The particles making up the alluvium decline in size in the down-pediment direction, but as shown by Dury's (1970) study of New South Wales surfaces, the sorting of the particles does not improve. This result suggests that the large particles are not transported by fluvial processes but merely weather in place (at least in the ancient landscapes of Australia), so that the youngest pediment surfaces close to the mountain front have the largest particles.

Pediments occur in all climatic regimes (King 1953), but they develop most prominently in drylands, and literature on all major deserts includes descriptions of them. In drylands there are three major types: apron pediments, terrace pediments, and pediment domes. Apron pediments extend from the bases of mountain masses and cover large percentages of the local landscape. As the pediments enlarge through poorly understood erosion processes at the piedmont junction, the mountain masses decline in size (Lustig 1969). Terrace pediments develop along major river courses (Tuan 1962). Pediment domes result when there are no intervening mountain masses between pediments which join in a broad upland surfaces, as in portions of central Australia and southern Africa. While in some cases lithology may control pediment angles (as in Australia: Mabbutt 1966), in other areas pediment form is independent of rock type (in the United States: Mammerickx 1964).

Pediments may exist coincidentally at several different levels, perhaps reflecting the influence of radical climate changes that adjusted formative processes, tectonic activities that changed relative base levels, or hydrologic changes that caused entrenchment of the base-level stream. In Arizona, Bryan (1926, 1936) suggested that these changes caused flights of pediments dating from Tertiary times along major rivers. Sequences occur on a variety of rock types and bevel structures in South Australia (Twidale 1967), Egypt (Butzer 1965), and Chile (Cooke 1964).

5.2.2 Origin

Theories attempting to explain the origin of pediments fall into four basic groups. First, parallel retreat of slopes making up the mountain front might leave an ever-enlarging pediment at its base. Lawson (1915) envisioned weathering and fluvial erosion of the mountain front as the primary mechanism, with base level changes determining the geometry of the pediment and piedmont junction. A variety of American workers provided subsequent modifications of the hypothesis. The weaknesses of the general concept, as identified by Ruxton and Berry (1961) in their study of pediments in the Sudan, include failure to account for the

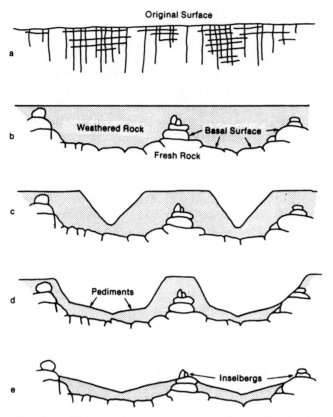

Fig. 5.9 a–e. Possible sequence of events leading to the development of pediments through weathering and stripping processes leading to the development of inselbergs in Uganda. (After Ollier 1960)

existence of the abrupt break in slope at the head of the pediment, and inability to elaborate on the role of interior mountain drainage basins.

A second set of explanations for the origin of pediments focuses on the role of drainage basins. Lustig (1969) pointed out that the fluvial processes associated with mountain drainage basins represent the most effective erosion forces in the system, and that as the erosion of mountains proceeds, production of a pediment is predestined because of nearby stable base levels in the basins. In this view, the mountain masses are interfluves between basins, and the streams that accomplish the work have smooth "graded" profiles from the interiors of the mountains onto the pediment. It may be that the striking differences between mountain slopes and pediment are not especially relevant to the overall explanation of the landscape.

A third group of hypotheses also focuses on the role of streams, but emphasizes those operating on the pediment. Johnson (1932b) hypothesized that as streams cross the pediment from mountain to basin they move back and forth laterally. According to the hypothesis (first advanced by Gilbert 1877), this lateral planation results in undercutting the edges of mountain masses and produces the sharply defined piedmont junction. Bryan (1922) partially accepted this explanation in his investigations of central and southern Arizona, but subsequent researchers have pointed out its fatal flaw. Lustig (1968, p. 191), for example, characterized the required movement of streams in the hypothesis as "truly absurd" and commented on the improbability of streams issuing from the mountain front and then turning nearly 90° in order to laterally plane the pediment next to the mountain front.

Australian workers have been the main proponents of a fourth group of hypotheses for pediment formation centered on weathering. Mabbutt (1966), for example, argued that pediment development in stable shield areas resulted from differential or compartmented weathering of near-surface materials. Fluvial erosion then strips the erosion products and exposes the former weathering fronts and boundaries (Fig. 5.9, Ollier 1960). Mountain masses represent more resistant zones, while pediments and basins are located in the less resistant areas. Lustig (1968) found this explanation acceptable for the fault-block Basin and Range Province of North America, and Thomas (1966) used it to explain the landscapes of northern Nigeria.

5.2.3 Fluvial Processes on Pediments

On time scales spanning several millions of years, pediments reflect variation in fluvial processes. In Brazil, for example, alternating humid and semiarid periods throughout the Quaternary produced a series of age-specific terrace pediments and terraces (Table 5.3, Bigarella and Andrade 1965). In Arizona, Tuan (1962)

Table 5.3. Quaternary stratigraphy and pediments in eastern Brazil

Geomorphic feature	Process	Climate, tectonics	Time period
Low fluvial terraces	Dissection and sedimentation	Humid with dry phases, uplift	Holocene
Paleo-pavement	Mechanical morphogenesis	Semiarid	Wisconsin
None	Dissection	Humid, Uplift	Sangamon
Pediment	Mechanical morphogenesis	Semiarid	Illinoisan
None	Dissection	Humid, Uplift	Yarmouth
Pediment	Mechanical morphogenesis	Semiarid	Kansan
None	Dissection	Humid, Uplift	Aftonian
Pediplane	Mechanical morphogenesis	Semiarid	Nebraskan

Note: Adapted from Bigarella and Andrade (1965), who indicate that the conclusions are preliminary.

correlated pediment, basin, and alluvial fan surfaces along the San Pedro River in a complex history of fluvial adjustments.

Modern fluvial processes on pediments are the outgrowth of a long mutual dependency between the smooth profile of the form and the actions of runoff. Some researchers have used sheetflow as a possible process explanation for the development of pediments (e.g., Davis 1938), but it cannot occur unless the form exists first. The numerous rills and small channels on pediments conduct fine particles downslope and probably play a significant role in the transport of debris across pediment surfaces (Bryan 1922). Relatively shallow depths of flow and gentle gradients insure that flows are unlikely to develop sufficient shear stress to move large cobbles and boulders which are left to weather in place. Large channels issuing from mountain fronts usually construct alluvial fans at their exit points, but if there is not enough sediment from the mountain mass to build fans, the channels may entrain and transport materials from the surface of the pediment. Filled channels reported on pediments attest to the locational instability of these streams.

Schumm (1962) investigated miniature landforms that mimicked mountain masses with surrounding pediments. His data from the White River Badlands, South Dakota, showed mutual adjustment of forms and processes when comparing conditions on the "mountain" slopes as opposed to the "pediment" surfaces. The mountain slope analogs were about 40° and had surfaces of fractured clay. The pediment slope analogs were about 6° and had smooth silt-veneered surfaces. The counteracting influences of slope and roughness balanced each other, so that depth and velocity of overland flow was approximately the same for each slope type. The applicability of Schumm's results to larger, more complex systems is problematic because of scale differences. The presence in many cases of coarse particles introduces an additional element of roughness to the pediment surfaces. The badland analogs are more useful in illustrating a tendency toward dynamic equilibrium across slopes of radically different characteristics leading to a remarkable consistency of water transfer in badland terrain. Complicating factors might be introduced to the simple systems during further experimentation.

5.2.4 Hazards

The piedmont zones near desert mountains have surface expressions that permit the identification of pediments from aerial photography. Pediments have relatively high drainage density and lack the radial drainage of alluvial fans and the concentration of flow found on alluvial slopes (Rhoads 1986b). Pediments represent areas of potential problems for human construction activities because of possible hazards from floods, sedimentation, and channel instability. Measurements of water discharges in the myriad of small channels crossing most pediments are rare, and prediction of possible flood magnitudes must be based on channel dimensions. An additional complicating factor is the presence of sheetflow that might result from the overflow of numerous small channels. Rahn (1967) found that near the mountain front sheetflow is rare on pediments, but on the downslope areas it is more common. The depth of sheetflow on pediments and similar surfaces is not great (probably less than 60 cm), but with the addition of street patterns and drains added to the surface, sheetflows in the undeveloped portion of the pediment concentrate quickly.

Because pediments have dispersed, small channels, transportation lines built across the pediments are subject to destructive discharges at many points. One solution to the problem is to construct wing dams upslope from the rail, highway, or canal line and to collect the runoff from a broad area and to concentrate it at a single point. The water can then move across the transport route through a culvert or under a bridge. Along the Colorado River Aquaduct, a canal conducting water from the Colorado River to Los Angeles, California, wing dams collect runoff from pediments and fans and funnel it to artificial channels that cross over the top of the canal.

The sediment transported by water across pediments also poses threats to human activities. Clogging of drain ways, culverts, and narrow bridge openings by sediment and debris during floods causes damaging backwater effects and increased overbank flows. Because natural channels on pediments are relatively small, they frequently fill with sediment that diverts their flows to new channels, leading to significant instability for channel locations. Planning agencies attempting to predict where to construct underpasses and culverts for channel flow cannot rely on the present locations of the channels on a multi-decadal basis. Efforts at stabilizing the natural channel locations through engineering works are difficult because of sedimentary fillings. The best solution for developments on pediments is probably to maintain as many of the natural channels as possible, stabilize the largest ones, and prepare for periodic sediment-clearing efforts.

5.3 Alluvial Fans

Like pediments, alluvial fans serve as transfer systems for materials eroded from mountain masses and destined for deposition in adjacent basins. Drew (1873) apparently was the first to apply the term alluvial fan to the cone-shaped deposits of stream sediments at the mountain-basin boundary where the streams that form them issue from narrow valleys (Rachocki 1981, p. 8). If the quantity of material available for deposition is great enough or if the depositional process has continued for long enough, adjacent fans may coalesce, forming an alluvial apron along the mountain front.

A hybrid landform is the fanhead valley (Scott 1973), which has a wide, relatively flat floor partially filled with alluvium. The fanhead valley surface connects directly downvalley with alluvial fan surfaces, but in the valley it is more restricted than the fan in the adjacent basin area. Investigations in the Ventura, California, area by Rockwell et al. (1984) showed that the degree of development of fanhead valleys is tectonically controlled. They occur most often in mountain masses with slow rates of uplift.

Alluvial fan literature is voluminous in comparison to the amount of space that fans occupy in drylands (Bull 1977, provides a useful literature review). Gilbert (1875) and Dutton (1880) defined the role of alluvial fans in the Colorado Plateau of the southwestern United States as temporary storage sites for erosion

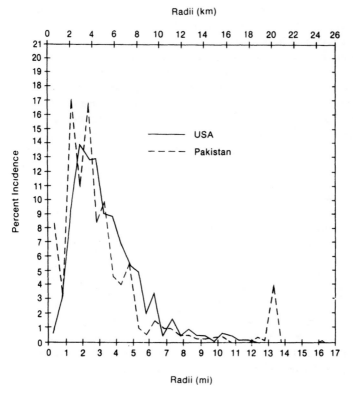

Fig. 5.10. Magnitude and frequency of 934 alluvial fans in the southwestern United States and Pakistan showing skewed distributions. (After data by Anstey 1965)

debris. Davis (1905) and Johnson (1932a) included fans as components of the arid geographic cycle of erosion, and Bryan (1922) recognized them as an integral part of the sediment transport system in semiarid southern Arizona. A few descriptive papers bridged the gap between these early works and an explosion of interest in alluvial fans beginning in the 1960s. By 1980 there were more than 100 papers on the subject (Rachocki 1981, p. 8).

5.3.1 Definition, Description, and Distribution

Individual alluvial fans are variable in size depending on their age and supply of sediment available for their construction. They range in size from tiny forms in badlands covering a few cm² to the fan of the Kosi River, India, which covers more than 10,000 km² (Gole and Chitale 1966). Anstey (1965) found that in the southwestern United States and in Pakistan the modal radii of fans was in the 1.6–3.8 km range (1 – 2.4 mi) range (Fig. 5.10). Fans in drylands tend to be larger than those in humid regions, probably because continually flowing streams in humid regions remove materials relatively quickly from the distal parts of fans.

Fan areas are directly related to the basin areas that supply their materials according to the function

$$A_f = c A_d{}^n, \tag{5.1}$$

where A_f = area of the fan, A_d = area of the drainage basin above the fan apex, and c, n = constants (Denny 1965). The constants vary according to geologic materials of the supply basins (Bull 1975b, pp. 123–125), vegetation, basin slope, and c climatic regime (Table 5.4). Given consistent units, larger values of the coefficient c reflect more rapid erosion in settings of equal age (Hooke and Rohrer 1979). Irrespective of units of measure, the exponent is usually slightly less than 1.0. Explanations for the magnitude of the exponent include storage of material on footslopes and along channels in the contributing basin. If fan size were measured more correctly as a volume instead of as an area, the exponent would equal 1.5 in an isometric relationship of the form of (5.1). If internal storage in the mountain channel network were a factor in the relationship, the exponent would be slightly less than 1.5.

Although the long profiles of alluvial fans are simple concave downward curves in the abstract, their field expression ranges across a continuum from simple curves to several profile segments. In drylands, their average slope is similar to that of pediments, often ranging between less than 1° to almost 10° at the apex of small fans. The long profiles of simple fans have negative exponential forms reflecting the profiles of channels that carry and then deposit sediments on the fan. The profiles of these channels usually extend along channels up into the mountain source areas. In a situation analogous to Schumm's findings that

Table 5.4. Coefficients and exponents for the power function relating basin area to fan area [Eq (5.1)]

Area, reference	Coefficient	Exponent
Central California, Bull (1964)	2.1	0.98
Central California, Bull (1964)	0.96	0.98
Death Valley, California , Denny (1965)	0.5	0.8
Northern Nevada, Hawley and Wilson (1965)	0.74	0.98
Southwestern USA, Hooke (1968)	0.42	0.94
Southwestern USA, Hooke (1968)	0.24	1.01
Southwestern USA, Hooke (1968)	0.15	0.90
Iran, Beaumont (1972)	1.18	0.946
Ventura, California, Rockwell et al. (1984) (mountains subject to uplift)	3.84	0.55
Ventura, California, Rockwell et al. (1984) (mountains not subject to uplift)	0.59	0.8
North Carolina, Mills (1982)	0.227	0.529
North Carolina, Mills (1983)	0.38	0.76
Costa Rica, Kesel (1985)	0.921	1.01

Note: Most data assembled by Kesel (1985).

mountain slopes and pediment slopes adjusted themselves to produce the same hydraulic conditions, alluvial fan slopes reflect adjustment to the available energy for debris transport. Inverse relationships between fan slope on one hand and fan area, drainage-basin area, and discharge on the other insure a constant amount of available force for the work of transportation (Hooke 1968).

Slope of the fan is also related to the relative relief of the basin which supplies its material (Melton 1965):

$$S = 6.69 [H (A^{-0.5})]^{0.88}, \tag{5.2}$$

where S = upper fan slope, H = basin height (difference between highest and lowest elevation in the basin), and A = basin area. Rugged basins with higher values for $[H (A^{-0.5})]$ shed more coarse debris which results in steeper fans (Denny 1965; Bluck 1964). In a global comparison, Kesel (1985, p. 458) found that fans in arid and tropical humid regions have similar gradients. He pointed out that previous literature on the subject concluded that arid fans were more steep than their humid region counterparts, but that this conclusion may have rested on humid region examples drawn mostly from glacial outwash rivers which may not be analogous.

Segmented fans occur because material may build up a cone with a smooth profile only to have the process interupted by climatic, tectonic, or internal system changes which initiate incision of the stream at the fan head. Stream processes excavate material from a fan-head trench and transport it to the perimeter of the fan, where it forms a new deposit with a slope more gentle than that of the original fan (Denny 1967).

When the downward concave long profile is combined with the general cone-like shape, alluvial fan morphology may be generalized by the function

$$Z = P + SR + LR^2, \tag{5.3}$$

where Z = the elevation of any point on the fan surface, P = elevation of the apex of the fan, S = slope of the fan at the center, R = radial distance from the apex to the point, and L = half the rate of change of the slope along the radial line (Troeh 1965). Bull (1968) confirmed the utility of the model with measurements of fans in California.

5.3.2 Materials

Particles on the surfaces of arid-region alluvial fans decline in size as distance from the fan apex increases according to an exponential decay function:

$$y = a e^{-bx}, \tag{5.4}$$

where y = particle diameter, e = base of the natural logarithm system, x = distance from the apex, and a,b = empirical constants. The rate of distance decay, the value of b, varies from one fan to another depending on the frequency and efficiency of fluvial processes and the nature of the source material. Low rates of change over distance are products of debris flows (Lustig 1965, in the American southwest; Beaumont 1972, in Iran; Boothroyd and Nummedal 1978, in humid regions). The distribution of particle sizes also responds to reworking of the sediments (Denny 1965), and the combined action of debris flows and fluvial events reduces the correlation between distance and particle size (Bluck 1964). Particle sorting also improves in the downfan direction, is better developed in humid-region fans than in those in arid regions, and is subject to the same disruptions as particle size (Kesel 1985).

Subsurface deposits in alluvial fans, unlike those on pediments, may extend to considerable depth. Depths of accumulation are greatest where tectonic activity continually lowers the basin level relative to the mountain mass during fan construction. The deposits terminate sharply at the apex-source and interfinger in a complex fashion with basin fill materials (Bull 1977). The various beds are of a variety of thicknesses and include fine and course particles in poorly defined zones (Leeder 1982, pp. 138–141). Lag deposits of coarse particles are common.

Mudballs may also appear in fan deposits. Bull (1964) reported that mudflows transport them across California fans, while along wadis in the Jordean Mountains and in southern Israel bank collapse produces masses of material that become mudballs and that subsequently are entrained by fluvial processes (Karcz 1969). On experimental fans, Rachocki (1981, pp. 78–80) found that mudballs in transit gained material and weight by accretion. Individual mudballs moved in a step process, occasionally becoming lodged in depressions in the channel floor or fan surface. Rachocki found that during transport the mudballs became armored with a surface layer of relatively coarse particles, a feature common to mudballs in ancient fan deposits in the Papago Hills near Phoenix, Arizona.

5.3.3 Origins

Alluvial fans originate where confined streams issue from mountain fronts onto relatively open basin floors. It is possible that early in the development of the fan there is a reduction in slope as the stream passes from the mountain area to the basin area, but as the fan structure increases in size, the break in slope disappears (Wasson 1974). Many investigators have stipulated that a break in slope along the stream channel at the mountain front is the reason for alluvial fan construction (e.g., Allen 1971; Blissenback 1954; Kadar 1957), but field surveys show that stream gradients cross the mountain-basin boundary without change (Bull 1964; Denny 1965).

A more likely explanation for the initiation and continuing development of fans is a combination of channel widening and channel migration. As the channel issues from the mountain area, confining rock banks are no longer a limiting factor, and channels usually (but not always) become wider in the fan area. Channels that are restricted to single threads in the mountain area become braided channels with many threads on the fan, while velocity of flow decreases on the fan resulting in deposition. Channel migration is possible in the basin-fan area, and generally not possible in the constricted mountain valleys. The braided channels distribute material in the available space in the fan zone, while no such space is available in the mountain area.

Segmented fans, those with slopes that consist of several separate components rather than one smooth curve, result when individual fans coalesce and "stack" their deposits one on top of the other, or when the location of deposition changes on a single fan. In the latter case, deposition may begin at the apex early in the history of the fan, but as the accumulation builds up, the main channel may conduct sediment some distance away from the apex before beginning deposition. When this occurs, a new spatial arrangement of deposited materials occurs, all related to the newly defined apex (Denny 1967).

Alluvial fan deposition is closely related to the tectonic activity of the associated mountain masses and basins. Basin and range dryland areas in the southwestern United States, northeastern Mexico, the Atacama Desert of Chile and Peru, and the deserts of Pakistan and Iran have extensive alluvial fan development. Bull (1964) explained segmented fans along the edge of the San Joaquin Valley, California, as products of the relative uplift of the mountain masses which initiated new regimes of fluvial activity in fan development. Denny (1967) concluded from evidence on fans in California and Nevada that activity on faults parallel to the mountain front caused incision that moved the locations of erosion and deposition on fan surfaces. In Death Valley, California, Hooke (1967) found that fans on the western side of the valley had been incised as a product of eastward tilting of the valley floor. Fans in northern Spain are frequently associated with fault scarps (Heward 1978), where uplift renews erosion on the upthrown block.

5.3.4 Modern Processes

Modern processes on alluvial fan surfaces include a series of activities that have a particular spatial arrangement. Near the apex fan-head trenching is likely to occur, in mid-fan areas debris flows and sieving are common, and in distal areas braided stream processes are likely (McGowen 1979), though the process areas are not entirely exclusive. Researchers have explained the entrenched channels that are common near the apex by three approaches: internal system adjustments, tectonic influences, and climatic change. The entrenchment might result from simple variation in flood discharges common in drylands (Denny 1967, p. 85; Beaty 1970), or from alternative over-steepening and subsequent erosional establishment of more gentle gradients such as those proposed for gully erosion by Schumm and Hadley (1957). Eckis (1928) suggested that as erosion reduces a stable mountain mass, the sediment supply eventually will decline, and trenching of the fans would occur without other influences. Abrupt tectonic activity might initiate increased rates of erosion in the mountain streams and at the fan apex (Bull 1964). Climatic changes could accomplish the same result by altering sediment supply and/or the fluvial regime (Lustig 1965).

Debris and mudflows occur when weathered material on hillslopes of the source basin flushes quickly through the channel system to the fan in a pulse of liquid with a high content of solid matter. Debris flows carry large cobbles and boulders, depositing most of them near the apex of the fan (Beaty 1963) but rafting a few of the large particles considerable distances downfan (Cooke and Warren 1973, p. 183). The debris-flow deposits have poor sorting and almost no internal sedimentological structure, but elongated fragments usually align themselves

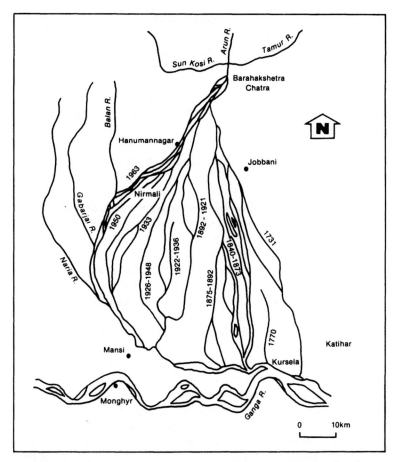

Fig. 5.11. Channel locations on the Kosi River alluvial fan, India, 1731–1950, showing lateral migration across the fan from east to west. (After Gole and Chitale 1966)

parallel to the direction of flow (Fisher 1971). Mudflows contain finer materials, usually sand size or smaller (Bull 1968), and frequently terminate in mid-fan areas by creating lobes a meter or so in height (Bull 1964). Unlike debris flows, mudflows create relatively uniform, fine-grained sedimentary deposits that are homogeneous. They may be up to 6 m (20 ft) thick (Blissenbach 1954). Debris and mudflows have relatively smooth upper surfaces that may mask more irregular fan surfaces upon which they rest (Wasson 1977), but subsequent erosion by relatively sediment-free flows may dissect the deposits, a process reported from direct observation on fans of the White Mountains of California and Nevada by Beaty (1963).

Hooke (1967) identified sieve deposits on fans as materials deposited by water flows that are abstracted from the fan surface by transmission losses. The flows seep into the porous materials of the fan, leaving behind sedimentary materials that once were in fluvial transport. Lobes of sieve deposits occur occasionally in mid-fan locations where they are easily recognized (Wasson 1974), but in comparison to other deposits they are relatively rare (Bull 1977).

The most common channel type encountered on alluvial fans is a braided arrangement. The braided channels of dryland fans occur because of the lack of lateral confinements by resistant materials and because of the high variability of flows. In the American Southwest depositional activity by braided channels may occur only once in several decades (Beaty 1970, 1974). Beaty (1970) dated the fans in the White Mountains of California and Nevada using a volcanic ash layer of known age and determined that the fans are at least 700,000 years old and still growing. Because water flows on dryland fans are discontinuous, sediment deliveries and transport across the fans are also discontinuous. Those sediments that are reworked by multiple events may experience sorting to a much greater degree than materials delivered in a single debris-flow or mudflow event. Braided stream deposits in alluvial fans include cross-bedded active channel deposits and fine-grained channel fill materials (Selley 1976, p. 265).

Braided channels on fans are inherently unstable and eventually migrate across the entire surface of the fan. On the Kosi River alluvial fan near Mansi, India, the main channel migrated laterally about 112 km (70 mi) between 1736 and 1963 (Fig. 5.11; Gole and Chitale 1966). Gradual avulsion has produced the channel migration irrespective of floods and tectonic processes (Wells and Dorr, 1987). The Kosi River fan is a wet fan (with a perennial stream), so this migration has been much more rapid than migration on large dryland fans. Kesel (1985) has shown that the differences between processes on dryland and humid-region fans (dry and wet fans) are substantial because of the episodic nature of process operation in drylands. Although Rachocki (1981 p. 51–55) has observed sheetflow on experimental alluvial fans, it is not likely on large alluvial fans in dryland areas. Vegetation and large particles on the fan surface prevent sheetflow from forming, and the cone-shaped geometry of the fan is not conducive to sheetflow. Overflow between many small channels in the distributary network of the distal parts of fans may give a short-lived appearance of sheetflow, but it probably does not behave as a major geomorphic agent as outlined by McGee (1897).

5.3.5 Hazards on Alluvial Fans

In his unique analysis of geomorphic hazards in the Los Angeles, California, area Cooke (1984, p. 50) identified five specific hazards associated with alluvial fans: unpredictable individual flows, variations between flows, debris flows which cause sedimentation problems, channel migration and avulsion, and long periods of inactivity which mask the hazards. Although Los Angeles provides the most striking example of conflict between economic development and the natural hazards associated with alluvial fans, it is not unique. There are 30 cities of over one million people each in dryland environments (Cooke et al. 1982).

Channel flows through fanhead valleys and across alluvial fans change character from relatively clear flows to debris and mudflows and then back again within a single event. In 1934 in La Canada Valley, California, for example, debris flows carried materials eroded from hillsides denuded by fires through channels in fanhead valleys and alluvial fans. In the same event, the relatively clear water flows followed the debris flows and were intensively erosive because they were not carrying sediment equal to their capacity. To further complicate the single event, there were as many as 15 pulses of material, and flow velocities varied 100 per cent (Eaton 1936). As with most dryland fluvial sediment transport, wave or pulsating activity on fans appears to be common.

Scott (1973) measured changes in the fanhead valley of Tujunga Wash, southern California, that resulted from flooding events in 1969. Extensive net scour reached 6 m (19 ft) while net sedimentation in some locations exceeded 10 m (33 ft). Widespread scour and fill results in channel mobility and avulsion in this particular example.

Alluvial fan hazard areas also experience variation between events in that some floods are dominated by clear water, while others are mostly in the form of debris or mudflows. In southern California this inconsistency develops because debris accumulate in the upstream basin between some events, supplying large quantities of material that move downstream as a debris flow or mudflow upon saturation (Troxell and Peterson, 1937). If the next event occurs soon after the first and no debris have accumulated from slopes in the source basin, the flow will be relatively clear and erosive. In semiarid hill and mountain lands, fire plays an especially important role in the process because denuded slopes contribute unusually large quantities of material to channels (Smalley 1971).

Debris flows pose special hazards that are common to fan apex areas but that diminish as distance from the apex increases. Urban development that extends to the fan apex allows debris and mudflows to extend farther from the source area because the impermeable surfaces associated with urbanization prevent transmission losses of water that otherwise would quickly stall the flows (Rantz 1970, p. 10). The flows are often laminar but become turbulent in constricted areas or over rough surfaces, and as they cross fan surfaces of less than 20 per cent they usually decelerate (Campbell 1975). Debris flows may transport boulders several meters in diameter with destructive consequences, and the general flow may be highly erosive (Sharpe and Nobles 1953). Mechanical damage to structures, sedimentation, and erosion of designed channels are common hazards associated with debris flows.

Sedimentation associated with debris, mud, and water flows in fan environments is probably the most serious problem facing planners and developers of fan environments in drylands. The problem is endemic to the entire range of dryland climates from the best-known issues in southern California (reviewed by Cooke 1984) to extremely arid areas. In southern Israel and Jordan, for example, Schick (1971a, p. 147) concluded that sedimentation problems associated with alluvial fan flooding probably affected highway maintenance on an annual basis. Even relatively clear water flows may transport large-caliber debris: in 1934–1935 flooding from the San Gabriel Mountains, southern California, transported boulders weighing 10 tons and sprayed gravel-sized particles onto the roofs of houses (Chawner 1935, p. 259).

Control of the amount of debris issuing from the source area is possible by the construction of dams designed to store the sediment (Guy 1970). These structures fill with sediment during flood periods but between floods (when their

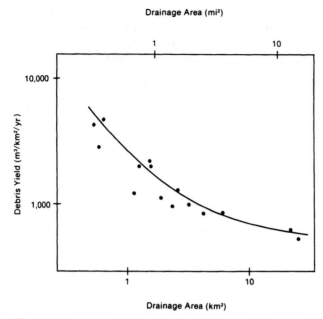

Fig. 5.12. Average annual sediment yield from catchments without debris basins in the Los Angeles area showing the general decline in yield per unit area with increasing total catchment area. (After Scott and Williams 1978)

capacity is reduced by 25 per cent or more) they can be dredged to restore their retention capability. In the Los Angeles area these "debris basins" are constructed to accommodate huge masses of materials: early structures could handle inputs of almost 30,000 m^3km^{-2} from their upstream watersheds, though later models could hold the maximum recorded yield of 59,000 m^3km^{-2} (Cooke 1984, p. 168). Tatum (1963) recommended even larger structures. The amount of debris expected on a unit area basis depends in part on the size of the drainage basin because of decreasing unit yields with increasing area (Fig. 5.12, Scott and Williams 1978).

Clear water flows define a series of hazard zones on the surfaces of alluvial fans. The areas at greatest risk are those adjacent to the major active channel and any fan surface that is near the apex and likely to receive large quantities of water from the mountain source area (Fig. 5.13). Fan toe areas are at moderate risk because of the distributary nature of the channel network and because over-bank flows from the many small channels may mimic sheetflow (Hogg 1982). Mid-fan areas away from the major channels have slight flood risk.

Application of numerical models provides more precise but perhaps less accurate methods for determining the distribution of flood hazards on alluvial fans. Dawdy (1979) provided a mathematical method for defining the 100-year flood zone on alluvial fans by combining concepts of hydraulic geometry, magnitude-frequency functions, and simplified fan geometry. Unfortunately the method requires assumptions that only single channels will be occupied by flood flows, that there will be a uniform distribution of channels on the fan, and that the

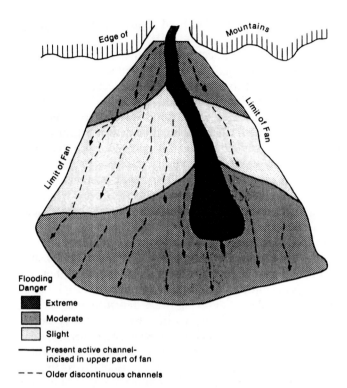

Fig. 5.13. Flood hazard zones on a typical alluvial fan based on data from the western United States and modified by experiences in southern Israel and Jordon. (After Kesseli and Beaty 1959 and Schick 1971a)

channels will not be braided. Since there are almost no fans that meet these assumptions, Dawdy's method is not likely to be successful (McGinn 1980). Magura and Wood (1980) developed simulation models for areas of flood inundation based on the HEC-II computer program (Hydraulic Engineering Center 1977), but the application of the program to fan environments where the flow is essentially without banks seems questionable, and no empirical confirmation of the method has been forthcoming.

Floods in the channels of alluvial fans also pose a dilemma for transportation route planning (reviewed by Schick, 1971a). Routes of communication and transportation frequently pass along the lengths of mountain fronts, and therefore must cross alluvial fans. Construction of the crossing at the fan apex insures that bridges, culverts, and related structures will bear the brunt of any flooding that occurs on the fan since all flows must pass across the apex. However, fan apex locations are safest from the standpoint of lateral channel stability (Fookes 1976). In mid-fan areas, profile irregularities are greatest, increasing construction costs. Fan toe areas have only small channels, but there are many of them, requiring many structures to carry the road, railroad, or communication line over the water courses. Since it is unlikely that all the courses will be used simultaneously, it may be possible to establish crossings over a few stabilized channels, while blocking the unused ones.

Channel instability on fans is problematical because when large amounts of debris enter the fan area, some channels may become blocked with sediment or trash from upstream. Flows then divert to new channels or reoccupy previously abandoned channels. In the floods of 1966 the southern Jordan town of Ma'am was partially destroyed and 70 people died when flood waters reoccupied an abandoned channel that had been sealed off by an earthen dam (Central Water Authority 1966). In 1969, local deposition apparently plugged main channels causing reoccupation of historically unused distributary channels on the fan of Tujunga Wash, southern California (Scott 1973). The result was widespread surface erosion, destruction of old alluvial terraces, headcut erosion due to locally lowered base levels, and extensive damage to areas where property owners had not recognized the potential hazard.

Perhaps the problem of lack of recognition of hazard areas on dryland alluvial fans is because most of the time they are completely dry. Scott (1971) and Giessner and Price (1971) found that historical observations indicated that destructive floods on alluvial fans in southern California have a recurrence interval of about 70 years. Minor damage such as that on highways of southern Israel and Jordan may escape public notice, and rapidly expanding urban areas may engulf hazardous alluvial fan surfaces where no one has witnessed water, debris, or mud flows. It is only costly disasters that capture the public attention, a situation common to other natural hazards (Burton et al. 1978). A second reason that hazard zones on fans are unrecognized is that surface features and soils do not reveal flood areas as readily as in humid regions where flood-plain materials are easily identified in the field (Cain and Beatty 1968).

The best defense against alluvial fan hazards is the identification of hazardous zones followed by administrative controls to insure that investment in high-risk areas is minimized. In his investigation of possible remedies for flooding of alluvial fans and pediments near the McDowell Mountains, Scottsdale, Arizona, Rhoads (1986b) suggested minimizing building densities, positioning structures away from channels, and instituting floodproofing measures. The generation of specific policies, planning efforts, and public education to control siting problems are ingredients of successful efforts (Guy 1970). Engineering efforts to control debris inflows, stabilize channels, and protect valuable property enjoy success, but only at a heavy price. Although not all its flood-control investment has been for protection on alluvial fans, the experience of Los Angeles is instructive. By 1974 the city had spent about $1.5 billion on flooding problems, with some programs being only marginally effective in a cost-benefit analysis (U. S. Forest Service 1974).

5.4 Channels and Flood Plains

After water and sediment make their way from highland areas across pediments and fans to lowlands, the dominant fluvial processes and forms are in alluvial channels on basin and valley floors. Some aspects of presently held fluvial theory for the explanation of these channel processes and forms are useful because some of the research that supports that theory is firmly grounded in dryland experience. Much of the early research on hydraulic geometry and braided channel behavior, for example, used semiarid field sites. Unfortunately, other aspects of fluvial theory, such as that pertaining to flood plains, derive almost exclusively from humid-region examples that are not directly transferable to the dryland cases. In many instances the difficulty in transfering concepts of channel behavior from humid to dryland areas lies in the underlying assumptions of continuous system operation with well-defined feedback mechanisms, assumptions that are not met in the dryland processes.

5.4.1 Channel Patterns and Controls

Limited reaches of rivers during short time periods of a few years are straight, meandering, or braided (Wolman and Leopold 1957). Straight channels consist of a single thread and have low sinuosity; meandering channels have sinuous single-thread courses; and braided streams have several intermingling channels that may be individually sinuous, but within an entire reach that is relatively straight. Over more extended periods of several days or weeks the same reach may switch between braided and meandering configurations in response to changes in discharge and be a compound channel (Graf 1987a). On still longer time scales of several decades, channel reaches may change configuration in response to climatic, tectonic, or human influences (Gregory 1977).

Sinuosity is the along-channel distance through a given reach divided by the straight-line distance connecting the beginning and ending points of the reach. Values range from 1.0 for a perfectly straight channel to values above 2.0 for highly sinuous examples. Valleys are also occasionally meandering and have measurable sinuosity. Sinuosity of a channel has two components (Mueller 1968): hydraulic and topographic. Hydraulic sinuosity results from the fluvial processes of the channel:

$$HSI = [100(TS - VS)]/[TS - 1)], \qquad (5.5)$$

where HSI = percentage of the total sinuosity ascribed to hydraulic forces, TS = total sinuosity, and VS = valley sinuosity. Topographic sinuosity is introduced because of the influence of meandering valley alignments on the stream:

$$TSI = [100(VS - 1)]/[TS - 1], \qquad (5.6)$$

where TSI = percentage of the total sinuosity ascribed to the influence of valley morphology.

Perfectly straight channels are not found in nature, but relatively straight reaches occur where faults or weathering of joints controls channel alignment. Irregularities in course alignment may occur as a result of the inherent tendency of flows to meander, as suggested by Gorycki's (1973) experiments with flows across hydrophobic surfaces, or they may merely be the result of the stream seeking the path of least resistance around topographic or sedimentologic irregularities that the stream is incompetent to overcome. Even in relatively straight

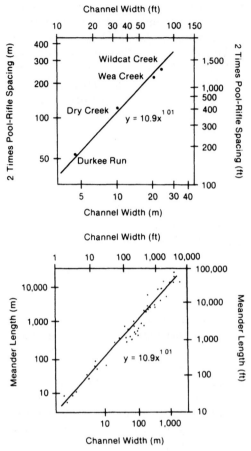

Fig. 5.14. Data showing the similarity between spacing of pool-riffle sequences and meanders showing close agreement. (After Keller and Melhorn 1973)

channels flow is inconsistent, with zones of higher and lower velocity. Yalin (1971) proposed that roller eddies partially explain this zonation, with the eddies spaced from each other at a distance of about 2π times the mean channel width.

Straight channels often develop pool and riffle sequences, where reaches with relatively shallow gradients and finer particles separate reaches with relatively steep gradients and coarser particles. In the East Fork River, Wyoming, for example, Andrews (1979) found pool and riffle sequences to be relatively stationary features of the stream. During low flows, velocity and shear stress are greater over the riffles, but during high discharges that form the channel, this arrangement is reversed and the pools become locations of scour and subsequent refilling (Keller 1971). In many rivers the spacing of pools and riffles is about five to seven times the mean channel width though in Dry Creek, an intermittent California stream, Keller and Melhorn (1973) found that pools varied greatly in size (Fig. 5.14). This particular spacing agrees well with Yalin's (1971) expectations that

Table 5.5. Relationships describing meandering stream geometry

Relationship	Reference
Rivers in flood plains	
Meander length and discharge	
$ML = 53.6\ Q_b^{0.5}$	Inglis (1949)
$ML = 29.70\ Q_b^{0.32}$	Agarwal (1983)*
$ML = 59\ Q_b^{0.48}$	Dury (1976)
$ML = 23\ Q_{1.5}^{0.62}$	Carlston (1965)
$ML = 1940\ Q_m^{0.35}\ M^{-0.73}$	Schumm (1968)*
$ML = 385\ Q_{2.33}^{0.48}\ M^{-0.74}$	Schumm (1968)*
Meander length and bankfull width	
$ML = 6.06\ W_b$	Inglis (1949)
$ML = 10.77\ W_b^{1.01}$	Leopold and Wolman (1957)
$ML = 12.15\ W_b$	Garde and Ranga Raju (1977)
$ML = 8.81\ W_b^{1.05}$	Dury (1976)
$ML = 15.0\ W_b^{1.17}\ D_{max}^{-0.51}$	Schumm (1972)
Meander belt width	
$MB = 153.4\ Q_b^{0.5}$	Inglis (1949)
$MB = 17.38\ W_b$	Inglis (1949)
$MB = 2.7\ W_b^{1.10}$	Leopold and Wolman (1957)
$MB = 18$ to $20\ W_b$	Dury (1976)
$MB = 0.476\ ML$	Agarwal (1983)*
$MB = 14.00\ W_b$	Garde and Ranga Raju (1977)
Sinuosity	
$P = 3.5\ (W_b/D_{max})^{-0.27}$	Schumm (1963)*
$P = 3.2\ (W_b/D_{max})^{-0.24}$	Williams (1984)*
$P = 0.94\ M^{0.25}$	Schumm (1963)*
$P = 0.99\ M^{0.22}$	Schumm (1968)*
Incised channels	
Meander Length	
$ML = 46.0\ Q_b^{0.5}$	Inglis (1949)
$ML = 7$ to $10\ W_b$	Dury (1976)
Meander Belt Width	
$MB = 102.1\ Q_b^{0.5}$	Inglis (1949)
$MB = 27.30\ W_b$	Inglis (1949)
$MB = 30.80\ W_b$	Garde and Raju (1977)

Notes: Compiled from Garde and Ranga Raju (1977, p. 460) and Williams (1984, p. 356–357). Symbol * indicates relationships most likely to be applicable to dryland rivers, though many relationships not tested in dryland environments. ML = meander length (m), $Q_{b,\ m,\ 1.50,\ 2.33}$ = discharge at bankfull, mean annual value, and various return intervals (m^3s^{-1}), W_b = bankfull channel width (m), M = weighted mean percent silt-clay in channel perimeter, MB = meander belt width (m), P = channel sinuosity.

spacing of forms created by roller eddies would be about 6.3 times the mean channel width. Bagnold (1960) has shown that such spacing minimizes energy loss due to bend curvature, although he used slightly different measures.

A possible variation of pool and riffle sequences characterizes steep channels in mountains, where the long profile may exhibit alternating step and pool features (Whittaker and Jaeggi 1982). The step faces may be vertical, while the pools may have reverse slopes resulting from plunge-pool erosion. The spacing of the steps and pools with a separation of about one to two channel widths is less than that for pools and riffles. As profiles of increasingly steep channels are considered, the steps are more prominent and their spacing is more regular (Judd 1964). Because of the plunging or tumbling flow in step-pool sequences, hydraulic processes are also dissimilar to those in streams with more gentle gradients (Beltaos 1982). Although relatively little is known about step-pool sequences and their operation, they are important components of dryland fluvial systems because they characterize many mountain water courses, and mountain areas cover a larger percentage of drylands than any other landform type. They are probably especially common in tectonically active areas as in the Dead Sea area of Israel and Jordan (e.g., Bowman 1977).

A regular spacing of pools and riffles also occurs in meandering streams. The connection is significant because the pools and riffles have particular locations along the channel, with riffles occurring between meanders and pools occurring at the outsides of the meander bends (Leopold and Wolman 1957). Discharge and sediment characteristics determine the size of the channel, spacing of pools and riffles, and ultimate size of the meanders. Table 5.5 summarizes the variety of relationships between morphologic characteristics of meandering streams.

The transportation of sedimentary particles through a meandering stream involves downstream and considerable lateral movement. During flow events large enough to entrain bed particles, sediment is scoured from pools and transported across intervening riffles so that overall the process is step by step, with individual particles starting and stopping. As the flow moves through bends, superelevation of the water surface on the outsides of the bends occurs because of centrifugal force (Leliavsky 1955):

$$z = [v^2 \, w]/[g \, r_c],\tag{5.7}$$

where z = magnitude of the superelevation, v = mean velocity of flow, w = width of flow, g = acceleration of gravity, r_c = radius of curvature of the bend.

Because of this superelevation and vortices generated by bank roughness (Einstein and Shen 1964), a secondary, cross-channel flow develops that moves material diagonally across the channel from the outside to the inside bank in meander reaches. A point bar deposit of relatively coarse particles develops on the inside of the meander complemented by a scour hole near the outside bank. A balance of longitudinal downstream forces and transverse lateral forces produces the surface form of the point bar (Bridge 1976, 1977):

$$\tan \alpha = (11 \, d \, \tan \varphi)/r_c,\tag{5.8}$$

where α = local surface slope of the point bar measured transverse to the direction of flow, d = depth of flow, $\tan \varphi$ = the dynamic friction coefficient of the sediment (usually somewhat less than 1.0) , and r_c = radius of curvature of the bend. By balancing the forces acting on individual particles with the surface configuration for the point bar and using concepts outlined by Rozovskii (1961), Richards (1982, p. 209) defined the equilibrium particle size at any location in the point bar system as

$$D = [24 \, \tau_x \,]/[g(\rho_s - \rho_w) \cos \alpha \tan \varphi], \tag{5.9}$$

where D = equilibrium particle size, τ_x = mean bed shear stress, g = acceleration of gravity, ρ_s = density of the sediment, ρ_w = density of the water, α = local surface slope, and tan φ = dynamic friction coefficient. Deposition of coarse particles in point-bar locations also results from flow separation at the meander (Carson 1986).

When the meander bends migrate, the coarse particles on point bars become buried by finer particles from other portions of the channel through lateral accretion to banks and by fine overbank deposits. The fining upward sequence typical of deposits of meandering streams therefore has coarse lag deposits and point bar deposits beneath coarse sands with cross stratification from dune bedforms, fine sand with ripple cross-laminae from lateral accretion, and finally silt and clay in laminar beds from over-bank deposition (Hickin and Nanson 1975; Jackson 1976).

Although the fining upward sedimentary sequence is based primarily on humid-region research, there is no reason to expect that the processes are different in drylands. Meandering streams in drylands are usually those perennial streams with source areas in mountain zones or those streams in semiarid settings with relatively dependable water supplies. In many of the world's drylands, streams have undergone changes from meandering to braided, so that the subsurface deposits do not reflect the processes of present surface streams.

Braided channels (with multiple subchannels) are wide and shallow relative to straight and meandering ones. Anastomosing systems are multi-channeled without lateral connections (for an Australian example, see Riley and Taylor 1978). Braided channels occur less frequently than single-thread channels on a global basis (Knighton 1984, p. 143), but in drylands braided channels are the most common forms. Braided channels develop under four circumstances that frequently occur in drylands.

First, braided channels are characterized by abundant bedload, and though the braided condition does not necessarily imply overloading by sediment, it represents a channel form especially well suited to transport of bedload. This efficiency results from the relatively straight course adopted by high-stage competent flows which fill the entire cross section. These flows can move directly downslope in a steep trajectory that produces greater shear stress on the bed than that possible in a more meandering course. Dryland areas provide large quantities of potential bedload material because of the predominance of mechanical weathering, which produces large particles.

Second, steep channel gradients are common for braided streams. If other major control variables are constant, an increase in slope results in a change from meandering to braiding in experimental streams (Schumm and Khan 1972), and most theoretical models based partially on empirical evidence indicate that increasing degrees of braiding accompany increases in slope (R. S. Parker 1976; Chang 1979). It may be that increases in slope represent only a partial explanation for braided channels, however, and that high amounts of stream power are the actual control. Since the definition of stream power includes slope, it may be that experiments and observations of the role of slope measured only one component of the more complex factor. Certain reaches of dryland streams are likely to have steep slopes because they must cross pediments or alluvial fans, forms that have gentle slopes compared to hillsides but which are steep in comparison with gradients on lowland rivers.

Third, highly erodible banks are a prerequisite to braiding because the initiation and growth of mid-channel bars requires that the boundaries of the channel change by erosion. Bank erosion may be controlled by the cohesiveness of the bank material, in part measurable as the per cent silt and clay content (Schumm 1960). Another important influence on bank stability is vegetation which may increase bank resistance several thousand times (Smith 1976). In drylands, banks are likely to be highly erodible because of the dominance of sand-size and larger particles (except in certain limited areas) and because vegetation is likely to be sparse.

Fourth, highly variable discharges promote braided channels but are not required because braided reaches may alternate with meandering ones in the same river at the same time (Hong and Davies 1979). If seasonal discharge fluctuations are large, braided channels can accommodate the variation without massive changes in form. In dryland areas large fluctuations in discharge are common even in the few perennial streams, so that braiding is more likely than in humid environments.

Exceptions to each of these four characteristics for braided channels may be found. Schumm and Khan (1972) could generate braided channels in laboratory settings by merely increasing the slope of the channel without altering the amount of sediment supplied. Miall (1978) pointed out that contrary to Schumm's (1968) conclusions that braided channels were characterized by bedload transport that differentiated them from meandering streams dominated by suspended load, some meandering channels have bedload transport as the dominant process. The braided channel of the Ganges River at Patna flows on a gradual slope of only 0.000066 (Wolman and Leopold 1957), so not all braided channels have steep gradients. Finally, streams crossing the Canterbury Plain of New Zealand have similar slopes, but the reaches of the Rangitata River that are sediment-poor do not braid while the other streams do, indicating that slope is not an over-riding control. Braided channels result from the interaction effects of several factors, but generally those factors are maximized in drylands.

Compound channels have two modes of operation: at low flow water occupies a single meandering channel while high flows occupy a wider "braided" channel (Graf 1987a). The braided channel has several subchannels of varying size but without a dominant one, while the compound channel has one subchannel which is clearly dominant over the others (Fig. 5.15). This meandering channel is nested inside the larger braided component in an arrangement also referred to as channel in channel (Gregory and Park 1974; Richards 1982, p. 266). When the meandering channel is filled nearly to capacity, it may have the ability to transport

Fig. 5.15. Vertical aerial photograph showing the compound channel of the Gila River west of Buckeye, Arizona. Cultivated fields mark the extent of terraces next to the channel system. Dense growth of phreatophytes lines the low flow channel with its meandering configuration. Less dense wild growth marks the flow lines on the braided channel component with its relatively straight course. (US Geol Surv photograph GS-VAJK-2–0370, May 22, 1961)

considerable material and to erode its banks. When flows exceed the capacity of the meandering low-flow channel, the system changes to a braided type and flows more directly down-gradient through the subchannels associated with the braided portion of the pattern. If the flow also exceeds the capacity of these subchannels, flow extends from one side of the compound channel system to the other and bed materials are completely mobilized. After the high flows recede, a new low-flow channel develops.

Compound channels represent an adjustment to a particular flow regime that is dominated by nearly continuous low flows with a few rare high-discharge events. Some dryland perennial streams meet this definition and develop compound channels naturally, usually in medium to large rivers with headwaters in

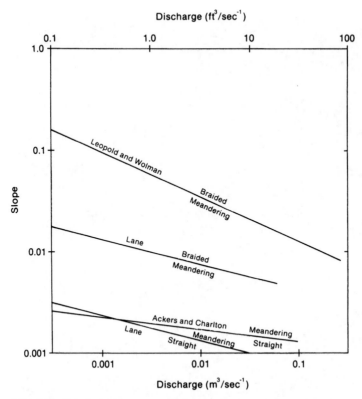

Fig. 5.16. Relationships among discharge, slope, and channel pattern as defined by various authors. Data from Lane (1957), Leopold and Wolman (1957), and Ackers and Charlton (1971). (After Schumm and Khan 1972)

mountain areas. The development of irrigation and flood control dams in many of the world's drylands produces compound channels as results of artificially modified discharges. The operation of many dams requires a continuous flow of water at a low level, which maintains the meandering low flow component of the compound system. Occasionally major flood discharges exceed the control capacity of the dams and their reservoirs and large flows maintain the braided component. As economic development of drylands continues, compound channels are likely to become more common as replacements for previously braided channels.

Straight, meandering, braided, and compound channels exist on a continuum where a variety of control factors operate to produce a range of forms. The most commonly identified control factor is slope or channel gradient which is combined with a second factor to define an information field within which observed channels may plot as points. A discriminant function separating meandering and braided (and sometimes straight) channels provides an expression of a geomorphologic threshold in such representations. The dual influences of discharge and slope are effective at discriminating among straight, meandering, and braided channels, though there is some difference of opinion as to the precise boundary between meandering and braided channels (Fig. 5.16; Schumm and Khan 1972;

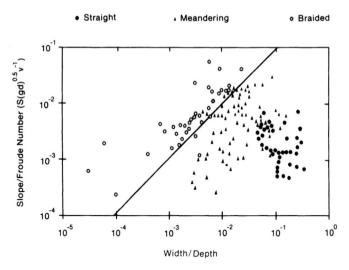

Fig. 5.17. Relationships among width-depth ratio, slope-Froude number, and channel pattern showing clear division between meandering and braided channels but less clear distinction between meandering and straight channels. (After Parker 1976)

Lane 1957; Wolman and Leopold 1957; Ackers and Charlton 1971). In an extensive investigation R. S. Parker (1976) found that in a field defined by the values slope/Froude number and width/depth the boundary between braided and meandering channels was clear, but the boundary between meandering and straight channels was poorly defined (Fig. 5.17).

If understanding of geomorphic systems is rooted in understanding of the relationships linking force and resistance (Chap 2), then it may be that the continuum of channel patterns is best seen in light of the relationship between stream power and the resistance offered by bed and bank materials. Brotherton (1979) maintained that straight, braided, and meandering channels would plot in separate compartments of a field defined by the transportability of bank particles on one hand and erodibility of particles on the other (Fig. 5.18). Transportability relies in part on the available stream power while erodibility is in part a measure of resistance. Andropovskiy (1972) successfully separated channel patterns from each other based on analysis of width and shear stress, but this is only a partial expression of the theme of force versus resistance.

Richards (1982, p. 215) used a catastrophe surface whose control factors were stream power and bank resistance and whose responding factor was sinuosity. The figure Richards produced is open to various interpretations because it shows braided channels having greater sinuosity than meandering channels and with the projection of the bifurcation does not correspond exactly to the location of the cusp. Graf (1987a) included compound channels in the fold of the cusp catastrophe surface to indicate that given certain force and resistance environments, such channels might be expected to have one of two possible states as measured by sinuosity or depth to width ratios (Fig. 5.19).

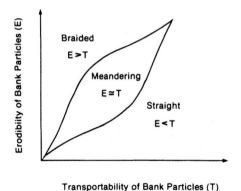

Fig. 5.18. Relationships among transportability of bank particles, erodibility of bank particles, and channel patterns. (After Brotherton 1979, p. 221)

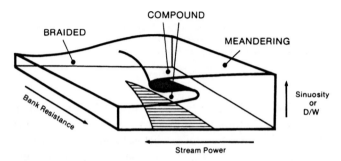

Fig. 5.19. Catastrophe theory representation of the relationships among stream power, bank resistance, and channel patterns. Compound channels lie in the fold or bifurcation set, two possible states of the channel exist depending on the magnitude of the discharge. (After Graf 1987a)

5.4.2 Channel Change and Recovery

The processes by which channel change from one type to another occurs are floods, changes in sediment inflows, and vegetation changes. Floods are almost always the forcing factors which convert meandering channels to braided configurations. In the western United States, floods were responsible for dramatic changes in channel patterns from meandering to braided on several streams of varying size (Table 5.6). The experience of the Gila River in eastern Arizona and western New Mexico was typical (Burkham 1972). Prior to the late 1890s the channel was narrow (a few tens of meters in some reaches) and meandering, but in 1905 a series of massive floods swept all traces of the meandering channel away and produced a braided channel more than a kilometer wide in some reaches. Dense phreatophyte growth and sedimentation narrowed the channel, expecially after 1940, and in the early 1980s the stream had a compound appearance,

Table 5.6. Representative changes in channel width along rivers in the western United States

River	Change	Time
Canadian River, Oklahoma	0.8 – 3.2 km	Flood in 1906
Rio Salado, New Mexico	15 – 168 m	1882 – 1918
Red River, Texas-Oklahoma	No change	1874 – 1937
Red River, Texas-Oklahoma	1.2 – .0.8 km	1937 – 1953
Cimarron River, Kansas	15 – 366 m	1874 – 1942
Cimarron River, Kansas	366 – 168 m	1942 – 1954
Platte River, Nebraska	1,161 – 111 m	1860 – 1979
S. Platte River, Colorado	790 – 60 m	1897 – 1959
N. Platte River, Wyoming	1,200 – 60 m	1890 – 1977
Gila River, Arizona	45 – 90 m	1875 – 1903
	90 – 610 m	1903 – 1917
	610 – 61 m	1917 – 1964
Salt River, Arizona	No Change	1868 – 1980
Fremont River, Utah	30 – 400 m	Flood of 1896

Notes: Sources include Hefley (1935), Bryan (1927), Schumm and Lichty (1963), Eschner et al. (1983), Hunt et al. (1953), Burkham (1972), Nadler and Schumm (1981), Graf (1983a).

apparently returning to its meandering geometry of almost a century before. Schumm and Lichty (1963) noted a similar series of events for the Cimarron River in southwestern Kansas.

In the cases listed in Table 5.6, change in channel pattern occurred without significant deepening of the channel. Occasionally pattern changes accompany incision and removal of large quantities of valley alluvium. The Fremont River near the Henry Mountains, Utah, experienced entrenchment of up to 7 m during a flood event in 1896 (Hunt et al. 1953; Graf 1983a). At the same time a braided channel 1.5 km wide replaced the original meandering 30 m wide channel (Fig. 5.20). Remnants of the meandering channel are now isolated on the surface of the terrace that once was the valley floor. Similar pattern changes and excavations occurred on the Paria River of southern Utah and northern Arizona (Hereford 1986).

The adjustment from braided channels back to meandering ones is much slower than the meandering to braided change (Fig. 5.21). Braided channels fill slowly with accumulated sediment. First a transitional form appears, the compound channel, and finally the meandering single channel dominates the system. Some observers refer to this process as "recovery," implying that the meandering form is indicative of some elusive state of environmental good health. Either the braided, compound, or meandering form might be a temporarily stable form, but in dryland rivers the change from one form to another is expectable and one condition may not be more likely than the others. Many fluvial sedimentary deposits reflect numerous changes in pattern over long time periods (Karl 1976).

The precepts of catastrophe theory are especially well suited to describing channel changes. If a single point on the surface of the catastrophe shown in Fig.

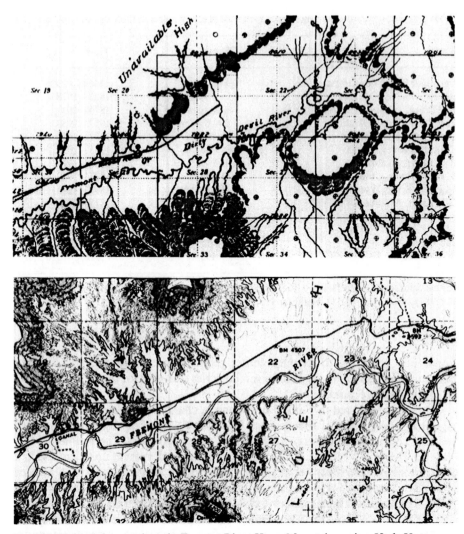

Fig. 5.20. Channel changes along the Fremont River, Henry Mountains region, Utah. Upper map from U. S. General Land Office Survey by A. D. Ferron of Township 28 South, Range 9 East of the Salt Lake Meridian, surveyed January 1883. Lower map from U. S. Geological Survey topographic map, Factory Butte, Utah, from aerial photography made in 1953. The river changed from a meandering to a braided pattern as a result of a flood in 1896 (Graf 1983a). Note especially the changes near the boundary of survey sections 22 and 23

5.19 represents the condition of a river reach at a particular time, system change is represented by moving the point about on the catastrophe surface. The path by which the point moves, determined by various combinations of stream power and bank resistance, can result in abrupt changes if the path crosses the fold area. Movement of the point might also depict relatively gradual change if the point moves in a path that avoids the fold. Observations on the Gila and Cimarron rivers suggest that the dramatic increases in stream power (and destruction of resistant

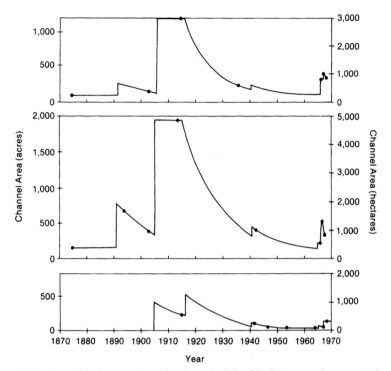

Fig. 5.21. Width changes along the channel of the Gila River, southeastern Arizona, showing the rapid expansion that occurred during floods followed by the much slower narrowing process which has been periodically interupted by renewed expansion. (After Burkham 1972)

channel-side vegetation) that accompany floods cause the point representing system conditions to move across the fold from right to left in Fig. 5.19. The results are abrupt reductions in sinuosity and depth/width ratios that characterize the change from meandering to braided. Later, reductions in stream power during flows much less than those of the large flood and increasing bank resistance from reestablished vegetation return the point to its original position by movement around the back of the fold in Fig. 5.19.

Quantitative description of the catastrophe surface and the conditions of the channels it represents is not yet common, but the general principle suggests that human management of the process is possible. Such management already occurs as an unintended byproduct of irrigation withdrawals which reduce flows and available stream power. Eschner et al. (1983) document consistent reductions in channel width of the Platte River in Nebraska between 1860 and 1979. The most dramatic change occurred near Cozad, Nebraska, where the channel width declined from 1,161 m (1,870 ft) to 110 m (177 ft) with gradual adjustment from braided to meandering geometry (Fig. 5.22). The channel changes have been produced by depletion of flows by irrigation efforts in upstream reaches where the changes have been greatest (Fig. 5.23; Karlinger et al. 1983).

When braided streams make the transition to compound and ultimately to meandering configurations, the narrowing process is accomplished by expansion

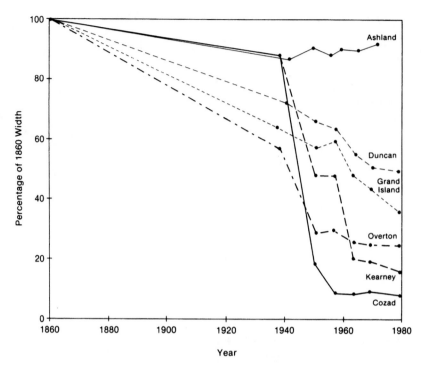

Fig. 5.22. Reductions in channel width of the Platte River, Nebraska, 1860–1979, based on maps and aerial photographs. Ashland is the site farthest downstream from irrigation diversions which deplete streamflows. (After Eschner et al. 1983, p. 30)

of bars and islands, stabilization of macro bedforms, and the attachment of bars and islands to banks by infilling of channels that once separated them. Karlinger et al. (1983) showed that in the Platte River, Nebraska, the stabilization of macro bedforms was critical to the narrowing process in a large braided channel. When sufficiently large flood flows were available to move large amounts of sediment through the system, macro bedforms in the channel appeared as waves of material several orders of magnitude larger than ripples and dunes. Crowley (1981b) showed that the macroforms are not the equivalent of dunes, but represent features with ramplike slopes facing upstream and fall faces oriented downstream. The macroforms migrate downstream under flood conditions at rates of up to 24 m (77 ft) per year (Crowley 1981a). Reductions in flood flows result in the stabilization of the macroforms and their conversion to islands that ultimately become attached to banks, resulting in channel narrowing. Nadler and Schumm (1981) showed that in the South Platte River, Colorado, the infilling of secondary channels led to rapid reductions in total channel width.

5.4.3 Horizontal Instability

Instability in channels resulting from changes in discharge, sediment load, and riparian vegetation may be manifest in horizontal migration of the channel form.

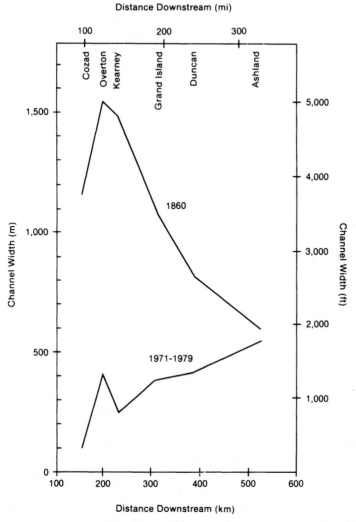

Fig. 5.23. Reductions in channel width of the Platte River, Nebraska, 1860–1979, based on a map for 1860 and aerial photographs for 1971–1979. (After Karlinger et al. 1983, p. 3)

Erosion on one bank and filling on the opposite bank results in lateral translation of the system. This process occurs in meandering streams with the general downvalley movement of the entire meander system, change in shape of individual meanders, and cutoffs (e.g., Hooke 1977, 1979, 1980). The dryland Fremont River of the 1880s shown in Fig. 5.20 was probably subject to extensive migration, especially in the reach between survey sections 22 and 23, where the form appears to be unstable.

Braided (and compound) channels are much more common in drylands than meandering ones, and horizontal instability is also present in braided forms.

Braided channels erode their banks by undercutting and thus enlarge themselves; the channel migrates if filling occurs opposite the eroding bank. On large alluvial fans, avulsion, or the plugging of channels with sediments, results in radical changes in the location of active channels (e.g., Gole and Chitale 1966). Rates of channel migration are highly variable and depend on magnitudes of flows and resistance of materials, but Table 5.7 provides some examples.

Table 5.7. Example rates of lateral channel migration

River	Period	Rate (m/yr)
Larami River, Wyoming	1851 – 1954	0.30
Missouri River, Nebraska	1883 – 1903	76
Ramganga River, India	1795 – 1806	80.5
	1806 – 1883	4.3
	1883 – 1945	4.0
Colorado River, Arizona/California	1858 – 1883	244
Kosi River, India	1736 – 1950	540

Notes: Data from Ackers and Charlton (1970) and Gole and Chitale (1966).

The deterministic prediction of the location and direction of movement of dryland channels by cutting and filling is not possible given the level of present scientific and engineering knowledge (e.g. Simons, Li and Associates 1981). A reasonable alternative is to adopt a probabilistic approach whereby observations of past channel behavior are combined with the assumption that past conditions will continue produce a statistical prediction. Even if predictions are not required and an accurate characterization of past channel changes is the objective, a spatial statistical summary provides valuable insight to variations in channel locations (following discussion based on Graf 1983a, 1984).

Locational analysis of channels depends on division of the available valley-floor surface into equal-sized subunits that are rectangles, squares, triangles, or hexagons (all forms that, given the same size and shape, exhaust the complete space with no gaps). Square cells are easiest to use and can provide a grid system with an arbitrary origin. The x- and y-distance from the origin in standard distance units or in number of cells identifies the location of each cell. Channels occupy some cells in the valley floor map, while those cells outside the channel have some probability of becoming a channel cell as erosion progresses. This probability of occupation or erosion in a given time period is a function of the magnitude of hydrologic events providing energy for the channel, the resistance of materials underlying the cell, and the distance of the cell to the channel. This method of analysis is similar to that used to assess spatial autocorrelation as outlined by Cliff and Ord (1981).

Assuming constant material properties beneath the surface of the valley floor and a system of square cells arranged in a rectangular coordinate system, the

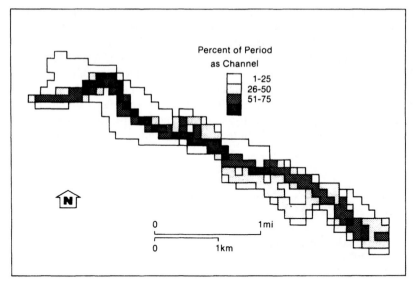

Fig. 5.24. Channel location probabilities for a reach of Rillito Creek, north of Tucson, Arizona. The values represent the percentage of time in the 107-year period of 1871–1978 that each cell was occupied by the main channel. The middle and downstream (left) portions of the channel were more stable and therefore have a concentration of cells with high values. The upstream portion was more unstable with a more even distribution of probabilities. (After Graf 1984)

probability that a cell will become occupied by the channel through erosion is

$$P_{i,j} = a \, (d_l)^{b_1} \, (d_u)^{b_2} \, (\sum_{t=1}^{n} r)^{b_3}, \tag{5.10}$$

where $P_{i,j}$ = probability of cell occupation by channel through erosion, d_l = lateral distance from the cell to the channel, d_u = upstream or along valley distance from the cell to the channel, r = return interval of annual floods experienced in each year t of the total record of n years, and a, $b_{1,2,3}$ = empirical constants (Graf 1984). Historical data from maps and aerial photography can provide definition of the empirical constants if (5.9) is converted to linear form by log transformations and solved through least squares regression methods.

Transition matrices summarize the locational changes of channels and provide input to the regression analysis. As an example, Rillito Creek near Tucson, Arizona, shows channel change behavior typical of dryland braided channels in shallow arroyos (Table 5.8). During the period 1871–1912, 75 per cent of the cells were located within 200 m of the original channel were invaded by channel migration. Fewer of the cells more distant from the channel experienced erosion in an illustration of rapid distance decay of the process. The distribution of total numbers of observations in the matrix is a function of the configuration of the channel, with meandering streams producing a relatively even distribution of observations throughout the matrix.

The probability of occupation through erosion recorded in Table 5.8 is a summary of past events, but if combined with similar tables from a variety of time periods, the resulting probabilities defined by (5.10) provide a reasonable estimator for future events. Data from Rillito Creek for the period 1912–1978 showed that distance and magnitude of events accounted for over half the variation in probability of erosion. Coefficients and exponents for the function were higher for 1912–1937 than 1937-1978, reflecting the greater erosional activity of the channel in the earlier period and increasing engineering efforts at channel stabilization in the later period (outlined by Pearthree 1983). The probabilities calculated for individual cells of the map also provided input for a collective map showing the spatial distribution of erosion probabilities over simulated future periods (Fig. 5.24).

Table 5.8. Matrix for probabilities of erosion and location for cells outlining Rillito Creek Valley, Arizona, 1871 – 1912

Lateral distance (m)	Upstream distance (m)			
	100	200	300	400+
100	P = 0.26	P = 0.16	P = 0.11	P = 0.34
	n = 46	n = 19	n = 9	n = 76
200	P = 0.00	P = 0.75	P = 0.18	P = 0.25
	n = 0	n = 8	n = 11	n = 85
300	P = 0.00	P = 0.00	P = 1.00	P = 0.16
	n = 0	n = 0	n = 4	n = 75
400	P = 0.00	P = 0.00	P = 0.00	P = 0.18
	n = 0	n = 0	n = 0	n = 45

Note: P = probability of occupation of a cell with the given lateral and upstream distances to the channel (number of cells occupied divided by total number sampled), n = total number of cells sampled. Data from project reported in Graf (1984).

5.4.4 Flood Plains

From the geomorphologic perspective a flood plain is an alluvial surface next to a channel, separated from the channel by banks, and built of materials transported and deposited by the present regime of the river (Wolman and Leopold 1957; Kellerhals et al. 1976, pp. 820–821). In other disciplines, definition of the term has slight though important differences. In hydrology, the flood plain is the surface next to the channel that is inundated once during a given return period (Ward 1978), so that to the hydrologist the flood plain might be underlain by bedrock instead of alluvium. Engineering approaches define the flood plain in a similar manner, especially those using standard computer programs to determine water-surface elevations (Hydraulic Engineering Center 1976). Again, the geomorphic history of the surface does not play a role in its definition. For the pedologist, flood plains may be defined by their soil characteristics, though these

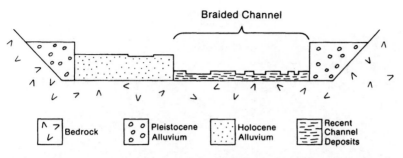

Fig. 5.25. Schematic cross section of the Agua Fria River near Sun City, Arizona, showing a typical dryland braided river without a flood plain in the normal geomorphologic sense of the term. The braided channel is confined between terrace edges. Not to scale

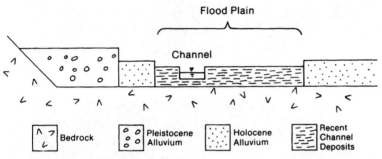

Fig. 5.26. Schematic cross section of the Gila River near Safford, Arizona, showing a typical dryland compound river that is in transition from braided to meandering configuration. The flood plain is an abandoned portion of the original braided channel. Not to scale

diagnostic characteristics vary from one climatic region to another (U. S. Soil Conservation Service 1975).

There is even little agreement on the correct spelling of the term. The U. S. Geological Survey, U. S. Soil Conservation Service, and geologists prefer the two-word form "flood plain." The present text uses the two-word form because it is preferred by the standard earth-science glossary (Bates and Jackson 1980, p. 235). When used as a modifier this form becomes hyphenated, as in "flood-plain sediments" (Bishop et al. 1978, p. 235). The hyphenated form as a noun, as in the phrase "on the flood-plain," occasionally appears in British English and was the first form of the word introduced to the scientific literature by Geikie (1882, p. 382). Geographers often use the single-word form (Goudie 1985, p. 187) which is also preferred in American English (Merriam Company 1977, p. 438). Continuing confusion in use of the term is illustrated by recent major publications in fluvial geomorphology. One uses "flood plain" (Richards 1982) while the other employs "flood-plain" (Knighton 1984); a prominent paper on the subject by Lewin (1978) uses "floodplain."

Flood-plain construction occurs in two primary ways: vertical and lateral accretion (Collinson 1978; Happ 1975). In vertical accretion, over-bank flows deposit fine-grained materials in horizontal beds with some lag deposits of more

Fig. 5.27. The transition from a braided to meandering channel as exemplified by the San Juan River near Aneth, Utah, 1928-1982. Note the phreatophyte growth which has invaded and stabilized the surface of sediments forming a flood plain where portions of the braided channel were once located. *Upper view*, 1928, photo by H. E. Gregory, University of Utah, Marriott Library Special Collection, photo P2000,9-1950(775). *Lower view*, 1982, photo by author

coarse materials from splays or exceptional flows. In meandering streams, cutoffs result in abandoned channels filled with very fine materials. In lateral accretion, coarse materials occur in cross-bedded forms as the stream deposits coarse particles along its margin while eroding the opposite channel side. Meandering and braided streams are capable of producing both processes of flood-plain development, though over-bank deposits are rare in braided configurations.

In drylands, meandering streams construct flood plains as in humid regions, but in drylands such streams are relatively rare. Braided channels, more common in drylands, are not often associated with flood plains in the geomorphologic sense. They frequently occupy the entire available space between low terraces, leaving no room for horizontal surfaces that are activated by present regime processes of the river as demanded by the definition of flood plain (Fig. 5.25). When braided channels make the transition to meandering, during the intermediate step of compound configurations, many of the original minor channels of the braided system become filled with sediment, so that strictly speaking a new flood plain develops between the terrace walls (banks of the original braided channel) and the banks of the new meandering (or low flow) channel (Fig. 5.26). As the transition from braided to meandering channel becomes better defined, the surface next to the remaining active channel takes on all the characteristics of a flood plain as the meandering channel migrates across the old floor of the braided system, smoothing the surface and reworking the deposits (Fig. 5.27).

5.4.5 Near-Channel Hazardous Environments

Although the major hazard from water in the relatively unconfined flows on alluvial fans and broad flow zones of basin fill surfaces is inundation, the major hazard from confined flows in dryland rivers is erosion. Graf (1984) estimated that in the next 50 years losses along Rillito Creek from lateral channel erosion will be five times greater than losses from flood water inundation. In the foregoing section lateral instability of river channels appears as an expected feature of dryland fluvial systems, but because the channel migrations occur rarely on a human time scale and because between episodes of lateral translation the channels may be nearly dry, public awareness of the hazard is usually low. As population growth and attending economic development of drylands continues to bring more people into contact with dryland channel hazards, evaluations of processes and implementation of mitigation measures become increasingly important. Fundamental research to support such measures is woefully lacking.

In addition to increased research efforts, a number of practical alternatives are available to deal with dryland channel hazards. Engineering modifications of the channels in order to stabilize their locations are possible in some cases, but the resulting disruption of system operations may result in unforeseen consequences and costs that may exceed the benefits (Keller 1976).

A more likely route to success in dealing with hazardous environments susceptible to channel instability is to identify the hazardous areas and then avoid them. The landscape provides numerous clues in the form of surface configuration and subsurface materials that inform on the likely surface processes (Schick 1971b; Graf 1987b), but these indicators must be accurately identified and then mapped. Cooke, Brunsden, Doornkamp, and Jones employed this approach in their analysis of flood hazard areas near Suez in Egypt (Egypt Ministry of Housing and Reconstruction 1978). In another example, the alignment of the proposed Super-Indus Highway in Pakistan has two obvious alternatives, the more desirable of which crosses the apexes of alluvial fans where the braided channels are most likely to be locationally stable (Griffiths 1978). In Albuquerque, New Mexico, Heggen and Leonard (1980) and Lagasse (1985) note that deterministic predictions of arroyo development are the best basis for the development of planning zones designed to avoid the hazard, though such predictions are not now possible.

Once mapping of hazardous areas associated with channels is complete, regulation of human use of the environment is required to make use of the information. Although erosion hazard zones associated with channels are fairly well known in the area of Tucson, Arizona, meaningful management of the problem is difficult because of the lack of public acceptance of zoning practices (Saarinen et al. 1984). The city of Los Angeles, California, probably has the most advanced system wherein geoscience professionals generate hazard maps which are subsequently included in formal local controls on land use (Cooke 1984).

5.5 Entrenched Channels

In addition to horizontal instability, many channels in drylands exhibit considerable vertical instability through entrenchment. The excavation of valley floor alluvial deposits by channel erosion results in the development of incised channels and the conversion of the original flood plain into terrace remnants. The process occurs on all continents with dryland environments, but has attacted the most attention in the American Southwest where the term arroyo refers to the resulting forms. The term arroyo has been in the formal written Spanish language since at least 775 with the meaning river or stream bed, but was first applied in the English language scientific literature by Dodge (1902). In American English the term refers to "a trench with a roughly rectangular cross-section excavated in valley-bottom alluvium with a major stream channel on the floor of the trench" (Graf 1983b, p. 280). In the United States the term gully refers to entrenched channels in V- or U-shaped cross sections excavated in colluvium, frequently on hillsides (Murray et al. 1933, Vol. 4, p. 505). In other parts of the world gully is the more common and all-inclusive term. Schumm et al. (1984) use the term gully for valley-side features and entrenched channel for the features developed on valley floors. In many dryland regions the majority of stream channels are entrenched.

5.5.1 Entrenchment Process

Entrenched channels in drylands are of two fundamental geographic types: discontinuous and continuous. Discontinuous gullies develop in small channel systems (less than 10 km² drainage area) as gradient adjustments. Early analyses in dryland areas revealed the dynamics of the discontinuous gully (Bryan 1927; Thornthwaite et al. 1942, pp. 92–94). As deposition occurs in headward portions of some channel reaches gradients steepen and channel erosion ultimately occurs, removing sediment from the upper reaches by erosion and depositing it in lower reaches to create a new more shallow gradient (later works provided elaboration: Strahler 1952b, 1956; Schumm and Hadley 1957; Mosley 1972). Thus the discon-

Fig. 5.28. Historical and modern views of Johnson Canyon, southwestern Utah, showing the excavation of an extensive arroyo system in a previously undissected alluvial valley floor. *Upper view*, 1872, photo by J. K. Hillers, U. S. Geological Survey photo and Field Records Library, Denver, Gregory photo 1272. *Lower view*, 1984, by author. The arroyo in the foreground is a tributary of a larger feature that appears at the right side of the view

tinuous gully has a headcut in the channel at its upper end and an alluvial accumulation at the downstream end. Occasionally, erosion might progress so far as to link successive discontinuous gullies along a single drainage line, resulting in a continuous form with a gradient similar to the original surface (Leopold et al. 1964, pp. 448-453), though the frequency with which this final step occurs is not clear.

In larger channels entrenchment may reach sizable proportions (Fig. 5.28). Cooke and Reeves (1976) found in southern California, for example, that San Timoteo Creek has entrenched nearly its entire length of more than 30 km (18 mi). In southern Utah, Gregory (1950) found that Kanab Creek had excavated a cavity in alluvial fill that was more than 20 m (70 ft) deep. The dramatic continuous arroyos on valley floors may rarely develop from discontinuous forms, but more often the recession of headcuts during flood periods produces them. Historical accounts of erosion throughout the southwestern United States describe such events and recount the upstream migration of headcuts as much as 10 m (30 ft) high at rates of several tens of meters per hour.

5.5.2 Causes of Entrenchment

Channel erosion is obviously the result of the disruption of some previous state of near equilibrium by alterations in the amount or rate of delivery of water, sediment, or both to the channel system. The voluminous literature on channel entrenchment contains three common themes for causal mechanisms that might result in such changes: land management, climatic change, and internal adjustments. For complete reviews see Cooke and Reeves (1976) and Graf (1983b).

The first hypothesis proposed to explain channel entrenchment in the American Southwest was related to land management, specifically overgrazing of vegetation by large numbers of livestock. Dodge (1902), Rich (1911), and many other early observers noted that entrenchment occurred at varying dates in different locations, but always after the introduction of Anglo-American cattle herds. The sequence of events envisioned by these researchers was that cattle removed much of the hillslope vegetation, which resulted in increased runoff and increased erosion along stream channels. They also observed that livestock disturbed vegetation on the valley floor and along stream banks, so that when increased runoff entered lowlands, resistance to erosion had also been reduced (Swift 1926).

A variation on this theme, later developed by Melton (1965) from sedimentological evidence, began with overgrazing on hillslopes, but with the important additional concept that the increased runoff from the slopes carried increased amounts of slope-derived sediment onto the valley floor. The newly arrived sediment buried vegetation or restricted its growth on the valley floor, while the large quantities of colluvium accumulating near valley side-slopes caused constriction of the channel in the valley center. Constriction of the lateral dimensions of the channel forced increased depths of flow and concomitant incision. This explanation, developed from evidence in southern Arizona, recognized the existence of broadly defined flow zones on undissected alluvial fills, but it appears unlikely that the explanation is applicable in areas other than those dominated by basin and range type topography and structure.

American governmental agencies accepted the overgrazing hypothesis in the 1920s and 1930s, and sponsored considerable research on the subject, leading to the documentation of much of the channel incision in major drainage systems (e.g., Cooperider and Hendricks 1937, for the Rio Grande system). The early focus on grazing led to the near-exclusion of research on alternative explanations, and

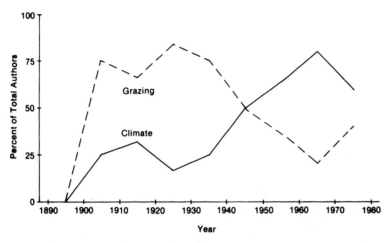

Fig. 5.29. Temporal trends in the publication of explanations for the origin of arroyos in the American Southwest. Note the early dominance of the grazing hypothesis which declined as more climatic data became available after about 1940. (Data from Graf 1983b)

public policies that attempted to control grazing and thus erosion were common in the 1930s. Because the strong link between grazing and widespread intensive erosion apparently did not exist, the programs failed to provide the expected degree of control over the erosion processes (Graf 1986).

The grazing hypothesis did not succeed completely because it failed to provide a complete explanation of all of the observations related to channel entrenchment. As Quaternary researchers began to develop more completely the sequence of earth-surface events of the last 10,000 years in the region, they found numerous episodes of erosion and sedimentation along channels (Bryan 1940, 1941). Based on palynological and archeological evidence, Hall (1977) found that in Chaco Canyon, New Mexico, a series of erosional and depositional episodes throughout the Quaternary were related to climatic changes. These pre-settlement periods of erosion were clearly not the product of overgrazing by livestock, and were likely to have been produced by climatic adjustments (Gregory 1917). Early settlement by Spanish and Mexican ranchers introduced large numbers of livestock in the 1700s without associated erosion (Denevan 1967).

As the overgrazing hypothesis gained adherents, a less numerous group of researchers emphasized the role of climatic change. After the 1930s, when more data became available and the limitations of the grazing explanation became apparent, climatic change became the dominant explanation (Fig. 5.29). The climatic hypothesis took several forms, explaining erosion as a product of increased precipitation, decreased precipitation, and variation in precipitation intensity.

The first proposal, that increased precipitation resulted in increased runoff and therefore increased erosion, was a logical extension of early reconnaissance work during the 1870s and 1880s in the American Southwest, when workers assumed that the surface conditions they were seeing were the product of increasing precipitation (Dutton 1882a). This period was coincidental with settlement by

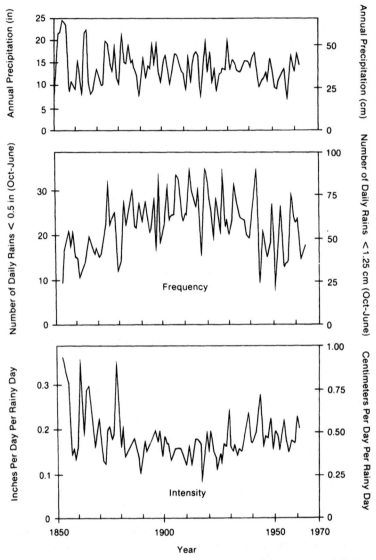

Fig. 5.30. Varying conclusions from the same precipitation record viewed from different statistical perspectives. Although the annual precipitation series appears to have no trends, the frequency and intensity series have a general peak and trough respectively between 1900 and 1920. (After Leopold et al. 1966, pp. 241–243)

Anglo-Americans and the introduction of their livestock, hence the potential for confusion of the evidence. Bryan (1922) at first accepted this scenario, but his subsequent geomorphic research in Arizona and New Mexico related to archeological materials caused him to modify his opinion and to suggest instead that when climate shifted from moist to dry conditions, vegetation cover declined, runoff (especially from occasional severe storms) increased, and erosion resulted.

Richardson (1945) believed that a shift in either direction in climate would result in erosion, and finally Haynes (1968) concluded that climate and erosion were linked, but that it was impossible to establish the nature of the link.

As more climatic data became available and geomorphologists became more adept at statistical analysis, the climatic hypothesis took on more sophisticated components. Annual precipitation as a control came to be viewed as oversimplified, and researchers investigated changes in the seasonality, frequency, and intensity of individual rainfall events (following an early suggestion by Visher 1913). Leopold (1951b), based on his work along the Puerco River of New Mexico and Arizona, suggested that periods of intense precipitation might be related to erosion through catastrophic runoff events. Others analyzed frequencies of rainfall or storm events (Cooke and Reeves 1976) or various combinations. Longterm, high quality records are still scarce even in the relatively highly developed American Southwest, but the available evidence indicates that the various components of precipitation do not necessarily all have the same temporal trends (Fig. 5.30). Further research is required before the nature of the connection between climate and resulting erosion can be specified in process terms rather than in statistical associations.

A compromise between the competing overgrazing and climatic change hypotheses is to accept both, with climatic change setting the stage for erosion and overgrazing "triggering" the final response by removing enough vegetation to move the entire fluvial system into the realm of disequilibrium. Huntington's world-wide geographical and anthropological work led to his early proposals on the connection (Huntington 1914a, b). Several subsequent workers developed the theme (e.g., Judson 1952).

A more likely marriage of the two hypotheses, however, is to relate them to the probable scale of their effects. Research on the Navajo Indian Reservation of the southeastern Colorado River Basin shows that on the scale of large basins (greater than about 26,000 km^2 or 10,000 mi^2 in area) variations in climatic conditions measured by a drought index explain large amounts of the variation in erosion (Graf 1986). On more restricted scales in the same area, overgrazing clearly controls erosion in areas less that a couple of km^2 (Hereford 1984). The areas intensively affected by overgrazing do not affect a large percentage of entire drainage basins, so grazing impacts are spatially limited. Both grazing and climate explain erosion, but they do so at different scales of analysis.

A third set of explanations for dryland channel entrenchment does not require external influences. In this perspective, successive episodes of erosion and deposition are the logical course of events, as gradients adjust to differential filling along the profile (as outlined by Schumm and Hadley 1957). These internal operations of the system are likely to affect small streams, but large river systems with drainage areas of more than a few tens of km^2 are probably too complex and have enough space to dampen the localized internal adjustments.

An all-encompassing theory for the initiation of channel entrenchment is probably not possible since the evidence that remains after entrenchment frequently does not include clues to the origin of the episode (Cooke and Reeves 1976; Heedee 1976). Broadly defined analyses that include erosion in large basins and that involve time and space changes in explaining channel filling as well as

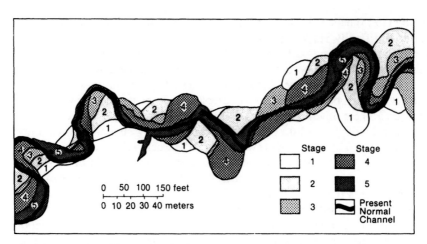

Fig. 5.31. A portion of the arroyo of Polacca Wash near Polacca, Arizona, showing the distribution of several erosion levels along the channel. (After Thornthwaite et al. 1942, p. 90)

channel erosion might lead to the development of a dynamic theory (Graf 1977a; Leopold 1978). Such a unified theory, as yet undeveloped, must account for scale variations, process reversals, internal adjustments, and external influences of climate and human activities.

5.5.3 Processes in Entrenched Channels

Debates concerning origin dominate the arroyo and gully literature, but of at least equal importance is the analysis of the operation of entrenched channels once they are established. Nickpoint migration is the process of entrenchment for many continuous arroyo systems, and after establishment, additional nickpoints may erode their way through the channel networks. The rate of nickpoint migration depends on the amount of flow available for erosion (Parker 1977; Schumm et al. 1984), and therefore on the drainage area upstream. Seginer (1966) found that in gullies

$$R = C \, A^{0.5}, \tag{5.11}$$

where R = average rate of headcut migration, C = a constant presumably related to soil and vegetation characteristics, and A = drainage basin area. The rate of recession therefore declines as the nickpoint moves upstream, a phenomenon observed in the laboratory by Parker (1977) and in the field by Graf (1977a) in a gully system in central Colorado. As outlined in Sec 2.3.4 above, the process may be described by a rate law employing a declining exponential function. Eventually the nickpoints "wash out" or decline in height by progressive smoothing of the long profile (Brush and Wolman 1957) or their movement stops for lack of water and associated stream power to perform the needed erosion.

Temporal and spatial variability of fluvial processes within entrenched channels produces a distinctive association of features. Because of internal adjust-

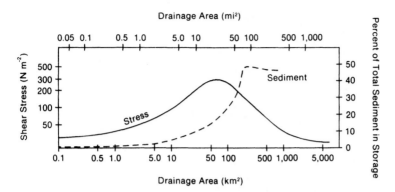

Fig. 5.32. The downstream distribution of shear stress in the streams of the Henry Mountains region, Utah, for the 10-year flood event showing a peak in the mid-basin areas where upstream drainage areas are 50–100 km² (19–39 mi²). (Graf 1983b)

ments, many entrenched channels contain a series of terrace-like steps contained within the arroyo and located below the pre-incision surface. Bailey (1935) found evidence of cyclic erosion throughout thousands of km² of the Colorado Plateau. Thornthwaite et al. (1942) noted similar evidence in Polacca Wash, northeastern Arizona, where the steps were created partially by discontinuous downcutting and partially by channel migration (Fig. 5.31). In Douglas Creek, Colorado, Womack and Schumm (1977) identified these steps as products of discontinuous, episodic downcutting. This episodic process is not ubiquitous as Hereford (1984, 1986) demonstrated using sedimentological evidence in the Little Colorado and Paria river basins.

Fluvial processes within entrenched channels also have particular spatial arrangements. Evidence from the Henry Mountains area, Utah, shows that when the entrenched channel is the product of headward erosion, the size of the excavation varies systematically with distance away from the place where the initial base-level change occurred (Graf 1982b). Discharge variations also played a role in determining size of the arroyo (see Sec 2.4.3 for a more complete discussion of the functions and statistical analysis).

Once created, the continuous arroyo influences the distribution of fluvial energy (Graf 1982a). In the smallest tributaries, there is insufficient water to generate deep flows and high amounts of stream power. In medium-size streams there are ample water and restricting arroyo walls that generate relatively deep flows and high values of stream power. In the downstream reaches where large rivers have more over-bank space for flood waters, the depth of flow is less than might otherwise be expected and values of stream power decline. The result is a distribution of stream power within entrenched channel systems that first increases and then decreases in the downstream direction (Fig. 5.32).

For those reaches with drainage areas upstream of about 1,000 km² (385 mi²) or less, stream power increases in the downstream direction in entrenched channels (left side of Fig. 5.32). Based on historical surveys in the Henry Mountains region, Graf (1983a) showed that when the stream systems were aggrading and were not entrenched in the 1880s, stream power trends were the reverse, and

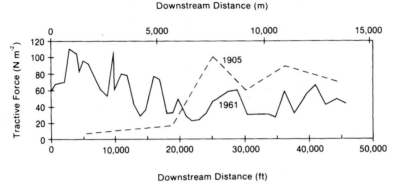

Fig. 5.33. The downstream distribution of stream power along the main stem of Walnut Gulch, southern Arizona, for 1905 and 1961. In 1905 the stream was generally aggrading in its upper reaches because the largest particles required about 45 N m⁻² for motion. In 1961, after an erosion episode in the 1930s, deposition was occurring in the middle and some lower reaches. (After Graf 1983, p. 650)

declined in the downstream direction, a circumstance that led to extensive deposition and meandering stream configurations. Stream power decreased in the downstream direction because channels were narrow, and as discharges increased in the downstream direction, over-bank flooding depleted channel flows and reduced stream power. Transmission losses may also have contributed to reduced discharges and power in the downstream direction. Vanney (1960) observed similar downstream transmission losses and effects on channel flows in the northern Sahara. In southern Arizona an erosion episode that created an arroyo network in Walnut Gulch resulted in an altered downstream distribution of stream power (Fig. 5.33).

When the distribution of stream power changes during flood events, fluvial forms are altered. At lesser flows the forms determine the distribution of stream power. The geographic distribution of stream power may therefore be an important theoretical building block for explanations of dryland streams that switch from erosion to deposition and that have predictable distributions of forms.

Just as channels might be said to "recover" from excessive widening caused by large floods, so arroyos and gullies might be said to "recover" by eventually filling with sediment. The process in entrenched channels may be linked to width of erosion. As the cavity containing the channel widens through erosion, its width may become so great that flows spread out over the floor of the arroyo to such a degree that depths of flow required to erode additional material cannot be often achieved. Moderate flows may deposit materials on the arroyo floor and the channel, which once appeared to be a braided form, becomes a meandering one. Arroyo side slopes may develop talus slopes to hasten the filling process (Fig. 5.34). This reversal from erosion to deposition has been widespread in the southwestern United States (Emmett 1974; Leopold 1976; Graf 1983a), even along large rivers (Hereford 1984).

The hazards of entrenched channels are clear and have been experienced in most parts of the dryland world. The development of trenches destroys transpor-

Fig. 5.34. Historical and modern views of upper Kanab Creek, southwestern Utah, showing the change from an eroding system to one dominated by deposition. Note the establishment of a meandering channel in the more recent view and the conversion of the vertical side walls into slopes with talus from bottom to top. *Top view*, 1931, photo by R. W. Bailey, Utah State Historical Society, Salt Lake City, photo 264931. *Bottom view*, 1984, by author

tation routes, channel-side structures, and agriculturally productive flood plains. The sediment that is mobilized by the erosion fills downstream reservoirs and alters channels far from the sediment source. Like the narrowing of channels that have changed from meandering to braided status, the filling of arroyos and gullies is several orders of magnitude slower than the erosion processes that created them. Once the hazard develops it is likely to remain for many decades.

5.6 Valley Fills

Arroyos represent channels entrenched into alluvial fills. Although many researchers have expended great effort in investigating the distribution and dynamics of the arroyos, few have investigated similar aspects of the alluvium. Valley and basin fills in dryland river systems reflect the changes experienced by the channels that formed them, so that the deposits show signs of alternative erosion and deposition as well as switches from meandering to braided channel configurations and back again. In Chaco Canyon, New Mexico, for example, deposits record more than 7,000 years of channel change and switching behavior with different facies dated by radiocarbon and archeological evidence (Love 1983a).

The interaction of fluvial processes in drylands insures that a distinctive pattern develops in the spatial distribution of valley fills along channels. Narrow valley floors and steep gradients in small tributary streams insure that they store few sediments at any time, while along medium-sized streams adequate space is available for storage and the alternating increases and decreases in available stream power causes alternating periods of storage and evacuation of sediments from the mid-basin locations. Butcher and Thornes (1978) found in ephemeral channels in Spain that large quantities of stored material did not occur in small tributaries but rather at that point in the channel system where several tributaries had an opportunity to contribute sediments.

Graf (1987b) showed that shortage of maximum amounts of sediment in mid-basin locations was at a maximum in basins of the Colorado Plateau (Fig. 5.35). Stream-power variations over time in the same general area provided a mechanism for process changes from deposition to erosion (Graf 1983a). Earlier, Love (1983b) came to similar conclusions for Chaco Canyon, New Mexico. The general downstream trend of first increasing and then decreasing amounts of storage at measured cross sections occurs in basin and range topography (Graf 1983c) and even in humid mountains (Madej 1984). Most analyses of this type show that the cross sections with maximum storage have upstream drainage areas of $50-150$ km^2 ($19-39$ mi^2) though the precise reason for this maximum remains obscure and may be coincidence given the few studies completed at this time.

The deposition of sediments in dryland fluvial systems may take place simultaneously with erosion occurring elsewhere in the system (Malde and Scott 1977). This arrangement is especially true of discontinuous gully development, because as the headcut retreats by rapid erosion, equally rapid deposition is occurring on downstream alluvial fan areas. The result is a series of time-transgressive deposits (Patton and Schumm 1981). Pickup (1975) mathematically modeled this process based on observations of Australian streams.

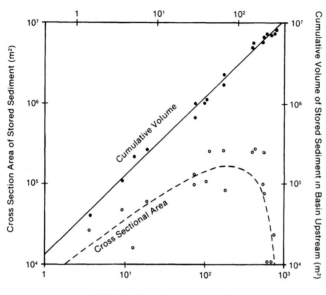

Fig. 5.35. The downstream distribution of stored sediments along White Canyon, Utah, showing the increase in the downstream direction followed by a decline. For cross sections in those reaches with drainage areas greater than about 10^3 km^2 drainage areas there is almost no sediment storage. (After Madej 1984, p. 32, and Graf 1987b)

Time-transgressive deposits need not always be the case. In some large streams (those with drainage basins several hundred km^2 in extent) nickpoint recession occurs very rapidly, either all within the same flood event or within the span of a very few years, so that subsequent depositional units are correlative over large areas and relate to the spatial distribution of stream power. Hunt et al. (1953) found this to be the case in the Fremont River system of the Henry Mountains, as did Hereford (1986) for the Paria River in southern Utah. It is possible that both modes of operation could occur in the same large basin, with time-transgressive deposits along small tributaries subject to discontinuous gully development. The step-by-step movement of sediments down these small tributaries might then supply materials for the major streams of the basin, which develop correlative deposits along their lengths as the large streams operate independently (in a hydrologic sense) from the local tributaries. Such an arrangement is to be expected in those drylands where the main stream is a throughflowing stream with water source areas in highlands that are climatically different from the dryland settings of the lower courses of the streams.

In comparison to the amounts of material stored in terraces, valley, and basin fills, the amount of material in transit at any one time or removed in a single erosion episode in typical dryland streams is relatively small. In a typical stream system in southern Arizona with a drainage area of about 150 km^2 (57 mi^2), an erosion episode in the 1930s caused channel entrenchment of widespread proportions, but it removed only about 15% of the total amount of unconsolidated materials beneath the channel (Graf 1983c). During the erosion episode, the annual sediment yield was about 18 times greater than during the post-erosion period when sediment yield and slope-sediment contributions to the channel system were about equal.

5.7 Three Landscapes

The landscape which faces the analyst in dryland settings has three components. First, there is a landscape of energy or force derived from the inputs of water from the atmosphere, a landscape largely invisible in a physical sense but nonetheless capable of definition and analysis. This landscape of force is highly variable in space and time and is susceptible to change under natural circumstances and to manipulation by human activities. Second, there is a landscape of resistance derived from the structure and materials of the earth. It, too, is variable from one time to another and one place to another and can be measured or mapped. When these two landscapes combine in earth surface processes and forms, a partly predictable suite of geomorphic environments results. The forms and processes have specific geographic patterns and specific histories of development that provide the foundations of fluvial geomorphologic theory to explain dryland rivers. As shown in this chapter, portions of that theory are known (especially aspects of the resistance), while many aspects are little explored (especially aspects of the force distributions). The next two chapters seek to explore the influences of vegetation and human activities that alter the three landscapes.

Part III   Modifications of Processes and Forms

6 Vegetation and Dryland Rivers

A direct and powerful link exists between vegetation and fluvial systems in dryland environments. Vegetation plays a pivotal role in determining the amount and timing of the runoff which ultimately supplies mass and energy for the operation of fluvial processes. Most analyses that assess the variability of sediment yield demonstrate that at the lower end of the precipitation scale (representing dryland conditions), small changes in the annual precipitation bring about major changes in vegetation communities and associated sediment yields (see Sec 4.3). In addition to occupying a central role in dryland fluvial processes, vegetation communities are susceptible to human manipulation through land management and are therefore avenues by which human activities may inadvertently or intentionally alter river behavior.

The purpose of this chapter is to review the degree to which upland and riparian vegetation interact with fluvial systems, and to assess the magnitude of adjustments that occur in river systems as a result of changes in basin vegetation. For upland vegetation, the impacts of wildfire and controlled burning are of special importance. For riparian vegetation, overt management, maintenance, and eradication have special implications for river channel change that complement discussions of channel behavior in the previous chapter.

In theory building for dryland rivers, vegetation has two inputs. First, from the perspective of explaining the hydrologic cycle, the movement of water through the biosphere implies that any theory attempting to explain water movement must account for the influence of vegetation. This book is concerned with river processes, and so is more concerned with the second input of vegetation in earth surface processes: that of a control variable that modifies the processes and forms associated with streams. The following pages emphasize this aspect of the vegetation/geomorphologic connection.

6.1 Upland Vegetation

For the sake of convenience, vegetation from the standpoint of the fluvial geomorphologist is of two types: upland and riparian. Upland vegetation occupies hillslopes and those portions of valley floors and alluvial fans that are removed from active channels. Riparian vegetation has close associations with channels. Dryland vegetation communities consist mostly of grass and shrub species, with some dense woody growth occurring in areas where the annual rainfall approaches about 50 cm (20 in). River-system operation results from the influences of factors in their drainage basins, so that the vegetation communities in those parts of the basin that are not semiarid, arid, or extremely arid can also influence dryland river processes. Of special importance are those forest communities that survive where the annual rainfall is slightly above 50 cm, because such communities are likely to be adjacent and immediately upslope from drylands areas.

6.1.1 Influences on Runoff

The definition of the effects of upland vegetation on runoff is at an early stage of development, primarily because of lack of data and the inapplicability of existing models to dryland areas (for a general review see Yair and Lavee 1985). Although instrumented watersheds have operated in some dryland areas for many years (in the United States, Mexico, parts of Africa, Israel, and Australia), the discontinuous nature of precipitation events prevents the development of a time series of sufficient length to characterize the major system input of precipitation accurately. As a result, most models that describe the operation of the integrated hydrologic system of drylands depend on statistical rather than deterministic approaches. Recent efforts of this type include those by Diskin and Lane (1972) for the American Southwest, Girard and Rodier (1979) for northern Africa, Cohen and Ben-Zvi (1979) for the Negev in southern Israel, and Pilgrim et al. (1982) for New South Wales in Australia. The necessary basis of these statistical approaches is the establishment of fundamental relationships between various system components, but even this aspect remains in initial stages (e.g., Shannan and Schick 1980, for the Negev).

Runoff processes in dryland areas are dominated by Hortonian runoff (over the surface after initial ponding), while shallow subsurface, saturation surface, or ground-water flows appear to be relatively unimportant (Yair and Lavee 1985). Given that vegetation (in addition to surface particles) contributes much to the hydrologic and hydraulic properties of the surface, it is logical to account for the vegetation in the surface runoff process. Moreover, because the vegetation is susceptible to human manipulation, it may be possible to intervene and partly control the runoff process by controlling the intermediate variable of vegetation.

Areas with sparse rainfall and little vegetation cover generate little runoff. Small amounts of rainfall imply that generally there is little moisture available, and without vegetation cover much bare ground allows rapid evaporation of the the little moisture that is available. In extremely arid regions with almost no plant cover, less than one per cent of the precipitation probably emerges as runoff, while in forested areas in mountains adjacent to and above drylands more than a quarter of the precipitation eventually appears as runoff. In an assessment of precipitation, runoff, and vegetation communities in the Rio Grande Basin of Colorado and New Mexico, Dortignac (1956) and Branson et al. (1981) found that a clearly defined positive relationship existed between precipitation and per cent of the precipitation as runoff (Fig. 6.1).

The management of vegetation for the expressed purpose of increasing water yield holds modest economic attraction for dryland watersheds. Removal of some vegetation generally increases the amount of precipitation that discharges from basins as water yields (Rowe 1948). At the heart of this approach is the concept

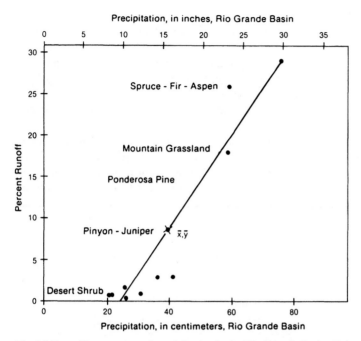

Fig. 6.1. Runoff as a percent of precipitation in the Rio Grande Basin, Colorado and New Mexico, showing the positive relationship and the variety of vegetation communities involved. (After Branson et al. 1981, p. 83, and based on data from Dortignac 1956)

Table 6.1. Evapotranspiration rates from upland vegetation likely to affect dryland fluvial processes

Vegetation	Season	Rate (cm day^{-1})	Reference
Spruce, Alberta	July	0.47	Storr et al. (1970–
Spruce, Colorado	July–Oct	0.34	Brown and Thompson (1965)
Spruce, Europe	–	0.43	Rutter (1968)
Fir, Canada	July	0.35	McNaughton and Black (1973)
Pine, Arizona	May–Aug	0.33	Thompson (1974)
Pine, Oregon	Aug	0.35	Gay (1971)
Aspen, Colorado	June–Sept	0.39	Rutter (1968)
Brush, Utah	May–Nov	0.17	Rowe and Reiman (1961)
Grass, Colorado	June–Sept	0.20	Brown and Thompson (1965)
Grass, Utah	May–Nov	0.12I	Rowe and Reiman (1961)

Note: Adapted from Rich and Thompson (1974, p. 5).

that vegetation transpires large amounts of moisture, and if the type of vegetation cover is changed to species with lower transpiration rates, the amount of water yield or runoff will increase (Table 6.1). Generally, the efficiency of vegetation changes in producing additional runoff increases as the annual precipitation increases (Fig. 6.1), so dryland settings are not as useful for this purpose as other areas (Hibbert 1971). When shrub- or forest-covered watersheds are converted to grass cover, runoff amounts may be increased from 2.5 cm yr^{-1} (1 in yr^{-1}) to as much as 22 cm yr^{-1} (8.6 in yr^{-1}) depending on the annual precipitation regime (Hibbert et al. 1975). In dryland ecosystems, clearing produces only minimal amounts of increased runoff, even in those systems with some original tree cover such as the pinyon-juniper forests of the American Southwest (Collings and Myrick 1966). Clearing of juniper-covered areas caused minor, statistically insignificant increases in runoff in Arizona (Wilm 1966).

Large drainage basins have a variety of ecosystems, and even if the majority of the basin has rainfall less than 50 cm (20 in), mountain areas will be likely to have the greater amounts and thus be candidates for vegetation manipulation. For example, the Colorado River Basin has a total area of about 65,000,000 ha (about 160,000,000 acres). The basin receives an annual average of about 14.2 in of precipitation but it produces only about 1.3 in of runoff. About 75 per cent of the

Table 6.2. Manageable vegetation and water yield for the Colorado River Basin

Vegetation	Area (ha x 10^6)	Precip-itation (cm)	Natural yield (cm)	Increase in yield (cm)
Alpine Grass	0.53	100	50–100	5.6–11.2
Mountain Grass	0.12	60–100	7.5–40	3.8–5.6
Subalpine Forest	6.07	70–75	30–40	2.5–7.5
Mixed Conifer	1.62	60–75	7.5–12	7.5–10.1
Aspen	1.34	50–100	7.5–50	12.7
Ponderosa Pine	3.04	40–60	7.5–10	2.5
Chaparral	1.42	50–55	2.5–10	2.5–12.5
Mountain Brush	1.34	40–60	2.5–15	2.5–7.5
Sagebrush	10.53	20–50	2.5–10	<2.5
Pinyon-Juniper	12.96	30–45	2.5–7.5	2.5–7.5

Notes: Does not include areas not susceptible to vegetation management. Table provides estimates based on measurements made in small areas only. Data from text of Hibbert (1979).

runoff comes from only 15 per cent of the basin, and about 37,600,000 ha (92,800,000 acres) are susceptible to treatment by reducing cover for increased runoff (Hibbert 1979). As shown in Table 6.2, the possibilities for increasing runoff in the Colorado River Basin lie mostly outside the dryland areas, so that such efforts would be spatially concentrated. They would be likely to produce only a fraction of a cm in increased runoff, but even this small amount would be an economically significant total. Whether the cost-benefit analysis would be positive is another unanswered question. In the case of treatment of pinyon-juniper in Arizona, water yield increases were statistically significant, but economically not worth the investment (Baker 1984).

Grazing alters plant communities and thus has a potential impact on hydrologic processes. Despite its general importance, the question of the impact of grazing on geomorphology and hydrology in drylands is little studied (Branson et al. 1981, p. 89), so many conclusions are preliminary and based on limited analyses. In small basins (those less than about 25 km^2 – about 10 mi^2) grazing can produce measurable effects in runoff. In carefully controlled conditions in Badger Wash basin in semiarid western Colorado, Lusby et al. (1971) found that runoff from a grazed watershed was 43 per cent greater than those that were not grazed. Similar results were derived from a study by Hanson et al. (1970) in semiarid South Dakota. In chaparral watersheds of semiarid Arizona grazing produced no increases in runoff (Rich and Reynolds 1963). The advantage of these works is that they used instrumented watersheds of limited sizes and where precise control of grazing was possible.

Grazing is not likely to alter the water yield from large basins (those greater than 2,500 km^2 – about 1,000 mi^2) because of the variety of environments in such basins. Grazing takes place in drylands only in subcomponents of the basin where plant communities that support grazing occur. Overgrazing, grazing so intense that plant communities are radically altered, is not likely to affect runoff significantly in large basins because overgrazing occurs in focused areas near water or salt sources and near human habitations. At very limited scales overgrazing undoubtedly causes increases in runoff, but at expanded scales the impact is dissipated by the complexity of the system involved. The role of widespread vegetation change through overgrazing and the resulting environmental changes continues to be an unresolved debate (e.g., Sheridan 1981).

6.1.2 Influence on Sediment Yield

Vegetation management might reasonably be expected to impact the amount of sediment supplied to channel systems because it influences the amount of water supplied. The complete removal of vegetation for construction purposes, for example, results in a flood of sediment that inundates downstream channels (Wolman 1967). In the semiarid region of Denver, Colorado, suburban development accompanied the nearly complete vegetation removal on over 50 per cent of a 15 km^2 (6 mi^2) basin; the released sediment enlarged flood plain areas by 270 per cent (Graf 1975).

Vegetation management to increase water yield as discussed in the previous section might also increase sediment yields, but the few data that are available suggest that the increase is minimal or that there is no clear trend (Table 6.3). Arnold and Schroeder (1955) suggested that the destruction of the herbaceous understory by grazing in pinyon-juniper areas caused increased rates of erosion, but offered no firm evidence. Dortignac (1956) found that erosion in the upper Rio Grande Basin was related to the condition of the plant cover, but other studies in nearby southern Utah showed opposite trends (Gifford et al. 1970). It may be that hydrologic changes that occur with vegetation manipulation are too small to affect erosion and sedimentation processes in any but the smallest drainage basins. At the local scale of a square kilometer or so, the erosion system may respond significantly (Table 6.4), but when larger areas are considered, the basin complexity may mask the effects of treatments on limited subareas.

Table 6.3. Rates of erosion from instrumented dryland watersheds with vegetation management

Watershed	Area (ha)	Vegetation	Period	Sediment yield (cm^3 m^{-2})
Salt River, Arizona:				
	1,491,800	Desert to alpine	1911–25	589.58
			1911–46	323.86
North Fork of Workman Creek, Arizona:				
	100	Mixed conifer	1939–58	2.77
		(Treated to increase water yield, 1958)		
			1959	15.92
			1960	42.90
		(Year of maximum daily rainfall, 1970)		
			1970	9.69
South Fork of Workman Creek, Arizona:				
	129	Mixed conifer	1939–1953	0.97
		(Fire destroyed 74% basal area, 1957)		
			1957	5,023.92[1]
			1957	96.88[2]
			1958–60	30.45
East Fork of Castle Creek, Arizona:				
	470	Pine-fir	1958	6.92
			1960	2.77
West Fork of Castle Creek, Arizona:				
	364	Pine-Fir	1958–60	0.00
		(17% clear cut in 1965–66)		
			Oct 1972	16.61
Fool Creek, Colorado:				
	289	Pine-fir	1952	15.22
		(Road construction completed in 1952)		
			1953	7.61
			1954–55	0.00
		(Timber harvest in 1956)		
			1956	12.46
			1957–58	19.38
			1959–65	3.46
Deadhorse Basin, Colorado:				
	270	Pine-fir	1955–65	2.77
Lexen Basin, Colorado:				
	124	Pine-fir	1956–65	2.08

Notes: 1) Sediment yield from the 24.3 ha burn site. 2) Sediment yield from the entire basin. Adapted from Rich and Thompson (1974, p. 8).

Table 6.4. Land-Use impacts on sediment production, per cent increases in production over "normal"

Land use	Basin area affected (%)	% Increase Mean effect	Local effect
Conversion of steep forest to grass (northern California)	14.8	470	2600
Forest and brushland fires (northern California)	5.3	230	1300
Brushland fires (southern California)	2.0	140	500
Poor logging (northern California)	1.4	126	1900
Unimproved roads (northern California)	0.6	124	900
Roads (Oregon)	0.3	192	1900
Recent logging (Oregon)	6.0	113	320
Bare cultivation (Oregon)	4.0	129	820
Eroding channel banks	8.0	224	1700

Notes: Data from Wallis and Anderson (1965) and Anderson (1949, 1954).

Grazing impacts on the production of sediment may follow the same precepts by being obvious in limited areas but not significant in large basins. The question of the specific nature of the connection between grazing practices and sediment yield responses is of obvious importance for scientific research and management applications, yet surprisingly little reliable information is available (Branson et al. 1981, p. 88). A recent state of the art review of the impacts of grazing contained only passing mention of sediment production (Moore et al. 1979), and review papers by Gifford (1975) and Smeins (1975) show that long-term precise measurements are generally lacking.

The most intensive study available may be that by Lusby et al. (1971; see also Lusby 1970) of several subbasins of Badger Creek Wash in semiarid western Colorado. This 13-year study included analyses of four pairs of watersheds with one member of each pair grazed and one ungrazed. Drainage areas ranged from about 5 ha (12 acres) to about 43 ha (107 acres). Over the 13-year period of instrumented record, sediment yield from the grazed basins was about 50 per cent greater than from the ungrazed basins (Fig. 6.2). Reduction in plant cover did not occur at the same time as the increases in sediment yield, and the exact mechanism for the increased sediment yield was not clear. The Badger Wash study is probably the best-documented study of its type, but the applicability of its results in other areas is limited because of the small size of the basins, the uniformity of the basins, and changes in grazing management practices during the experiment (Moore et al. 1979, p. 31).

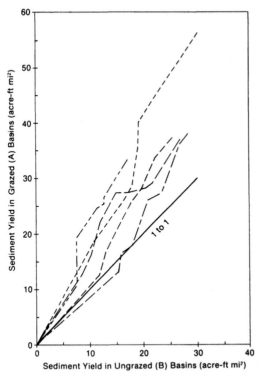

Fig. 6.2. The impact of grazing on small watersheds of Badger Creek Wash, western Colorado, based on 13 years of record showing the general increase in sediment yield. Values plotted on the horizontal axis are the yields for ungrazed basins while the values plotted on the vertical scale are for the matched grazed basins. (After Lusby et al. 1971, p. 46)

Dunford (1949) investigated the impact of grazing by following the temporal trends of sediment yield from several small plots in Colorado. He measured the sediment yields before grazing, and then after grazing with some basins excluded from grazing during both measurement periods as controls. The results showed a substantial increase in sediment production from the heavily grazed plots but only minor increases from those that experienced moderated grazing (Fig. 6.3). The problem with interpreting these results for the geomorphologist is that the research was for small components of hillslopes and not for drainage basins. Because of internal storage mechanisms associated with drainage networks, and because as yet there is no account of the spatial variability of grazing in watersheds, it is impossible to extrapolate the results to larger areas.

Additional instrumented studies of grazing impacts in drylands have shown a relationship between grazing and sediment production (Renner 1936; Croft and Bailey 1964; Dixon et al. 1977; Stephenson and Street 1977). Results are difficult to interpret because the studies included trends over time without accounting for climatic variations, and experimental areas were small.

At the opposite extreme of the scale from small plot studies is the analysis of entire large drainage basins for the relationship between sediment yield on one

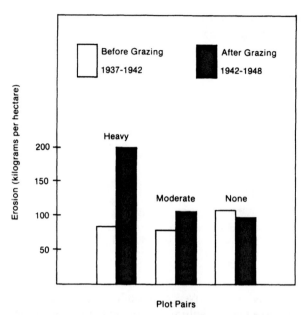

Fig. 6.3. The impact of vegetation management through clearing and replacement on runoff. In each case runoff increased after treatment, but the increments of increase for humid regions were much greater than for drylands. (After Branson 1976)

hand and various possible control factors on the other. As an example, Graf (1986) evaluated annual sediment yield from the Little Colorado River (69,837 km² – 26,964 mi²) and San Juan River (59,598 km² – 23,011 mi²) basins for the period 1930–1962 for which high quality data were available. The control factors were annual values for a drought index (which integrated the effects of temperature and precipitation) and annual numbers of livestock in the basins. Regression analysis showed that the variation in livestock numbers explained 1 to 5 per cent of the variation in sediment yield, while the climatic factor explained 38 to 66 per cent of the variation in sediment yield. The results strongly indicate that when very large basins are considered, climatic influences overwhelm the influences of grazing. Based on sedimentological analyses in the same region, Hereford (1984) reached the same conclusion. The major problem with this large-scale analysis is that the connections established are only statistical, and the exact mechanism by which climate influences the processes is not revealed.

6.1.3 Influence on Channels

The discharge of altered amounts of water or sediment from source areas subject to vegetation manipulation into dryland streams might be expected to produce channel adjustments, but this case has not been proven. The amounts of adjustment are relatively small, especially in comparison to changes associated with severe surface disruptions caused by construction or mining activities. Further-

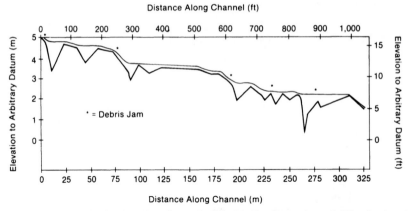

Fig. 6.4. The longitudinal profile of a reach of Prairie Creek, northern California, showing a series of pools and riffles with the impact of log steps which affect the pool and riffle spacing. (After Keller and Tally 1979, p. 181)

more, the adjustments experienced by dryland rivers as a result of flood events are likely to be larger than adjustments resulting from upland vegetation management.

A possible exception to this generalization is the role of log or debris steps that occur in medium-sized channels in mountain watersheds. These steps in longitudinal profiles result from logs and branches which fall into the channel and create a vegetative damming effect (Fig. 6.4). The step development is greatest in intermediate streams because in small streams sideslopes are so steep that the fallen organic debris orients itself along the channel length, while in larger streams the available stream power is great enough to remove the debris (Marston 1982). The significance of log steps in streams of forested mountain areas for dryland rivers downstream is of two types: dissipation of energy and storage of sediment.

The establishment of log steps converts a simple steep profile into a stepped profile with reaches of relatively shallow gradients. Heedee (1972, 1975, 1976b) reported that the riffles of organic debris and gravel bars dissipated more than half of the energy of log-step streams he investigated in Colorado and Arizona. In northern California, Keller and Tally (1979) deduced that the amount of energy dissipation was 18–60 per cent. In the mountains of Oregon conflicting results have emerged: for the Cascade Mountains Keller and Swanson (1979) reported a reduction of 30–80 per cent, and Swanson et al. (1976) reported 32–52 per cent. In the nearby Oregon Coast Range Dietrich (1975) found the reduction to be only four per cent and Marston (1982) 12 per cent. Marston suggested that the different results were due to different methods of accounting for the energy; in any case the steps dissipate some energy and their removal by land managers would alter energy conditions downstream in dryland rivers.

Log steps also perform an important role in the storage of sediment. Marston (1982) found that the amount of sediment stored in the steps was 123 per cent of the mean annual sediment discharge of the entire system. Megahan and

Nowlin (1976) and Swanson and Lienkaemper (1978) found that the stored sediment was ten times greater than the annual sediment yield. If this material were to be released through removal of the logs and other organic debris, downstream areas would receive significantly more sediment than the mean annual amount during a relatively short period of time (Beschta 1979) and the hydraulic geometry of the channels might change (Heedee 1972).

If log steps are removed, natural processes in the streams probably operate to establish new steps defined by boulder- or gravel-covered nickpoints and dominated by bedload transport. In mountain streams of Colorado and Arizona, Heedee (1981), for example, found that log and gravel steps operated in the same fashion in terms of gradient adjustment, but in the latter case bedload transport of sediment was common. In mountain streams in Arizona, the removal of log steps triggered the development of gravel bar/steps in 74 per cent of the previous locations of log steps (Heedee 1985). Upstream migration of a few nickpoints resulted in the general enlargement of the channel cross sectional areas by 6.2 per cent.

6.1.4 Impact of Fire

Wildfire and controlled burning affect water and sediment production from slopes and thus influence dryland channels downstream. Fire tends to be significant in semiarid regions where there is enough vegetation to provide fuel, and those areas with marked seasonality in the precipitation regime are more susceptible to catastrophic fires because they generate vegetation during a "wet" season and then develop fire-prone conditions during a "dry" season. In arid and extremely arid areas fires are relatively unimportant because fuel is generally lacking. Fires affect runoff processes because they cause soil surfaces to become water repellent, generating large amounts of runoff which in turn causes more rapid erosion than in pre-fire conditions (Foggin and DeBano 1971; Debano 1981). Heat from the fire burns surface litter (exposing the mineral layer beneath) and causes chemical and physical bonding of soil particles.

The fire-geomorphology connection is a circular one with a defined series of events referred to by Smalley and Cappa (1971) as the fire-induced sediment cycle. When a fire occurs in a semiarid brush-covered terrain such as that in the mountains of southern California, the flames destroy the brush and grass as well as the root net. The fire also destroys the protective layer of humus and chemically and physically alters surface soil properties. After the fire, a precipitation event with a return interval of as little as 1 to 2 years is sufficient to cause extensive sheet and rill erosion, processes which are accelerated by the absence of humus and which remove enough soil to retard the recovery of vegetation. Eventually, a precipitation event with a return interval of 5 to 10 years occurs, resulting in debris slides and the development of slide scarps, features that would not occur if dense vegetation were present.

The fire-induced sediment cycle continues when a series of years occurs with modest precipitation events. The brush begins to redevelop and slows the mass wasting processes, though debris slides may continue to expand and accelerated creep processes may be widespread. Finally, when the brush cover is completely

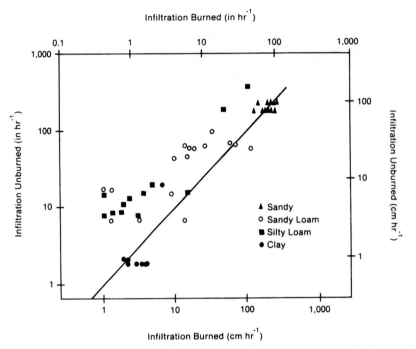

Fig. 6.5. The effect of fire on infiltration rates based on data from a range of soil types. Generally fire results in reduction of infiltration except for clay soils and a few sandy cases. (After Settergren 1967)

reestablished, the slopes become stable, as they are protected by expanding root systems and canopies with attending humus accumulations. Sediment production declines to relatively low values until the next fire occurs.

The effect of fire on hydrologic processes at and near the surface is generally to increase runoff. In semiarid Kansas, Hanks and Anderson (1957) found that in unburned watersheds 83 per cent of the precipitation that fell in a 11.3 cm (4.5 in) event was stored, while in burned watersheds only 40 per cent was stored. There is considerable spatial variability in the degree of changes because of differing responses by various soil types. In some California soils, little or no change occurs in the infiltration rates (Veihmeyer and Johnson 1944; Burgy and Scott 1952; Scott and Burgy 1956). In a few cases infiltration rates are higher after a fire (Scott 1956), but this is usually not the case, and most research documents significant reductions in infiltration (Rowe 1948). Settergren (1967) compiled results from ten previous studies and found that fire impacts on infiltration were least for clay soils and greatest for loams, with sandy soils demonstrating a variety of responses (Fig. 6.5).

Evidence from Australia and the United States suggests that the change in infiltration rates after fires is a product of chemical interactions (DeBano and Rice 1973). Plants and micro-organisms produce waxy substances which increase the water repellent properties of soils, but the presence of litter and vegetation provides countervening influences which result in moderate infiltration rates. Fire destroys the litter and vegetation, but the waxy materials (aliphatic carboxylic acid, pentose, and hextose) and their influence becomes dominant (Morris and Natalino 1969).

The decrease in infiltration rates and increase in runoff rates from burned areas frequently results in increased sediment production (Rowe et al. 1954). Rowe (1948) and many others (e.g., Cooke 1984) have observed increased sediment production from chaparral and brush watersheds in southern California, where the processes have been intensively studied. Increases in sediment production occur over a wide range of vegetation and soil types, many affecting dryland

Table 6.5. The impact of fire on sediment production from various upland vegetation types of the western United States

Vegetation	Soil	Increase sediment (t ha yr^{-1})	Reference
Pine-fir	Fine sandy loam	509.6	Helvey (1973)
Pine, oak	Sandy clay loam	9.1 – 256.2	Rowe (1941)
Pine-oak-fir	Loam/clay loam	79.4	Rich (1962)
Brush	–	19.3	Shantz (1947)
Grass/meadow	–	6.8	Branson et al. (1981)
Chaparral	Coarse	2.5	Pase and Lindenmuth (1971)
Fir	–	2.0	Brown and Krygier (1971)
Fir	Sandy loam	2.0	Megahan and Moliter (1975)
Fir-spruce	Silt loam	0.2	DeByle and Packer (1972)
Pine, oak	Gravel loam	0.0	Biswell and Schultz (1957)
Pinyon-juniper	Sandy loam	0.0	Buckhouse and Gifford (1976)

Note: Adapted from Branson et al. (1981, pp 138–139).

river systems (Table 6.5). Slope characteristics exert strong control on the degree of the sediment production response to fire (Fig. 6.6), and the timing of post-fire storms plays an important role: storms that occur soon after the fire cause larger amounts of erosion (Doehring 1968).

Management of upland vegetation results in some small hydrologic and sedimentologic responses by dryland rivers that drain the uplands (Hibbert et al. 1974). These small adjustments are likely to concern the manager only if the river system is close to a disequilibrium condition. Upland vegetation management through controlled burns and clearing by cutting or chemical applications was viewed by water managers in Arizona as a major new source of water supply (Barr 1956), but subsequent research using instrumented watersheds and controlled experiments failed to demonstrate the large increments expected (e.g., Brown 1971). The cost of increasing water yields exceeded by two to three times the cost of water from other sources, and the effort was abandoned (C. L. Smith 1972, pp. 80–81).

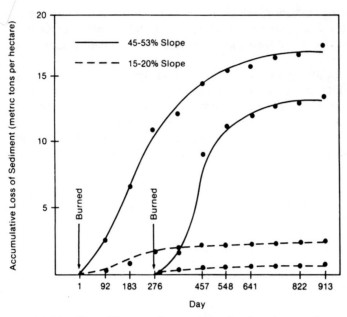

Fig. 6.6. The effect of fire on sediment production from very small watersheds in central Texas showing the importance of slope angles in the response. (After Wright et al. 1974)

6.2 Riparian Vegetation

Riparian vegetation grows in or near river channels and has strong interdependency with channel processes and near-surface water supplies. The interactions among the biotic, hydrologic, and geomorphic systems take on added importance in riparian areas because human activities frequently focus on riparian zones. In drylands, irrigable land is frequently concentrated on flood plains which also serve as sites for urban development. Planned, engineered adjustments imposed on one system of the riparian environment often have unintended consequences for the other associated systems. The following sections define the connections of riparian vegetation to the hydrologic and geomorphologic systems along dryland rivers with special emphasis on the impacts of human manipulations.

6.2.1 Hydrologic Connections

In dryland settings, the most important riparian vegetation species are phreatophytes (from the Greek meaning well plants), plants with tap-root systems that connect directly with the ground-water table (Robinson 1958). Because of this direct connection, what affects the ground-water table affects the phreatophytes and vice-versa. Early researchers viewing the dense growth of phreatophytes assumed that there was withdrawal of large quanities of water from the subsurface, altering the local hydrologic budget (Gatewood et al. 1950). The precise definition of the amounts of water involved became a question of economic significance in the 1950s, because if elimination of the phreatophyte vegetation could prevent loss of water from the subsurface, the moisture "salvaged" could be used for human enterprises.

Researchers developed three common methods for assessing the amount of moisture likely to be transpired by phreatophyte vegetation: lysimeters, tents, and water balance approaches. Lysimeters, the oldest and most common approach to assessing the evapotranspiration of plants, are tanks large enough to contain a column of soil with its associated plant community (Turner and Halpenny 1941). These tanks may range in size from the dimensions of a common flower pot to canisters several meters in diameter and 6 m deep. Experimenters arrange the tanks so that the soil surface is coincidental with the natural ground surface. Weight scales measure the amounts of water added to the tanks to supply plant growth. The difference between the water added and that which remains after the measurement period represents the amount of evapotranspiration. Lysimeters provide direct measurement of the process in a controlled environment, but they exclude accurate simulation of the horizontal movement of water in natural circumstances and represent highly simplified analogs to exceedingly complex systems.

An alternative approach to measuring evapotranspiration is the use of plastic tents which cover the plants to be analyzed and their associated soil surfaces (Campbell 1966). Pumping systems supply air to the interior of the tent and measuring devices provide constant indications of the amount of water vapor entering the tent for comparison with the amount exiting. The advantage of the tent system is that it provides direct measurement of evapotranspiration under natural hydrologic conditions. The disadvantages of the tent approach are that the scale of analysis is restricted, periods of measurement are usually brief and not necessarily representative of usual conditions, and the tents alter the micro-climatic environment of the plants.

Finally, the water balance approach provides analysis of entire hydrologic systems by assessing the input and output of moisture in entire drainage basins or in reaches of large rivers (Culler et al. 1982). Stream gauges provide data for surface flows, observation wells for ground water, and precipitation gauges for rainfall. Mass balance calculations for extended periods of several weeks or months provide indications of the quantity of water assumed to be accounted for by evapotranspiration. The general systems concepts inherent in the approach make it desirable from the standpoint of the hydrologist or geomorphologist, and the scale of analysis is similar to the scale used for related research. The problem with the water balance approach is that huge quantities of water make up all the components of the system except for evapotranspiration, which is about the magnitude of the measurement error.

The amount of moisture lost from the ground-water reservoir through phreatophytes is highly variable, depending upon the species involved, climatic conditions, and the soil conditions. A review of measured rates shows that even for the same species in the same location, different researchers reached different conclusions (Table 6.6). The data also indicate that there are considerable differences among the various species in terms of evapotranspiration. As was the case with upland vegetation, this perceived difference gave rise to efforts in the

Table 6.6. Evapotranspiration rates measured for riparian vegetation in the western United States

Reference	River	Method	Rate cm yr^{-1}
Tamarisk *(Tamarix chinensis)*[1]:			
Gay and Fritschen (1979)	Rio Grande, NM	Lysimeter	28.8
Gay and Fritschen (1979)	Rio Grande, NM	Water Balance	30.0
Sebenick and Thames (1967)	San Pedro, AZ	Tent	33.5
U. S. Bureau of Reclamation (1973)	Rio Grande, NM	Lysimeter	87.1 - 112.5
U. S. Bureau of Reclamation (1973)	Rio Grande, NM	Lysimeter	100.6 - 230.1
Hylckama (1974)	Gila, AZ	Lysimeter	101.6 - 203.2
Turner and Halpenny (1941)	Gila, AZ	Lysimeter	121.7 - 155.2
Blaney (1961)	Pecos, NM and TX	Various	131.1 - 182.9
Blaney (1961)	Pecos, NM	Various	150.1 - 159.8
Gay and Hartman (1982)	Colorado, AZ	Water Balance	166.4
Gatewood et al. (1950)	Salt, AZ	Various	211.8
Horton and Campbell (1974)	Gila, AZ	Various	213.4
Muckel (1966)	Western US	Various	213.4 - 279.4
Blaney (1961)	Gila, AZ	Various	219.5
Cottonwood *(Populus fremontii)*[2]:			
Muckel (1966)	Western US	Various	152.4 - 182.9
Robinson (1958)	Western US	Lysimeter	158.5 - 246.9
Blaney (1961)	San Louis Rey, CA	Various	158.8 - 232.4
Blaney (1961)	Gila, AZ	Various	182.9
Gatewood et al. (1950)	Gila, AZ	Various	182.9
Willow *(Salix gooddingii)*[3]:			
Robinson (1958)	Western US	Lysimeter	76.2 - 134.1
Young and Blaney (1942)	Rio Grande, NM	Lysimeter	77.5
Blaney (1961)	Santa Ana, CA	Various	114.3
Young and Blaney (1942)	Santa Ana, CA	Lysimeter	133.9
Muckel (1966)	Western US	Various	137.2
Mesquite *(Prosopis spp)*:			
Gatewood et al. (1950)	Gila, AZ	Various	100.6
Tromble (1977)	Walnut Gulch, AZ	Water Balance	36.5
Baccharis *(Baccharis* spp):			
Gatewood et al. (1950)	Gila, AZ	Various	143.3
Turner and Halpenny (1941)	Gila, AZ	Lysimeter	78.7 - 132.1
Saltgrass *(Distichlis stricta)*:			
Muckel (1966)	Western US	Various	24.4 - 121.9
Young and Blaney (1942)	Santa Ana, CA	Lysimeter	34.0 - 108.7
Young and Blaney (1942)	Rio Grande, NM	Lysimeter	46.0 - 117.9
U. S. Bureau of Reclamation (1973)	Rio Grande, NM	Lysimeter	48.3 - 84.1
Greasewood *(Chenopodiaceae)*:			
Muckel (1966)	Western US	Various	5.3 - 41.2
Robinson (1970)	Humbolt, NV	Lysimeter	7.4 - 19.6*

Table 6.6. (continued)

Reference	River	Method	Rate cm yr^{-1}
Wildrose *(Rosaceae):*			
Muckel (1966)	Western US	Various	41.7
Robinson (1970)	Humbolt, NV	Lysimeter	1.0– 8.2*
Rabbitbrush *(Compositae):*			
Muckel (1966)	Western US	Various	61.0
Robinson (1970)	Humbolt, NV	Lysimeter	2.8– 19.8*
Bermuda Grass *(Cynodon dactylon):*			
Young and Blaney (1942)	San Bernardino, CA	Lysimeter	71.6– 77.5
McDonald and Hughes (1968)	Colorado, AZ	Lysimeter	185.4
Russian Olive *(Eleagnus angustifolia):*			
U. S. Bureau of Reclamation (1973)	Rio Grande, NM	Lysimeter	53.6–130.1
Alder *(Alnus oblongiflolia):*			
Muckel (1966)	Western US	Various	152.4
Wet Meadow Grass:			
Muckel (1966)	Western US	Various	45.7
Arrowweed *(Tessaria sericea):*			
McDonald and Hughes (1968)	Colorado, AZ	Lysimeter	243.8
Quailbush *(Atriplex lentiformis):*			
McDonald and Hughes (1968)	Colorado, AZ	Lysimeter	111.8
Four-wing Saltbush *(Atriplex canescens):*			
McDonald and Hughes (1968)	Colorado, AZ	Lysimeter	96.5
Sacaton Grass *(Sporobolus airoides):*			
Blaney (1961)	Pecos, NM	Various	122.2

Notes:
1. Other studies producing rates not directly translatable to comparable form are Tomanek and Ziegler (1961), Tromble (1977), Campbell (1966), and Decker et al. (1962).
2. Other studies with nontranslatable rates are Tomanek and Ziegler (1961) and Tomanek (1958).
3. Other studies with nontranslatable rates are Tomanek and Ziegler (1961), Tomanek (1958), and Robinson (1970).
* Indicates seasonal data only.

American Southwest to change riparian vegetation communities to forms likely to conserve water.

Because of the large water "losses" associated with tamarisk (Table 6.6) this plant became the target for many experiments at control, removal, and replacement (Gatewood et al. 1950; Robinson 1965). Attempts to control phreatophyte vegetation by mechanical destruction, chemical applications, and fire resulted in

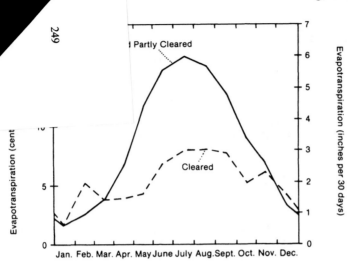

Fig. 6.7. The reduction in evapotranspiration rates resulting from operations on the Gila River that removed phreatophytes. When revegetation of the test reach occurred, reduction in evapotranspiration was minimal. (After Culler et al. 1982, p. 23)

considerable expense but little in the way of definable increases in available water. Only three significant projects offer useful indications of the utility of radical riparian management. Rowe (1963) cleared riparian vegetation from a 354 ha (875 acres) drainage basin in the San Gabriel Mountains of southern California and measured streamflow changes for comparison with a 300 ha (740 acre) control watershed which had been left undisturbed. He found streamflow yields substantially increased in the cleared watershed, but the scale was too small for extrapolation to major river systems. Bowie and Kam (1968) found that streamflow increased by about 50 per cent along a 2.4 km (1.5 mi) reach of Cottonwood Wash in western Arizona after a clearing project.

In a large scale, 8-year analysis of the water balance of a 24 km (15 mi) reach of the central Gila River, Arizona, Culler et al. (1982) attempted to resolve the debate of how much water would be salvaged by phreatophyte removal and replacement with grass. Using 414 budget periods of a few days each, they concluded that clearing the phreatophytes temporarily reduced evapotranspiration (Fig. 6.7). Bare flood-plain and channel surfaces do not represent vegetative equilibrium conditions, however, and when replacement vegetation covered the cleared areas, no water savings were apparent. Manipulation of riparian vegetation to alter aspects of the hydrologic cycle for increased water supplies was not generally effective for upland vegetation, and presently available evidence suggests a similar conclusion for riparian vegetation.

6.2.2 Impacts on Channels

Natural and artificially induced changes in riparian vegetation impact stream-channel processes because the vegetation affects the resistance of banks to

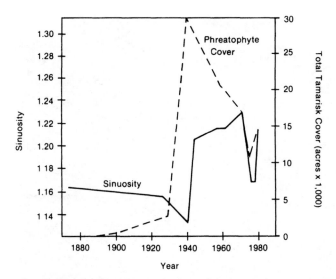

Fig. 6.8. Relationship between density and coverage of phreatophyte growth and channel sinuosity on the Gila River, central Arizona. The channel changes occurred mainly during flood periods, so that there is a lag between changes in vegetation and corresponding changes in channel sinuosity. (After Graf 1981, p. 1093)

erosion and because it induces sedimentation in channels and on flood plains. Hadley (1961) showed that the growth of tamarisk affected channel erosion and sedimentation on small streams in arid northern Arizona. Graf (1978) found that in the Colorado River system major throughflowing channels experienced an average width reduction of 27 per cent when tamarisk became firmly established on bars and beaches after about 1930. The width reduction took place because the vegetation anchored otherwise highly mobile sediments and protected them from erosion, whereas before the growth of the vegetation instability was common. Bank vegetation may increase the resistance to erosion by a factor of several thousand (Smith 1976), a circumstance typical of streams in all environments (e.g., Zimmerman et al. 1967).

Although most studies indicate that the causal relationship is from vegetation to sedimentation, Everitt (1980) viewed the causal relationship as operating in the other direction, with sedimentation occurring first and thus creating a habitat favorable to new growth. Both arrangements probably occur in most dryland settings.

The growth of phreatophytes in the channels of dryland rivers fills channel capacity with biomass and leaves little space for flood waters. The result is increased overbank flooding, which poses seriously expanded flood hazards (Graf 1980a). As flood waters divert from established channels, new channels appear and are in turn the site of new colonization by phreatophytes. The process represents a positive feedback loop because the vegetation growing in the channels induces sedimentation by providing increased hydraulic roughness. In simulations of hydraulic conditions in channels with dense phreatophytes, the U. S. Army Corps of Engineers uses Manning roughness coefficients of 0.1 or greater to

Fig. 6.9. The mismatch between a channel with straight reaches (light strip) resulting from phreatophyte clearing operations on the Gila River near Arlington, Arizona, and the meandering channel (narrow dark line) that developed in response to gradient and water-sediment adjustments. Dense phreatophyte growth (dark patches) occupies the previously braided channel components. (U. S. Geological Survey aerial photograph GS-VAJK 2 0492, May 22, 1961)

account for the impact of the vegetation. Streamlined accumulations of sediment develop behind stems and trunks, and eventually the stream deposits large amounts of sediment in the once clear channel. Smith (1981) found that tamarisk thickets caused as much as a meter of sedimentation in the channel of the Gila River, Arizona, during five floods.

The exclusion of flood waters from phreatophyte- and sediment-choked channels and the relocation or development of new active channels results in significant planimetric changes in some dryland rivers. Channel sinuosity increases as the density and extent of phreatophyte coverage increases (Fig. 6.8). When phreatophyte growth became dense in the Gila River, Arizona, in the early 1950s, sinuosity of the channel increased from a low of 1.13 to a high of 1.23 (Graf 1981). This increase appears small in numerical terms, but it implied significant erosion of channel banks and destruction of near-channel property to accommodate the more extensive meander pattern. Adjustments in pattern and attending sedimentation resulted in a gradient reduction from 0.0012 to 0.0006.

Recognition of the increased flood hazard from dense phreatophyte growth in channels may prompt environmental managers to clear the vegetation in the channels in an attempt to promote efficient discharge of flood waters (Graf 1980a). Unfortunately, attempts at establishing clear channels have resulted in straight clear reaches instead of the slightly meandering or compound channels likely to develop under natural conditions without the phreatophytes (Fig. 6.9). The channel instability that occurs in the newly designed forms could be reduced if managers were to install meandering low flow channels with wider zones for high flows as recommended by Keller (1975).

6.2.3 Exotic Vegetation – Example of Tamarisk

Exotic plant species are those that are planted by humans in areas where the plants originally did not grow under natural circumstances. Intercontinental transfers of vegetation species have been common over the last several centuries in attempts to increase agricultural productivity, provide erosion protection, and for ornamental purposes. Drylands have been candidates for exotic plantings in the search for species that are drought-resistant and because natural migration of seeds may have been difficult between widely separated dry environments (Adams et al. 1979). Exotic plantings may also be accidental (with accidental incidents probably outnumbering intentional ones – Ridley 1930), unexpectedly releasing new species to the natural environment where the exotics may effectively compete with native vegetation or occupy otherwise empty ecological niches. In semiarid grasslands of North America and New Zealand, for example, European colonists introduced grazing activities and new plants that created entirely new biotic communities in a few decades (Clark 1956). The significance of these vegetation changes for dryland rivers is that the upland vegetation in part controls water and sediment supplied to the channels while riparian vegetation influences channel processes directly. Introduction of exotic vegetation may therefore result in changes in fluvial processes.

On a global basis there are about 8,000 species that might be considered weeds, undesirable plants that impose economic costs on agriculture and environmental management (Holm et al. 1979, p. vii). Most of these species are exotic to at least part of their present ranges. Each species has a particular spatial distribution, often clustered at the sub-continental scale. Hastings et al. (1972), for example, found that in the Sonoran Desert of Mexico and the United States spotty distributions characterized most exotics. The following paragraphs outline the nature and impact of tamarisk, an exotic phreatophyte in the Sonoran Desert. Tamarisk serves as a representative example. Other plants that have expanded ranges resulting from human manipulation with possible implications for fluvial systems in the region include mesquite (*Prosopis juliflora*), Utah juniper (*Juniperus oteosperma*), big sagebrush (*Artemisia tridentata*), creosote bush (*Larrea divaricata*), snakeweed (*Gutierrezia* spp), burroweed (*Haplopappus* spp), and cholla (*Opuntia*) (Harris 1966).

Fig. 6.10. Phreatophyte growth (mostly tamarisk) in the channel of the Gila River south of Arlington, Arizona, showing the density of growth and complete coverage of what was once a braided channel. Arrow added by photographer to show direction of flow at the time of the photograph. (U. S. Army Corps of Engineers photo, 1949)

Tamarisk is a tree or shrub native to Mediterranean drylands that has several difficult to distinguish botanical forms. It consists of long, thin branches, scale-leaf structures that resemble pine needles, lavender or pink flowers, and an extensive tap-root system that enables it to operate as a phreatophyte. Early workers identified several subspecies including a coastal salt-tolerant *Tamarix galicia*, a five-stamen *Tamarix pentandra*, and a massive tree-forming *Tamarix aphyla* (McClintock 1951; Horton 1962). Present usage suggests that the general term *Tamarix cheninsis* encompasses most wild-growing tamarisk in the Sonoran region (Baum 1967). The popular name "saltcedar" recognizes the high salt-tolerant characteristics of the plant and its wispy green foliage, but because the plant is unrelated to the cedar family, "tamarisk" is more accurate.

Botanists in the United States imported tamarisk seeds and seedlings from southern Europe in a seed exchange program in the mid-1800s, and by 1852 southern California nurseries sold the plant as an ornamental (Robinson 1958). Tamarisk escaped cultivation and invaded moist sand environments along Sonoran Desert rivers, especially in the United States, by the end of the century (Christensen 1962). Tamarisk competed effectively in its new range because it occupied a previously empty ecological niche on sand bars and along channel margins, and it grew more rapidly (as much as 3 m – 10 ft per year) after rooting than its natural competitors (Horton 1962, 1964). Once established, tamarisk resisted the effects of drowning during flood periods (Horton et al. 1960; Warren

Fig. 6.11. Historical changes in vegetation in the channel of the Gila River near Geronimo, Arizona, showing the growth of tamarisk between 1939 and 1982. *Upper view*: 1939, U. S. Army Corps of Engineers Photo. *Lower view*: same location, 1982, photo by author

and Turner 1975). The plant spread most rapidly between about 1900 and 1940, and its range expanded along the trunk streams of the Colorado River system at a rate of about 20 km (12 mi) per year (Graf 1978). By the 1980s tamarisk occurred throughout the Sonoran Desert river systems in varying degrees of density, but in some places it colonized braided streams and filled channels with vegetation (Fig. 6.10).

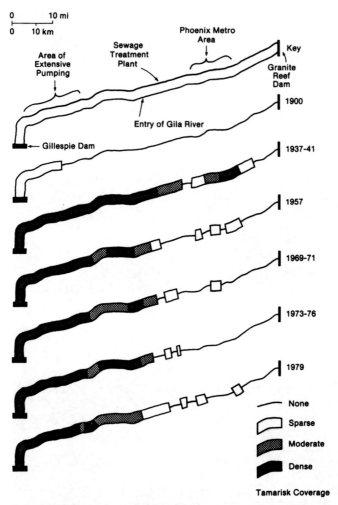

Fig. 6.12. The temporal-spatial variation of tamarisk cover in the channel of the Salt and Gila rivers, 1900–1979, showing the progressive spread upstream and subsequent decline in the post 1941 period as a result of ground-water pumping. Rises in local water tables in response to recharge by floods increased cover in the 1976–1979 period

The geomorphologic impacts of tamarisk growth in the Colorado River system are those discussed above for phreatophytes. It alters channel dimensions through induced sedimentation and changes in bank resistance (Hadley 1961; Graf 1978; Smith 1981). The result is smaller, more unstable channels than under earlier conditions. Small ephemeral streams do not have tamarisk growth because as a phreatophyte the plant requires access to ground water, and many small tributaries are elevated too far above the water table. Intermediate-sized streams that have periodic flow may be virtually overwhelmed with tamarisk in some areas (Fig. 6.11), and in some reaches the banks of the Colorado River present a seemingly inpenetrable wall of tamarisk.

Tamarisk distributions along river channels are variable in time and space. During the period of its initial range expansion, the plant typically began at one end of a channel reach and spread along the length of the channel. Subsequently, the distribution typically became patchy as environmental controls altered the original distribution (Fig. 6.12). The range of the plant along a single stream may oscillate from time to time in response to changes in the local hydrologic system.

Potential controls on the variation in range of tamarisk include surface flows and dams, floods, irrigation and sewage effluents, ground water, and sedimentation. Because tamarisk transpires large amounts of moisture and depends on moist surfaces for seedling establishment, the availability of surface water might affect its growth. During the period of initial spread of the plant in the Colorado River system in the period of about 1900–1940, surface flows declined, however, so that the trends of tamarisk coverage and surface water availability were in opposition to each other. Floods may create fresh seed beds for tamarisk and be especially important if they coincide with the peak seed production period of mid-spring (Dietz 1952), but this timing does not occur in most areas of the Colorado system. Floods may play an indirect role in encouraging tamarisk growth by ground-water recharge.

At a local scale the discharge of irrigation runoff and sewage effluents into otherwise dry channels encourages the growth of tamarisk near the outfall. General regional density does not respond to variation in such discharges, but tamarisk thickets are likely to be located at and near the water sources which also recharge ground-water supplies.

The availability of ground water appears to be the primary determining factor in explaining the spatial and temporal distribution of tamarisk (Graf 1980a). Because the plant is a phreatophyte it is effective in tapping the ground water directly, so that when this source is close to the surface, dense growth results. If the ground-water table declines slowly, tamarisk roots may grow to follow – in one example to a depth of 17 m (55 ft) (Gatewood et al. 1950). If ground water is pumped for irrigation purposes and declines rapidly, tamarisk density on the surface becomes less, and eventually the plant may survive only as isolated individuals.

Areas of rapid sedimentation may provide seedbeds for tamarisk and contribute to expansion of its range (Harris 1966). At the time tamarisk invaded the Rio Grande system in New Mexico, the installation of several reservoirs aided its growth by creating new deltas composed of moist sandy surfaces. On a local scale these additional ecological niches are significant, but on a regional scale they affect only limited areas and do not explain, as does the variation in ground water, the later decline of the plant in some areas.

The supposed heavy use of ground water by tamarisk and the flood-control problems associated with its growth in channels prompted environmental managers in the American Southwest to mount a campaign of eradication in hopes of establishing clear channels. Clearing efforts usually employed mechanical destruction of the plants, burning, or applications of herbicides (Hollingsworth et al. 1973; Horton 1960; Hughes 1966). Mechanical destruction of the plants was virtually impossible because of their long tap-root systems. Although heavy machinery could remove the crowns and trunks of the plants, regrowth occurred almost immediately. Root plows, devices that cut the stems below the soil surface, provided marginal improvement. Tamarisk also proved to be remarkably fire-resistant, and even after nearly complete combustion of the above-ground portions of the plant, regeneration from protected root systems was rapid. Application of herbicides alone or in conjunction with cutting or burning was the most effective method attempted, but within 2 years tamarisk had returned to the sites involved.

Undesirable consequences of each clearing method became apparent in reduced habitat for game birds (Anderson et al. 1977), and the loss of green belts along rivers seemed to represent a decline in general environmental quality (York 1979). Except for preliminary work by Burkham (1976), the geomorphologic impacts of clearing remain little studied. Finally, value of the small amounts of water "salvaged" by the process was probably less than the cost of clearing and maintenance, even without accounting for the suspected environmental impacts.

Tamarisk was joined in its invasion of Sonoran river courses in the 1960–1980 period by Russian olive (*Elaeagnus angustifolia*), a silver-leafed tree that also is a phreatophyte. The spread, behavior, and geomorphologic impacts of Russian olive have yet to be investigated, but its growth is generally not as dense as tamarisk. If recent history serves as a guide, it will not be the last exotic invasion of the region, and cumulative effects may become substantial.

6.3 Conclusions

Vegetation is intimately linked to fluvial processes in upland areas by influencing runoff and sediment production, but in a scale-dependent way. For small areas the connection is strong, and must be accounted for in the development of theoretical explanations of the operation of fluvial systems. In large drainage basins vegetation changes made by human activities have much less of an impact, and climatic forces (working in part through vegetation) become the dominant influence. In riparian settings, vegetation is linked to surface water, ground water, human management, and geomorphic processes in a direct fashion, so that what influences one influences all.

Fluvial geomorphologic theory does not yet adequately account for the role of vegetation, and few geomorphologists have investigated the vegetation-fluvial process connection. Laboratory experimentation may contribute to understanding of hydraulic processes around plant stems and in channel flows, while field and historical research are much needed to provide data on system changes. Field-scale experiments with intentional manipulation of controls may become less common than in the past because of financial limitations and environmental impact concerns. Fluvial theory for dryland rivers will remain incomplete without an explanation for the affect of natural and human-related variation in vegetation.

7 Direct Human Impacts on Dryland Rivers

Human activities indirectly influence fluvial processes and forms in dryland rivers through intended and inadvertent changes in vegetation, as outlined in the previous chapter. Human activities also directly influence river behavior through direct intervention in fluvial processes through control works, mining activities, development of urban areas, and the establishment and maintenance of agricultural activities. This chapter provides an examination of each of these activities and their fluvial consequences in dryland settings.

7.1 Upstream Impacts of Dams

Of all the direct impacts of human activities on dryland rivers, the most significant is the construction of dams or barrages. Whether designed for storage of irrigation and urban water supply, flood control protection, development of hydroelectric potential, or recreation usage, the erection of a dam across a water course causes an array of adjustments in the fluvial system. The impacts that extend upstream from the dam are mostly the result of induced sedimentation and the attending adjustments that channel processes make. Impacts downstream from the dam are more complex and less well understood, but they may reach farther downstream than the sedimentation impacts extend upstream. The following section explores the upstream impacts by analyzing the adjustments in processes caused by dams, the resulting deposits in the reservoir area, and finally the impacts on stream processes above the limit of the reservoir.

7.1.1 Impacts on Processes

From the geomorphological perspective, the effect of the closure of a dam is to raise the local base level from its former position on the channel bed to a new elevation coincident with the water level in the newly created reservoir. In the following discussion this new base level appears as a constant, but usually it fluctuates from time to time in response to varying inflows. In dryland areas the additions of water to reservoirs usually occurs as a result of infrequent storms (for small systems) or seasonally (for large systems), so that on a time scale of a single year the reservoir level may vary substantially. Additionally, reservoir operation may cause radical fluctuations in water level, with management objectives dictating the pattern of changes over time. For example, a reservoir being managed for flood control purposes must have a relatively large, empty reserve of storage space

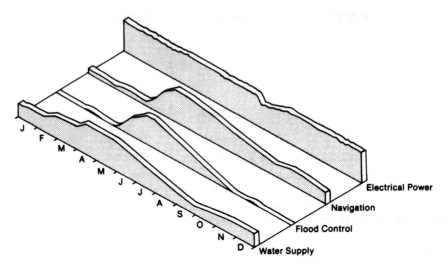

Fig. 7.1. Pool elevation changes in a hypothetical reservoir resulting from different management strategies: management for an assured water supply or for recreation purposes with a constant high reservoir level to permit a relatively constant supply for withdrawl, management for flood control with draw-down of the reservoir to allow storage space for flood inflows, management for downstream navigation with releases designed to maintain a constant discharge in the river below the dam, management for electrical power generation with the reservoir kept as full as possible at all times to insure adequate water for discharge through turbines during peak demand periods. (After Gardner 1977)

to accommodate unexpected inflows, but if a reservoir is managed for irrigation storage, as much water as possible is retained. Power generation requires the periodic release of water and resulting drawdown of the stored water, while recreationists prefer high water levels in the reservoir for as much of the time as possible. If the same structure is to perform all of these functions, trade-offs among the objectives insures that none is perfectly met and that water level fluctuations are complex (Fig. 7.1).

In drylands, natural dams and reservoirs may result from landslides, construction of alluvial fans across trunk streams by more active tributaries, flash flood deposits from similar sources, and falling sand dunes that enter from the side and form temporary plugs. Examples include the 1971 flood in the southeastern portion of the Sinai, which flushed sediments from the tributary Wadi Mikeimin into the main trunk stream, Wadi Watir (Lekach 1974). The sediments formed a fan covering 6,200 m^2 (66,667 ft^2) that blocked the main stream for almost 2 years before a flood on the trunk stream destroyed it. Falling dunes and flash floods from a tributary constructed a dam across Lake Canyon in southeastern Utah that survived for several centuries (Lipe et al. 1975). Begashibito, Shato, and Reservoir canyons in northeastern Arizona are rock-bounded gorges that falling sand dunes have divided into segments, with marshes and lakes developed between the sand blockages (Gregory 1917, p. 137).

Natural and artificial dams create lakes of two geomorphic types, basin and channel. Basin lakes or reservoirs have inflow but no constant surface outlet and are common in fault block mountain and basin terrain. Waters entering the basin reservoir do not flow through the impoundment, so the movement of water and sediment in the basin is dictated by the location of inflow entry points, thermal variations in the reservoir, and surface winds. Bottom topography frequently forms itself into a relatively smooth saucer shape through sedimentation. Channel lakes or reservoirs have consistent inflow and outflow and partially fill pre-existing valleys, canyons, or gorges. Currents in such impoundments are strongly influenced by the movement of water from input to output points, while the influences of winds and thermal characteristics play lesser roles. Bottom topography in channel reservoirs exerts strong influence on locations for deposition, with the deepest portions of the pre-lake valley filling first. Irregular bottom topography persists in such reservoirs, unlike the basin cases.

Within the channel reservoir, normal channel processes give way to a new set of processes that are part fluvial and part lacustrine (reviewed by Annandale 1987). At the upstream end, a dissipating jet of flow issues from the entry point of the river in a common delta setting (Thakur and Mackey 1973; Church and Gilbert 1975). The inflow distributes coarse particles in a delta that has a shape modified by bottom topography into a long narrow body of material on the reservoir floor. As the inflow distributes itself into the reservoir, it carries finer materials far from the input point where they may slowly settle onto the bottom if water residence time in the reservoir is long enough. Otherwise, as the water escapes the reservoir through outlet works it carries some suspended sediment with it.

Turbidity (or density) currents are of special importance in sediment transport processes in reservoirs (Howard 1953; Kulesh 1971). Sediment-laden water entering the reservoir has a density slightly greater than that of the reservoir water. The new water with its sediment load flows rapidly down the face of the delta materials at the upstream end of the reservoir and may flow for long distances along the floor of the reservoir, even to the base of the dam. Turbidity currents may also flow across a thermocline in the lake waters, a sharply defined transition zone between warmer surface waters and cooler deep water (Sturm and Matter 1978). The flows are changeable in nature, depending on conditions surrounding the lake or reservoir (Weirich 1986b), and in some cases are capable of eroding fine particles from the lake bed (Weirich 1986a).

Observation of turbidity currents in deep ocean basins and experimentally in laboratories provides reasonable descriptions of their behavior (reviewed by Allen 1970, pp. 188–210). The more dense fluid flows downslope at the bottom of the surrounding medium and forms a moving mass with a well defined head and tail. The head is usually 1.5–2.0 times the height of the tail, although the ratio may be less when the depth of the turbidity flow and the depth of the medium approach each other. Keulegan (1957) showed that the velocity of the turbidity flow is:

$$U_h = 0.7[(\rho_a - \rho_f)/\rho_a)gh]^{0.5}, \tag{7.1}$$

where U_h = velocity of the turbidity flow, ρ_a = density of the ambient fluid, ρ_f = density of the flow, g = acceleration of gravity, and h = thickness of the head of the flow. As the head moves forward, mixing with the ambient fluid occurs causing complex mingling of clear and turbulent fluid at the forward boundary (Simpson 1972; Middleton 1966).

Turbidity currents derive their forward motion from the difference in density between themselves and their surroundings, so their behavior is different from that of normal channel flow. A positive feedback mechanism develops whereby suspension of sediment causes motion which in turn results in turbulence and increased suspension, a process Bagnold (1962, 1963) defined as autosuspension. Autosuspension is dampened only by a change in slope or density which reduces velocity of the flow. Pantin (1979) defined the condition for autosuspension as:

$$(e_x \, S \, U_h)/V_g \geq 1, \tag{7.2}$$

where e_x = an efficiency factor, S = bed slope, U_h = flow velocity (Panton used velocity of the suspended load, which is approximately equal to flow velocity), and V_g = fall velocity of the suspended grains. The efficiency factor is the ratio between the power needed to transport the suspended load and the power expended against bottom friction.

Additional processes that may be significant in reservoirs include landsliding and shoreline processes. When the reservoir first forms after the natural or artificial closure of a dam, rising lake waters inundate previously dry terrain. As surface materials absorb moisture from the lake, those still above the water line may become destabilized through increased weight and increased lubrication. This circumstance is especially likely in dryland areas, where sand dunes may have surfaces at or near the angle of repose, so that minor moisture additions have great impacts on stability. The filling of Lake Mead behind Boulder Dam in Nevada and Arizona resulted in widespread landsliding with slumping in falling and climbing dunes along the shore (Smith et al. 1960). When Glen Canyon Dam, Arizona, created Lake Powell on the Colorado River system in 1962, many sand dunes along the lower portions of canyon walls slipped into the reservoir as rising lake waters reached their footslopes (U. S. National Park Service 1977a). Two decades later, sand slopes along the shorelines of the lake were still unstable, in part as a result of lake-level fluctuation which alternatively wetted and dried foot slopes.

In the largest reservoirs, shoreline processes more commonly associated with marine coasts modify the fluvial-lacustrine processes. Longshore currents may be established by prevailing wind systems, and the construction of deltas contributes materials that ultimately may be reworked into bars and spits. Perhaps the best known dryland example of such features is that of Lake Bonneville, Utah, the forerunner of the modern Great Salt Lake. Delta, bars, and spits formed in the Pleistocene lake, which was hundreds of meters higher than the present lake, are now dry features, which allowed Gilbert (1890) to deduce many of the near-shore processes usually associated with oceanic coasts.

A final process of significance in reservoirs is evaporation. If the lake has no outlet, lake size is a function of the interplay between inflow and evaporation, an arrangement occasionally used to deduce the paleoclimatic regimes of the lake area (e.g., Snyder and Langbein 1962, for Spring Valley, Nevada). Even in lakes with substantial amounts of throughflow, evaporation may stimulate the precipitation of materials to the reservoir floor and may significantly reduce the downstream flow. The Colorado River system probably loses 15 per cent of its discharge through evaporation of moisture from its reservoir surfaces.

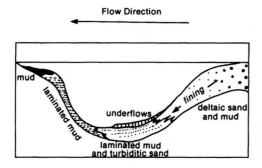

Fig. 7.2. Distribution of sediments in a hypothetical channel reservoir. (After Sturm and Matter 1978 and Leeder 1982)

7.1.2 Reservoir Deposits

Reservoir processes create a series of deposits, including those of delta, turbidity, and evaporite materials, that are useful indicators of rates of geomorphic activity (Fig. 7.2). Delta deposits at the upstream end of the reservoir assume characteristic shapes and distributions. The delta usually consists of a series of topset beds that are gently sloping in the downstream direction. Currents carry most materials across these topset beds farther into the reservoir. At the downstream end of these beds most of the coarse materials are deposited in a series of steeply dipping foreset beds. Fine particles remaining in suspension continue on their journey to settle in deep water (Jopling 1963). The accumulation of delta materials proceeds so that most rapid construction occurs through additions to the foreset bed area with much lesser additions to the topset beds (Fig. 7.3; Bogardi 1978, p. 583). Deltas observed in reservoirs through bathymetric surveys show that this idealized model is fairly close to reality in many cases. In the Govindsagar Reservoir behind Bhakra Dam, India, the topset beds have somewhat irregular upper surfaces, while in Lake Powell above Glen Canyon Dam, Arizona, tributary inputs and a steepening bedrock line in the upstream direction alter the expected delta form in profile (Fig. 7.4).

Based on data from Indian reservoirs, Miraki (1983) used simplified geometric and mathematical descriptions to characterize delta accumulations in reservoirs. Using the geometry illustrated in Fig. 7.4, the upper slope of the topset beds is related to the age of the reservoir and the initial slope of the channel floor:

$$S_u/S_o = K \, (T/T^*)^{-0.08}, \tag{7.3}$$

where S_u = slope of the upper surface of the delta, S_o = slope of the original channel floor, T = year of the age of the reservoir, and T^* = the anticipated number of years required to fill the reservoir with sediment (see further definition below). The constant K for the Indian data is 0.34, though the general applicability of the value remains to be tested. Also using the geometry of Fig. 7.4, the same independent variables may be used to define the slope of the foreset beds:

$$S_d/S_o = J \, (T/T^*)^{0.20}, \tag{7.4}$$

where S_d = slope of the foreset beds downstream from the peak of sedimentation

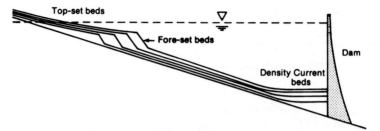

Fig. 7.3. Arrangement of the surface of sediment in a hypothetical channel reservoir. (After Bogardi 1978)

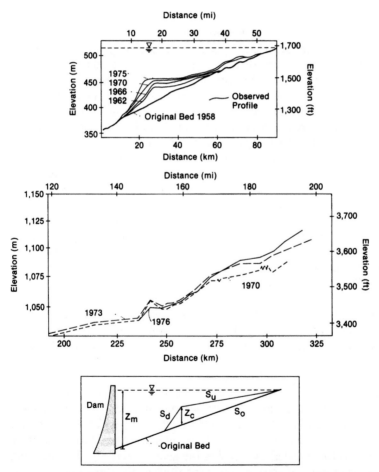

Fig. 7.4. Longitudinal profiles of reservoir deltas. *Upper*, delta in Govindsagar Reservoir, redrawn from Garde and Ranga Raju (1977). *Middle*, Colorado River delta in Lake Powell, redrawn from U. S. Bureau of Reclamation (1976) and Lazenby (1976). *Lower*, idealized geometry of a reservoir delta after Miraki (1983)

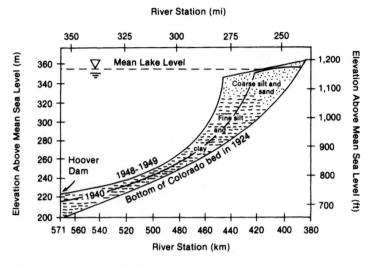

Fig. 7.5. Distribution of sediments in Lake Mead, Nevada, Utah, and Arizona. (Based on U. S. Geological Survey data)

and the other variables are as defined for (7.3). The constant J for the Indian data is 3.85.

The exponents in (7.3) and (7.4) are instructive concerning the changes in delta morphology with increasing age. The upper surface of the topset beds becomes progressively flatter through time, as evidenced by the negative exponent, but the change is of small proportions, as revealed by the small magnitude of the exponent. The foreset beds become more steep with advancing delta and reservoir age, with the change occurring rapidly in comparison with changes in the geometry of the topset beds. The empirical data for the Govindsagar and Lake Powell reservoirs (Fig. 7.4) appear to broadly support the generalization. Because of the simplified geometry and previous functions, the depth of deposition at the crest of the delta is:

$$Z_c/Z_m = M (T/T^*)^{0.285}, \tag{7.5}$$

where Z_c = height of the crest of the delta, Z_m = maximum depth of reservoir water at the dam, and other variables as defined for (7.3). The constant M is 0.717 for Miraki's (1983) Indian data.

Turbidity or density currents carry fine-grained sediments downstream beyond the delta foreset beds and distribute materials throughout the reservoir. Each flood event which empties water and sediment into the reservoir results in turbidite deposits of massive fine-grained beds which may extend from the delta foot to the dam face. The deposits of density currents may be of relatively uniform thickness over most of their length, but the fine deposits may be appreciably thicker near the dam face which checks the downslope movement of the more dense water and forces deposition. The example of Lake Mead behind Boulder Dam, Nevada, is typical with its thickest accumulation of sediments occurring at the upstream and downstream ends of the reservoir (Fig. 7.5).

Internally, deposits of turbidity currents are not uniformly fine. Ideally, the turbidite consists of coarse, partially graded particles on the bottom, overlain by parallel laminations, then cross laminations, silt and mud laminates, and finally a top massive bed of mud, all sloping downstream (Fig. 7.6; Walker 1975). Dzulynski and Slaczka (1958), von Rad (1968), and Wood and Smith (1959) provide examples of turbidites from ancient deposits. In many dryland reservoirs the vertical sequence of materials in turbidity current deposits simplifies itself into event-generated couplets that resemble varves (Lambert and Hsu 1979). Coarse materials occur on the bottom and fine ones on the top of each couplet that is related to a single flood event.

Event-generated couplets are especially likely in dryland reservoirs where most of the additions of water and sediment are the result of discrete events. Christiansson (1979) reported them from Imagi Reservoir, an artificial lake in semiarid central Tanzania with a drainage area of only 1.4 km² (0.5 mi²). Laronne (1984) found event-generated deposits in the Yatir Reservoir in an arid and semiarid portion of the Negev Desert of Israel. The artificial reservoir drained an area of 13 km² (5 mi²) for about 2.5 years before failure which permited exposure and dissection of the deposits by gully erosion. Laronne found 40 couplets of alternating beds of fine sand and silt/clay, with each couplet apparently related to one of the 20 recorded large storm events during the history of the reservoir. Large storms were those with greater than 10 mm (0.4 in) of precipitation per day.

Evaporite deposits are more common in basin reservoirs than in channel reservoirs, but because of the infrequency of flow in dryland streams, deposits in dryland channel reservoirs may have numerous beds of evaporites among delta and turbidity materials. Evaporite deposits may take the form of precipitates from solutions in lake waters or they may form by capillary action with evaporation from ground water below the dry surface of lake sediments. In either case the result is sediment rich in calcium, magnesium, and related compounds. In closed basins the various chemical constituents may have definable geographic distributions, forming concentric zones about the lowest point in the basin. Hardie (1968) and Hardie et al. (1978) found in the deposits of Saline Valley, California, that the sequence from low point upward and outward is salt pan deposits, spring travertine, gypsum/glauberite, glauberite/halite, halite, and finally gypsum.

7.1.3 Reservoir Sediment Budgets

Sediment inflow is highly variable from place to place and time to time in dryland settings. Perennial streams almost always have highly variable flows in drylands, and ephemeral streams contribute sediments to reservoirs on a discontinuous basis, so that annual averages are misleading in detail. Generally, those reservoirs storing drainage from very small basins receive relatively small amounts of sediment because there is not enough discharge to move large quantities of materials. Reservoirs storing drainage from intermediate-sized basins receive the largest input per unit area because they are located in those portions of drainage networks that have large amounts of sediment and large amounts of water concentrated in channels to transport the sediments to the reservoirs. Reservoirs

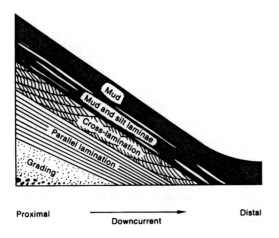

Proximal   ──────────────▶   Distal
              Downcurrent

Fig. 7.6. Generalized arrangement of sediments in a turbidity current deposit. This arrangement may be simplified in some reservoirs to event-generated couplets of fine over coarse beds. (After Allen 1970, p. 205)

with large basins upstream receive intermediate amounts of sediment on a unit area basis because although large amounts of sediment are available, much of it is stored along channels upstream from the reservoir. This arrangement appears in empirical data concerning drainage basin sizes and reservoir sediment inflows for the western United States (Table 7.1), and is a logical outgrowth of the spatial variation of stream power and sediment storage as outlined in Chapter 5 above.

The gross amount of sediment that enters a natural or artificial reservoir and that is stored there is the difference between the sediment inflow and the sediment released downstream from the dam or barrage. The ratio between inflow (Q_{si}) and outflow (Q_{so}) is the trap efficiency (T_e) of the dam and reservoir system:

$$T_e = Q_{so}/Q_{si}. \tag{7.6}$$

Most of the sediment lost from artificial reservoirs escapes in the form of suspended sediments carried through flood gates or penstocks, though if the dam is overtopped or spillways are used an additional amount of material leaves the system. Low diversion dams that serve only to direct low flows into canal intakes have almost no trap efficiency in drylands and even bedload moves across their crests during periods of high water. Stock watering reservoirs on small streams and natural dams often have trap efficiencies of nearly 100 per cent, at least for a relatively brief period of time.

The trap efficiency of reservoirs may be characterized by a number of methods that attempt to account for the relationship between the loading of the reservoir by incoming water and sediment and the ability of the reservoir to retard the downstream movement of materials and to increase the probability that the material will settle out in the reservoir area. Brune's (1953) method relied on data from 44 "normal ponded reservoirs" and produced a relationship between trap efficiency and the ratio between reservoir capacity and the annual water inflow

Table 7.1. Sedimentation data for example reservoirs in American dryland locations

Reservoir	Catchment area (km^2)	Reservoir capacity (ha m)	Reservoir deposition (ha m km^{-2} yr^{-1})
Mud Springs, Idaho	2.7	1.5	0.08
Camp Marston, California	4.1	5.4	0.32
Bennington Rago, Kansas	3.6	9.2	9.31
Kirk, Kansas	6.1	13.7	0.79
St. Marys, California	7.7	16.5	1.13
High Valley Ranch, Idaho	10.6	1.1	0.05
Gilmore, California	12.7	71.4	0.25
Baker, Montana	12.9	93.3	2.59
Mission, Kansas	20.1	228.5	6.79
Crystal Springs, California	31.1	3,595	3.23
Ericson, Nebraska	106	131.5	1.33
Emigrant Gap, Oregon	159	1,024	0.49
Morena, California	282	8,239	4.29
Muddy Creek, Colorado	394	2,088	1.54
Cold Springs, Oregon	482	6,134	1.88
Hodges, California	779	4,517	0.93
Sheridan, Kansas	463	53.8	0.22
Cucharas, Colorado	1,574	4,723	2.13
Sevier Bridge, Utah	2,821	30,850	1.36
Buffalo Bill, Wyoming	3,781	56,250	0.81
Arrowrock, Idaho	5,620	34,459	0.30
Piute, Utah	6,309	10,020	0.24
Black Canyon, Idaho	6,579	4,647	0.30
Guernsey, Wyoming	13,986	9,108	0.41
Roosevelt, Arizona	14,918	187,839	1.95
Seminoe, Wyoming	18,951	125,868	0.27
McMillan, New Mexico	32,634	11,229	0.25
John Martin, Colorado	44,237	86,599	0.69
Elephant Butte, N. Mex.	66,993	325,134	1.40
Mead, Nevada	43,4084	3,856,250	1.54

Notes: Conversions from original data in Gottschalk (1964, pp. 17 – 28, 17 – 29) assumed 85 pounds sediment = 1 ft^3.

rate (Fig. 7.7). The empirical relationship operates on the assumption that reservoirs which are large in comparison with the amount of annual inflow produce quiet waters and little throughflow of water, so that deposition of transported sediment is likely. When the capacity/annual inflow ratio is unity or greater the reservoir can hold over discharges from one year to the next, resulting in long residence times (Maddock 1948) and trap efficiencies near 100 per cent. Because of the limited data base for the Brune method, it is not especially instructive for those reservoirs that have extended dry periods or for desilting basins and flood-retarding structures (Borland 1971, p. 29/7). For other circumstances the method enjoys wide acceptance (Garde and Ranga Raju 1977, p. 400).

In geomorphologic applications, the rate of inflow to a reservoir is sometimes unknown, making application of the Brune method impossible. An alternative is

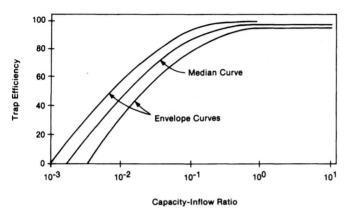

Fig. 7.7. Reservoir trap efficiency as defined by Brune (1953)

to assess trap efficiency as a function of the ratio of storage capacity to drainage basin area. Moore et al. (1960) defined a capacity/area function based on American data:

$$T_e = 100 \, [1 - (1/(1 + K \, C/A_d))], \tag{7.7}$$

where T_e = trap efficiency, C = capacity of the reservoir in ac ft, and A_d = drainage area in mi^2, and K is an empirical constant equal to 0.1. If capacity is in ha m and drainage area is in km^2, the constant K = 20.75. Trap efficiency exceeds 90 per cent in those reservoirs with values for the capacity/area ratio greater than about 100. Area-based relationships such as (7.7) are broad generalizations, however, and do not account for variations in runoff per unit area. In drylands, the runoff per unit area is relatively low, and trap efficiencies are probably higher than those predicted by the function.

Churchill (1948) developed a logarithmic relationship between trap efficiency and a sedimentation index defined by the ratio between the period of retention to the mean water velocity through the reservoir. Churchill defined period of retention as reservoir capacity (acre ft) divided by mean daily inflow rate (ft^3 s^{-1}), and the mean water velocity (ft s^{-1}) as the inflow rate divided by the average reservoir cross sectional area. Churchill originally used data from humid regions, but subsequent analysis by the U. S. Bureau of Reclamation added data from drylands that substantiated the utility of the approach (Borland 1971; Bube and Trimble 1986 made further important revisions). The relationship between sedimentation index and trap efficiency is therefore probably applicable to many dryland areas and is likely to be more successful than Brune's methods (Fig. 7.8).

Einstein (1965) developed an alternative method based on flume experiments for gravel bed streams, so that his method is applicable to channel reservoirs in systems dominated by coarse sediments. His empirical data indicated that

$$T_e = 1 - [1/(1.055 \, L \, V_s)]/[e^{(V\,d)}], \tag{7.8}$$

where T_e = trap efficiency, L = length of the reservoir or settling basin in miles, V_s = settling velocity (ft s^{-1}) of particles of a given diameter, e = base of the natural logarithm system, V = velocity of inflow (ft s^{-1}), and d = depth of water (ft).

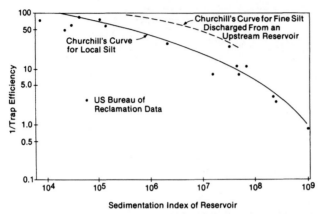

Fig. 7.8. Reservoir trap efficiency as defined by Churchill and modified by U. S. Bureau of Reclamation data from semiarid regions. (After Borland 1971, p. 29/8)

This method permits determination of trap efficiency for each particle size through different calculations for different values of settling velocity.

All relationships that predict trap efficiency from functions involving reservoir capacity, whether measured as volume or water depth, are not stationary; that is, they change over time. As the age of the reservoir increases, sedimentation reduces its capacity and thus alters the residence time of inflows. The capacity or depth term of the prediction functions becomes smaller, and the trap efficiency declines, resulting in the need for incremental calculations. Rate of sediment accumulation therefore becomes a critical issue in evaluating reservoir sediment budgets.

The rate of sedimentation in reservoirs is relatively rapid in the first years after dam closure, but declines with reservoir age. In many cases the period of rapid sedimentation is limited to less than 5 years after dam closure (Bogardi 1978, p. 581). Actual sedimentation rates are dependent on the rate of sediment inflow, particle sizes of the material, and reservoir location relative to the drainage network. These influences combine to make generalizations difficult to apply: Orth (1934) found that graphic curves relating time to the per cent of storage volume lost in reservoirs not only had different slopes (rates of loss) but that the curves had different shapes for various reservoirs. Shamov (1939) extended Orth's work and produced a family of curves relating the annual sedimentation rate to the capacity/drainage area ratio wherein upstream geomorphology and vegetation were important (Fig. 7.9). The volume of the reservoir relative to the annual inflow of water also appears to influence the rate of sedimentation, as suggested previously in the discussion of trap efficiency. Although there is no specific mathematical expression, preliminary data indicate that when the ratio of capacity to inflow is small (less than about 0.8), a small increase in drainage area results in a substantial increase in annual sedimentation rates. When the capacity/inflow ratio exceeds about 1.4, however, the rate of sedimentation is relatively insensitive to drainage basin area. The relationship in Fig. 7.10 is probably applicable to drylands as many of the points used in its establishment were from arid or semiarid environments.

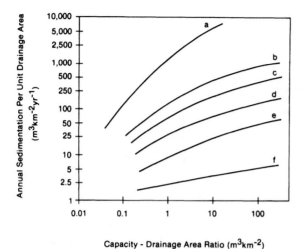

Fig. 7.9. Annual reservoir sedimentation rates defined by Orth (1934) and Shamov (1939). *A* headwaters and mountainous areas, *B* rivers draining areas with annual precipitation greater than 1000 mm (39 in) with flashy discharges, *C* similar to *B* but with extensive forest cover, *D* rivers draining hilly country with extensive forest cover, *E* rivers draining hilly or mountainous terrain with limited vegetation cover, *F* plains areas. Dryland reservoirs are most likely to plot near curve *F*. (After Bogardi 1978, p. 589)

Planners require information on the likely time period expected to elapse before filling of the reservoir with sediment, and the geomorphologist may require hindcasting capabilities given a filled reservoir and needing an estimate of filling time. In evaluating data from Indian reservoirs, Miraki (1983) found that the expected time required for complete filling of reservoirs was:

$$T^* = 2.21 \times 10^{-8} A_d^{0.38} P^{4.11} C^{-0.019}, \tag{7.9}$$

where T^* = expected time in years to complete filling of the reservoir, A_d = drainage area above the reservoir (km²), P = average annual rainfall (cm), and C = reservoir capacity (10^6 m³). Although the function is the product of limited data, the concept is a useful one that could be developed in other drylands.

7.1.4 Impacts Upstream from Backwater

The construction of a delta at the upstream end of channel reservoirs influences the gradient of the inflow channel upstream from the elevation of the reservoir water surface, but only to a limited degree. The channel gradient of the original stream connects to the gradient of the topset beds of the delta by means of a smooth curve, so there is some filling above the water level of the reservoir. In small reservoirs completely filled with sediment, this filling occurs only a short distance above the high water elevation (Eakin and Brown 1939, p. 139). Surveys by Leopold et al. (1964, pp. 261–266) on a variety of structures (mostly less than about 10 m high), all in arid and semiarid environments showed that the deposi-

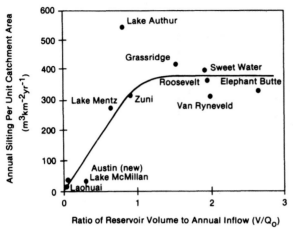

Fig. 7.10. Annual reservoir sedimentation rates defined by data predominantly from semiarid environments. (After United Nations 1953)

tion rarely extended more than a few hundred meters upstream from the backwater elevation.

The wedge of sediments forming the delta and the materials upstream from the backwater elevation has an upper surface with a gradient much less than the gradient of the original stream. Newly established gradients vary generally between 30 per cent to 90 per cent of the original gradients (Woolhiser and Lenz 1965; Leopold and Bull 1979), with a wide range of values in dryland rivers (Fig. 7.11). Preliminary data published by Leopold et al. (1964) indicated that coarser particles result in steeper gradients for the surface of the sediment wedge or topset beds of the delta, so that in streams carrying coarse materials the gradient adjustments extend farther upstream than for those cases with finer particles and more shallow new gradients.

The newly established gradient represents an arrangement that allows the stream to transport materials from the stream into the reservoir and is therefore adjusted to particle size. Once the equilibrium gradient is established, it is not likely to change, and continued accumulation of sediments in the upstream direction ceases. The difference in gradient between the topset beds or top of the sediment wedge and the original gradient of the stream probably represents the energy used by the original stream in overcoming hydraulic roughness in the original channel. The newly established channel is hydraulically more efficient and therefore transports its materials at a lower gradient.

In a few cases sedimentation above the limit of reservoir backwater may be substantial. If the original gradient is particularly gradual, any change in gradient extends for long distances. Elephant Butte Reservoir on the Rio Grande River, New Mexico, influences channel gradients beyond the backwater for several kilometers, necessitating the construction of levees to protect near-channel lands that would otherwise be inundated. Upstream structures may complicate the backwater effects of a given reservoir: Imperial Dam on the Colorado River has extensive upstream influences that interact with numerous dams above it (Lane

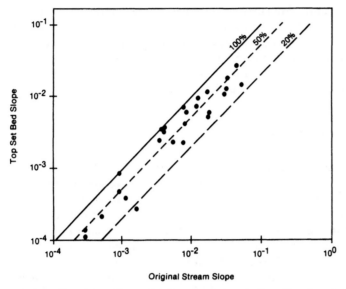

Fig. 7.11. Channel gradient reductions above dams in the arid and semiarid western United States showing a wide range of values. The *lines* indicate per cent reduction in gradient. (After Borland 1971, p. 29/5)

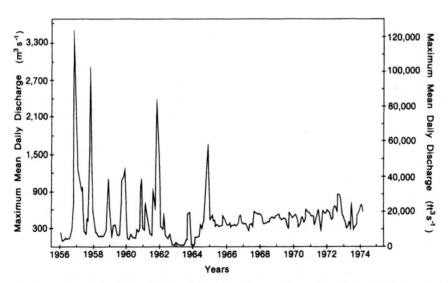

Fig. 7.12. Flow of the Colorado River below Glen Canyon Dam, closed in late 1962, showing the impact of reservoir operations on water releases to the previously undisturbed reach of the river downstream from the damsite. (After Pemberton 1976)

1951). Finally, the transition zone between original channel gradient and top of the sediment wedge or topset beds provides an ideal seed bed for phreatophyte vegetation. Increased growth of the vegetation may alter the hydraulic properties of the channel, induce sedimentation, and produce the need for development of a steeper gradient. The result is the extension of reservoir impacts further upstream than would otherwise be the case in the absence of dense vegetation.

7.2 Downstream Impacts of Dams

The downstream impacts of dams are in many cases less prominent than upstream sedimentation, but the downstream impacts often extend over greater distances and in sum may be of equal geomorphologic importance. Research on downstream impacts has not been as extensive as on upstream effects, but much of the investigative effort directed toward problems downstream from dams has considered dryland rivers (for reviews, see Fraser 1972; American Society of Civil Engineers 1978). The effects addressed in the following section include degradation and armoring of the channel flow downstream from dams, impacts on alluvial channels, impacts on nonalluvial channels (restricted canyons and gorges), and implications of dam-reservoir operations for vegetation.

7.2.1 Regime Changes by Dams

Dams influence the downstream geomorphic processes because they impose radical changes in river regime. Closure of a dam traps most of the sediment that once flowed through the river system, and releases relatively clear water into channel reaches that previously received significant quantities of incoming sediment. Outflow of sediment from the site of Glen Canyon Dam, Arizona, declined from a mean sediment concentration of about 1,500 parts per million before the dam to about 7 parts per million after dam closure.

In many cases, the newly constructed dam fulfills a flood-control objective by reducing peak flows and offering protection from inundation for downstream properties. Storage behind the dam of the flow otherwise contained in these peak discharges necessitates the gradual release of these waters over longer time periods. The result is that although the dam operations reduce peak flows, the operations also increase low flows. In some drylands this adjustment produces a continuous, more consistent flow instead of a flashy intermittent one.

Two cases on the Colorado River system illustrate the magnitude of the changes in river regime introduced by dams. On the Green River, Flaming Gorge Dam, Utah, reduced peak discharges from $510 \, m^3 \, s^{-1}$ ($18,000 \, ft^3 \, s^{-1}$) to about $170 \, m^3 \, s^{-1}$ ($6,000 \, ft^3 \, s^{-1}$) (U. S. National Park Service 1977b), and in some reaches more than 160 km (100 mi) downstream reduced the sediment discharge by 54 per cent (Andrews 1986). Further downstream, Glen Canyon Dam reduced mean annual floods from $2,440 \, m^3 \, s^{-1}$ ($87,000 \, ft^3 \, s^{-1}$) to only $760 \, m^3 \, s^{-1}$ ($27,000 \, ft^3 \, s^{-1}$) (Howard and Dolan 1981). The annual median discharge that increased from $210 \, m^3 \, s^{-1}$ ($7,500 \, ft^3 \, s^{-1}$) to $350 \, m^3 \, s^{-1}$ ($12,500 \, ft^3 \, s^{-1}$) reflected the impact of the dam on increased low

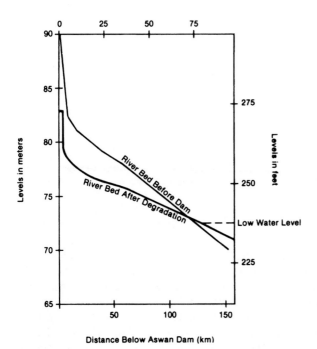

Fig. 7.13. Profiles of the Nile River below Aswan Dam, Egypt showing the reduction in gradient. The sedimentation wedge behind Esna Barrage influences the lower (right) reaches. (After Shalash 1974)

flows. The dam significantly reduced the variability of flow, as indicated by the flow records below the dam for the few years prior to and after the closing of the structure (Fig. 7.12).

7.2.2 Degradation and Armoring

Because reservoirs trap a large percentage of the pre-dam sediment load of the stream, the reaches of the stream below the dam experience flow that includes sufficient power to transport significant quantities of sediment but without the natural load. Fluvial processes compensate for the lack of in-stream sediment by eroding new materials from the bed and banks in a tendency to reestablish an equilibrium between ability to perform work and sediment load. Two direct effects of this adjustment are degradation of the channel bed and armoring of the bed surface.

Degradation may occur as a wholesale decline in the elevation of the bed over the entire reach of the stream below the dam, producing a relatively uniform lowering of the channel floor. The result is a lower elevation for the floor but no change in gradient. A second and probably more common arrangement is that greater amounts of degradation occur near the dam, with decreasing amounts in the downstream direction resulting in a decrease in bed gradient. The Nile River

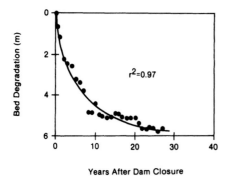

Fig. 7.14. Bed degradation on the Colorado River 1.1 km (0.7 mi) downstream from Davis Dam, Arizona and Nevada, showing the parabolic time function. (After Williams and Wolman 1984)

below Aswan Dam, Egypt, provides an example of the degradation downstream from a large dam that produced significant gradient changes over a 10-year period (Shalash 1974). In the Nile example, the next structure downstream, Esna Barrage, causes sedimentation to alter the profile further and lessen the gradient (Fig. 7.13). As with previously discussed cases, the influence of Esna Barrage extends only a short distance upstream from the normal low water level of its reservoir.

The effect of degradation extends to variable distances downstream from dams. Degradation impacts are measureable up to 300 km (180 mi) below Sariyar Dam, Turkey (Simons and Senturk 1976, p. 714). Below Glen Canyon Dam, Arizona, the degradation impacts extend only about 24 km (15 mi) downstream (Pemberton 1976). The downstream distance and the degree of the degradation below a dam depend on the nature of the water released from the dam, trap efficiency of the structure, geomorphic characteristics of the channel, size characteristics of the bed sediments, and controls on bed level such as bedrock sills or engineered structures that influence the erodibility of the channel. The downstream distance of the impact increases with time as indicated by direct observation (Stanley 1951; Makkaveev 1970) and the operation of mathematical models (Albertson and Liu 1957; Hales et al. 1970).

A number of approaches are available to quantify the process of bed degradation assuming that the result will be a final profile with decreased gradient. Mostafa (1957) combined basic sediment transport functions of Einstein and resistance as evaluated by the Shields function (Chap. 3) to define the slope of the newly adjusted channel:

$$S = [(0.06\ \gamma_s\ k_s)/Y]/[\gamma_f\ R] \tag{7.10}$$

where S = final gradient of the new profile, γ_s = specific weight of the submerged particles, k_s = characteristic length of the protuberances of the particles on the bed (usually assumed to be equal to the particle size D_{98}), Y = a value related to the original Einstein functions which ranges between about 0.8 and 1.7, γ_f = unit weight of water, and R = hydraulic radius.

Other approaches to the degradation problem have generally employed differential equations to account for sediment inflow and outflow from channel segments. Methods by Komura and Simons (1967) and Aksoy (1970) include

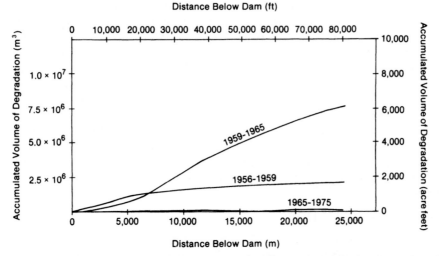

Fig. 7.15. Volumes of material eroded from the degrading channel bed downstream from Glen Canyon, Arizona, showing the high initial volumes of erosion contrasted with stable conditions after a short period of operation. The period 1956–1959 characterizes construction period for the dam, 1959–1965 the period of closure and impoundment, and 1965–1975 the period of normal dam operations

calculations for limited reaches in limited time segments and may be useful in applications where many data are available concerning channel and sediment properties.

Empirical studies of the rate of degradation in alluvial streams below dams show that degradation occurs rapidly at first, then progressively more slowly. Based on data from 12 cases in the dryland western United States, Williams and Wolman (1984) found the most useful time-related function was

$$D = t/[c_1 + (c_2\,t)], \tag{7.11}$$

where D = depth of degradation at a particular cross section (m), t = time in years since the beginning of the erosion, and $c_{1,2}$ are empirical constants. This hyperbolic function plots as a straight line on arithmetic scales if expressed in the form

$$(1/D) = c_2 + c_1(1/t), \tag{7.12}$$

where c_2 is the intercept and c_1 is the slope of the function. Standard regression analysis permits numerical determination of the constants. Application of the hyperbolic function showed strong agreement between the model and the data for some dams but not others where the degradation was episodic (Fig. 7.14).

Equations (7.11) and (7.12) suggest that the rate of removal of materials through degradation declines over time. Immediately after dam closure, the amount of mobilized sediments is likely to be substantial, but to decrease rapidly to a low, relatively steady rate. The rate of change depends on local conditions, but observations at Glen Canyon Dam, Arizona, indicate that relatively steady condtions there are likely within two decades (Fig. 7.15).

The massive removal of sediment from the bed of the channel downstream from a dam is often a selective process, with finer particles eroded and coarser particles remaining in place (reviewed by Hammad 1972). This process results in an armored condition, with the bed covered by particles that were mobile during high flows before the dam closure but that are too heavy to be removed by the new regime. The thickness of the armored layer is usually only one or two grains, but once the layer is established it inhibits further erosion. When bed material is relatively homogeneous in terms of particle sizes, armoring does not occur because there is no selective transport. Many sediment transport functions derive from conditions where similar particle sizes are the rule, so that modified functions are required to model the armoring processes. For example, Egiazaroff (1965) modified Shield's equation to account for the protective effects of the large particles:

$$\tau_c = [0.1]/[(\log 19 \, D/D_m)^2], \tag{7.13}$$

where τ_c = critical shear stress at which motion begins on the bed of mixed particle sizes, D = diameter of the particular grain, D_m = diameter of the mean particle of the mixture.

Gessler (1970) determined the particle size distribution of the final armor coating based on a probability density function describing the grain sizes in the mixture. In his model, the motion in the armored layer begins when the probability of grain movement for a given diameter exceeds 0.50. Komura (1971) proposed finite forms for Gessler's functions along with a simplified calculation procedure and nomograms for solution. Although experimental data are available for the armoring process (e.g., Harrison 1950; Little and Mayer 1972), deterministic predictive functions produce substantial error in applications, probably as a result of rigid assumptions required to simplify complex in-stream processes.

Empirical studies in India produced an additional set of functions to describe armor development below a dam. Ilo (1976) assumed clear water discharges in defining the following equation which can be solved for the armor coat particle size:

$$(1 - d_i/d_{ac})(d_{max}/d_{50i}) = 13.5 \, [1 - \exp \{-0.165 \, N_d \, ((L_d - x)/d_i)^{0.32}\}], \tag{7.14}$$

where d_i = a particle size in the original mixture, d_{ac} = particle size of the armor coat, d_{max} = maximum particle size, d_{50i} = mean particle size in the initial mixture, N_d = a degradation number related to channel and sediment conditions as defined below, L_d = length of the degraded reach, and x = downstream distance from the dam in the degrading reach. The degradation number is

$$N_d = [q_t \, t \, (S_{oi})^3)]/[\gamma_s \, D_i \, d_{50i}], \tag{7.15}$$

where N_d = degradation number, q_t = mean discharge in year t, t = age of the dam, S_{oi} = slope of the initial channel bed, γ_s = specific weight of the sediment, D_i = depth of initial flow, and d_{50i} = mean diameter of the original sediment mixture. The function is a product of limited data, but the general concept may prove useful in other cases.

7.2.3 Width and Pattern Changes

Diversion of flow from dryland rivers for irrigation purposes, channel manipulations for flood control purposes, and the introduction of exotic phreatophyte vegetation suggest that, in general, width and pattern changes are likely to be evident in such streams. Channels are likely to be more narrow than they were in the earlier decades, and channel patterns are more likely to be meandering or compound rather than braided, as was common prior to dam construction. The closure of dams contributes to these changes by further alterations of the river regime. Changes in channels described in Chapter 6 and related to hydrologic management are in large degree connected to dams. That dams have a major influence in the process is demonstrated by the case of John Martin Dam on the Arkansas River, in semiarid Colorado. The channel below the dam narrowed by 70 per cent after the dam closure in 1943, but upstream from the dam in reaches unaffected by backwater there were no changes. The clear separation of dam impacts and other effects is rarely possible, but the combined impacts extend downstream for several tens of kilometers.

Some reaches downstream from dams may widen as a result of the erosion processes associated with channel degradation. The widening extends as far downstream as degradation, and so is highly variable. Because erosion produces widening and deposition produces narrowing, it is difficult to combine both types of changes in the same theoretical or numerical models. Williams and Wolman (1984) found that a hyperbolic function described the temporal aspects of the changes:

$$W_1/W_t = c_3 + c_4 (1/t), \tag{7.16}$$

where W_1 = original width at the time of the closure of the structure, W_t = width in year t after closure, and $c_{3,4}$ are empirical constants. The constant c_4 is positive if the channel widens, negative if it narrows. The final width to which the channel is adjusting can be predicted using the constants in (7.16). The constant may be defined by standard regression techniques in the function which identifies the value of the hyperbola's asymptote:

$$W_r = 1/(c_3 + c_4/t) + 1.0, \tag{7.17}$$

where W_r = the final value of the W_t/W_1 ratio and other variables are as before. The data collected by Williams and Wolman (1984) from throughout the western United States and (7.16) indicate that changes in width and pattern are like those for degradation. They proceed rapidly at first, but at declining rates thereafter.

7.2.4 Impacts on Non-Alluvial Channels

When canyons or rock-bound gorges occur downstream from dams, the channel responses to dam closure are of a different kind than the changes experienced by streams flowing on broad alluvial deposits. Alluvium is unconsolidated so that channel dimensions, gradients, and patterns can respond quickly to changes in river regime imposed by dam closure. In canyons, however, channel shapes are

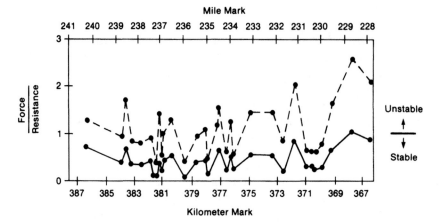

Fig. 7.16. Change in the stability of rapids in the Canyon of Ladore, Green River, Utah and Colorado. The stability ratio is the ratio between the amount of force exerted on the largest particle of each rapid by the largest flood of record (dashed line) and the largest flow after the closure of Flaming Gorge Dam upstream from the canyon (solid line). (After Graf 1980, p. 135)

mostly fixed by relatively nonerodible boundaries and the small amounts of alluvium present do not provide flexibility for system adjustment. Streams in canyons usually have alternating series of rapids with coarse particles and intervening pools floored by fine material (Laursen et al. 1976). During low flows, there is relatively little movement of bed material, though suspended materials continue their movement. At high flows much of the bed material is in motion, and the coarse particles move from one rapid to the next downstream.

The closure of a dam upstream from a canyon severely disrupts the movement of bed materials through the system because the dam eliminates the flood peaks, which under natural circumstances transported the coarse particles. The result is that the material in the rapids becomes stationary. If the rapid is located at the source of coarse materials brought down to the main channel by tributary streams, it continues to accumulate material, creating a longitudinal profile that is even more stepped than the natural one that characterizes canyon streams.

The closure of Flaming Gorge Dam on the Green River has had important impacts on the stability of rapids in Ladore, Whirlpool, and Split Mountain canyons downstream from the dam. An analysis of the balance of the force available in flows and the resistance to motion offered by the mass of the individual boulders in the rapids specified the impact (Graf 1980b). Before the dam was closed, boulders up to several meters in diameter formed the rapids. Under present hydrologic conditions, the boulders were so large that 62 per cent of the rapids were stable and even the peak flow of record was not large enough to move them. After the closure of the dam, the peak flow permitted by the structure was so low that 93 per cent of the rapids were stable (Fig. 7.16).

Fine materials, those that are sand-size and smaller, move through the canyon system under free-flowing conditions by a step process, moving from one pool to the next (Leopold 1969). In those places where canyon width permits, channel side bars in the form of beaches appear in reaches between rapids. These beaches represent temporary storage sites for sediment that is deposited during the waning phases of high flows. During low flow conditions, channel waters slowly erode the beach materials, returning them to the channel for continued downstream transport. The closure of a dam upstream from the canyon disrupts the alternating cycle of deposition and erosion because the dam eliminates the high flows and arrests the influx of new sediment from upstream. Consequently, the beaches receive no input, yet the low flow erosion continues unabated. The result is drastic reduction in the number and sizes of beaches along the canyon river.

Beaches along the Colorado River in the Grand Canyon, Arizona, are important components of the river management system because they are used by white-water recreationists as camping sites. Because the canyon is a National Park, it receives heavy usage, with about 14,000 boaters passing through the canyon each year. The stability of the beaches is therefore of economic interest in sustaining a multi-million dollar recreation industry.

Glen Canyon Dam has adversely affected the beaches in the Grand Canyon (Dolan et al. 1974). After closure of the dam in 1962, beaches began to decline in area and volume as low-flow erosion continued. The trend was not universal: some beaches remained unchanged, while others even expanded (Turner and Karpiscak 1980). In 1983 releases of water from the dam were larger than at any time since its closure, and for the first time in 20 years many of the remaining beaches were inundated. After the 1983 high discharges some beaches had enlarged through the addition of new materials (Bues et al. 1985), probably from channel pools. Modest contributions of sediment from undammed tributaries are apparently sufficient to prevent the complete elimination of the beaches.

7.2.5 Impact of Dams on Riparian Vegetation

The impact of dams on riparian vegetation, especially phreatophytes, is significant and direct. By manipulating the variability of flow and the balance of erosion-deposition downstream and by altering the depositional environment immediately upstream from backwaters, dams change preexisting environmental controls on near-channel vegetation. The deposition of deltas upstream provides new ecological niches for colonization where exotic species may have an unusual advantage (Harris 1966). Fluctuating water levels in reservoirs periodically reveal new margins for seedbeds, and some species are able to grow rapidly enough to establish themselves before the next period of high water. Tamarisk in particular is a phreatophyte commonly found in drylands that is capable of growing rapidly and withstanding long periods of inundation (Warren and Turner 1975). Reservoir areas may provide local niches that otherwise would not be available to the plant. It is possible that the Elephant Butte Reservoir on the Rio Grande, New Mexico, performed such a role in the spread of the plant through the river system.

In dryland alluvial rivers, the establishment of a dam may result in an increase in riparian and phreatophyte vegetation downstream. Dam operations convert near-channel areas that originally were subjected to destructive flood flows to relatively stable zones where vegetation may flourish. The growth of the vegeta-

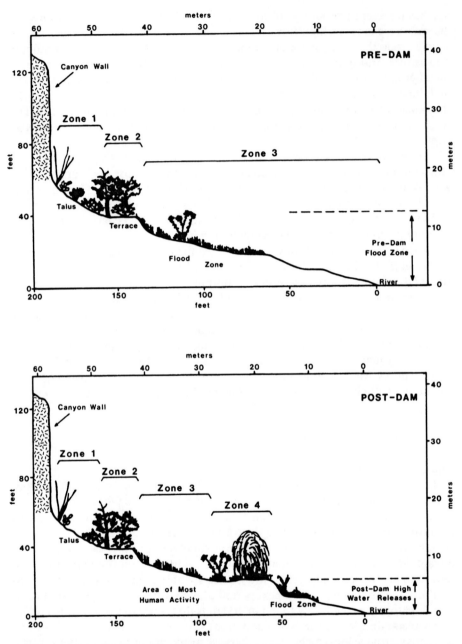

Fig. 7.17. Generalized changes in riparian vegetation along the Colorado River in the Grand Canyon showing the impact on geomorphic and vegetation systems of the reduced flood zone. The vegetation zones are 1) stable desert vegetation, 2) stable woody vegetation, 3) unstable zone, 4) new riparian vegetation, especially phreatophytes. (After an original by S. W. Carothers, U. S. National Park Service 1979)

tion encroaches on the channel and may contribute to the process of chan narrowing below dams by trapping and stabilizing otherwise mobile sediment. I many areas of the western United States, it appears that heavily vegetated areas represent prior channel locations so that the vegetation increase is proportional to the channel area decrease. This channel-vegetation invasion arrangement occurs along the Platte River below Kingsley Dam in semiarid Nebraska (Williams 1978) and along the middle Gila River below a series of dams in arid Arizona (Graf 1981). In these cases vegetation occupies as much as 90 per cent of the former channel floor. In an evaluation of 12 river reaches below dams in generally semiarid parts of the west central United States, Williams and Wolman (1984, p. 52) found that vegetation cover increased by an average of 57 per cent.

The connection between vegetation increase and flow regulation by dams on alluvial rivers is related to the elimination of destructive and destabilizing high flows, a factor that might in some cases aid in the expansion of range for exotic species. However, many other factors influence the spread of exotic species, including soil conditions, climatic conditions, availability of ground water, and human activities. Harris (1966) attributed the spread of tamarisk throughout the river systems of the American Southwest in large part to flow regulation by dams, but subsequent investigations showed that the plant spread with equal ease through dammed and undammed reaches and that other controls were more important than dams (Turner 1974; Graf 1978, 1980a). A reasonable conclusion is that dam operations contributed to the spread of the exotic, but that they were not solely responsible (Larmer et al. 1974).

In canyons and gorges below dams, the alteration in flow regime and adjustments in the stability of channel-side deposits also produces increases in vegetation cover. After only 13 years of operation, Glen Canyon Dam caused the conversion of the Colorado River in the Grand Canyon from a zone of sporatic vegetation to a stream lined on both sides with dense growth of native and exotic species (Dolan et al. 1974). Turner and Karpiscak (1980) relocated many historical photo sites, and their recent comparative photos clearly document the invasion of stabilized channel-side deposits by phreatophytes and riparian trees. The sequence of geomorphic surfaces next to the river is related to the sequence of vegetation, so that when the dam altered the river regime, it altered surfaces and plant communities simultaneously (Fig. 7.17).

7.3 Impacts of Urban Development

The development of cities in drylands usually occurs on flood plains, pediments, alluvial fans, bajadas (slopes of deposition downstream from pediments sometimes consisting of coalesced fan deposits), and bolsons (broad, nearly flat areas of deposition in the centers of basins). The construction and maintenance of urban structures disrupts the preexisting natural drainage and imposes an altered set of fluvial processes on the landscape. The purpose of the following section is to review the physical changes caused by urbanization, their impact on processes and forms, and possible modes of accommodation.

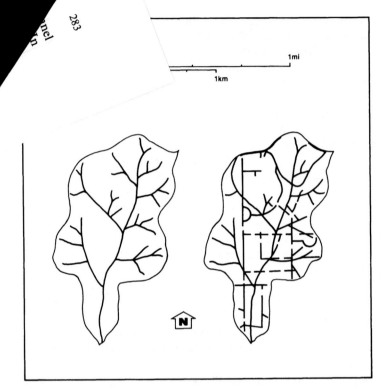

Fig. 7.18. The nature of network changes introduced by urbanization as illustrated by changes in a small Iowa drainage basin. Many of the additions to the network were by drains associated with streets. (From original data associated with Graf 1975)

7.3.1 Physical Changes Caused by Urbanization

Urban development brings about three types of physical changes on the surface that have implications for fluvial processes: exposed surfaces during construction, increased area of impervious surfaces after construction, and changes in channel networks. First, the construction of cities exposes surfaces for varying lengths of time to erosion without the impeding influences of vegetation. In nearly all cases, the first step in construction is to clear the surface of natural vegetation and to shape the surface using heavy equipment. This process breaks down resistant surficial crusts and breaks down particles and soil agglomerates, rendering them more susceptible to erosion than they were in their natural state.

When urban construction is complete, large areas of impervious surfaces replace natural surfaces that once covered the site of the city. Urbanization can bring about rapid changes: in central California some basins experienced almost a threefold increase in the amount of impervious area in a single decade due to urbanization (Harris and Rantz 1964). Leopold (1968) and Rantz (1971) estimated that within urbanized areas, about half of the surface is impervious. Roofs, streets, walkways, and other artificial coverings convert porous surfaces to ones which prevent purcolation of precipitation into the subsurface. The change is especially marked in dryland basins where the surface consists of highly porous unconsolidated materials. Data from Australia (Slatyer and Mabbutt 1964), the United States (Renard 1970), and southwest Asia (Schick 1971a) indicate that channel runoff occurs under natural conditions in drylands when the intensity of rainfall is at least 1 mm min^{-1} (0.24 in hr^{-1}) with totals at that intensity of at least 10 mm (0.4 in) (International Association of Hydrologic Sciences 1979). With the imposition of urban surfaces channel runoff occurs as soon as the surface is entirely wetted, so that channel flows occur quickly. Runoff volumes from any given total rainfall are also much larger for urban than for undeveloped surfaces.

Table 7.2. Channel network variables likely to be affected by urban development

Variable	Definition	Effect and impact of urbanization
U_c	Total number of links	Increases, causing decrease in lag and increase in kurtosis
l_e	Total length of exterior links	Increases, causing decrease in lag time and increase in kurtosis
l_i	Total length of interior links	Increases, causing decrease in lag time and increase in kurtosis
D	Drainage density	Increases, causing decrease in lag time and increase in kurtosis
λ	Length of exterior links divided by length of interior links	Decreases, causing increase in lag time and decrease in kurtosis
K_i	Length of interior links squared divided by mean interior contributing area per link	Increases, causing increase in lag time and increase in kurtosis

Notes: Density defined by Horton (1945), λ and K_i defined by Smart (1972). The influence of changes in λ and K_i is overwhelmed by the influence of the other variables.

The third physical change brought about by urbanization is a profound alteration in the network characteristics of the stream channels on the surface. On dryland bajadas and bolsons many channels are small (a few meters wide) and are easily eliminated during construction. After completion of urban construction, the drainage network is not simplified from its natural configuration, however, because designed structures add new links to the channel system. Spouts, gutters, conduits, storm sewers, streets, and designed channels represent extensions and additions to the natural network that conduct runoff from the newly increased area of impervious surface rapidly through the system (Fig. 7.18).

The addition of artificial links to natural channel networks changes almost all the basic measures of network topology. Graf (1977b) found that in a 7.7 km^2 (3 mi^2) drainage basin that experienced urbanization in about 25 per cent of its area, additions to the network increased drainage density by 50 per cent. Total channel length increased dramatically, as did the number of links in the network. The shapes of areas contributing to each link also changed, with the urban configuration having more compact contributing areas than the natural arrangement, thus speeding urban runoff into nearby channels. Additions of links also cut natural links into shorter segments, producing a more hydrologically efficient network. In each case, the network changes as measured by topologic variables were related to the percentage of basin area urbanized by a simple linear function. Table 7.2 provides a summary of the topologic variables that were altered to a statistically significant degree (0.05 confidence interval) by urbanization.

7.3.2 Fluvial Responses to Urbanization

Fluvial processes respond to urbanization with altered water and sediment discharges. With few exceptions, investigations into the process implications of urbanization have focused on humid regions and in any case empirical data are rare, especially in developing nations. Soliman's (1974) analysis of urbanization in the Nile Valley of Egypt is an exception. General principles likely to be applicable to dryland cities are available for further testing, however.

Changes in water discharges from urbanized watersheds are increased volumes of runoff, decreased lag time between precipitation and runoff, and increased flashiness of flow (or increased kurtosis) of the runoff hydrograph. Runoff volumes are particularly sensitive to the development of urban surfaces (Miller 1972), a generalization likely to hold true for drylands. In summarizing previous studies by Anderson (1968), Wiitala (1961), Martens (1966), Wilson (1966), Espey et al. (1966), and James (1965), Leopold (1968) illustrated the connection between urbanization and the expectable increase in discharges (Fig. 7.19). He found that in a typical basin of 2.6 km² (1 mi²) that if half the basin surface were urbanized and served by extensive storm sewer systems, the discharge from a given precipitation event after urbanization would be more than 2.5 times the discharge of the same event without urbanization. Carter (1961) found a fivefold increase in discharges. At the extreme, if the entire basin were urbanized and served by sewers, the increase would be more than sixfold.

Computer simulations using standard hydrologic assumptions and models predict that the changes caused by urbanization in small runoff events are different than the changes in large events. James (1965) and Rantz (1971) found that urbanization had a profound impact on the event with a 2-year return interval with 100 per cent urbanization causing a fourfold increase in runoff over the expected natural runoff. For the 100-year event the increase predicted for the completely urbanized basin was only about 2.5 times (Fig. 7.20). The ratio between an event of a given return interval and the annual flood declines as the area of impervious surfaces increases (Anderson 1970). In very large precipitation events, the natural ground surface would quickly become saturated and behave in a fashion similar to the impervious urban surface in the production of runoff.

The impervious surfaces and increased hydrologic efficiency of urban channel networks causes a decrease in lag time between the precipitation and resulting

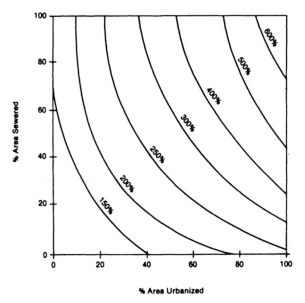

Fig. 7.19. The impact of urbanization on the mean annual flood from a drainage basin 2.6 km^2 (1 mi^2) in extent. (After Leopold 1968)

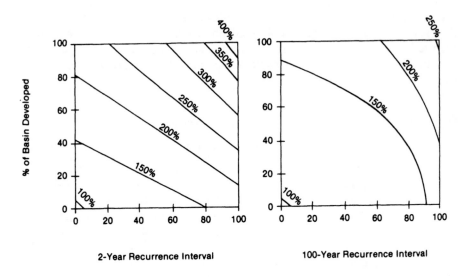

Fig. 7.20. The changing impact of urbanization on flood peaks of different return intervals showing the lesser impacts on large floods of longer return intervals. (After Rantz 1971)

runoff. Because larger volumes of runoff are involved in the post-urbanization case, the reduced lag time is accompanied by increasingly peaked hydrographs. Because many relationships between control and response variables related to flooding processes are in the form of power functions (e.g., Thomas and Benson 1970), lag time and kurtosis of the hydrograph in urbanizing watersheds are also likely to be related to basin characteristics by power functions:

$$t_{lag} = a_1 (V_x)^{b_1}, \tag{7.18}$$

$$K = a_2 (V_x)^{b_2}, \tag{7.19}$$

where T_{lag} = lag time (min), K = kurtosis (fourth moment about the mean discharge divided by the standard deviation), V_x = variable describing characteristics of the urbanizing watershed, and $a_{1,2}$, $b_{1,2}$ = empirical constants.

Solutions by standard regression techniques for (7.18) and (7.19) using data from the Iowa watershed showed that some network characteristics controlled only one of the responding hydrologic variables. For example, the total number of links exerted statistically significant control on lag time and kurtosis, but the total length of the exterior links (finger-tip tributaries) significantly explained changes only in kurtosis (Table 7.3).

The sequence of events in the construction of new urban areas and the exposure of surfaces to erosion produces a definable time series of sediment yield to downstream channels. Wolman (1967), based on data from humid regions, postulated a low but variable sediment yield during pre-construction periods, an extremely high yield during construction, followed by very low yields as hardened urban surfaces resisted erosion. In drylands there may be no fluvial erosion from construction sites because no precipitation events occur during the period of exposure. On the other hand, extensive erosion may occur if a rare event of large magnitude occurs during that period.

Table 7.3. Regression results for Eqs (7.18) and (7.19) applied to a small Iowa watershed

Independent variable	Dependent variable	Coefficient	Exponent	Correlation coefficient	Sign. level
U_c	Lag time	1475.50	−0.93	−0.95	0.02
	Kurtosis	92.94	1.43	0.99	0.01
l_e	Lag time	41.84	−0.42	−0.75	NS
	Kurtosis	5.35	1.20	0.96	0.01
l_i	Lag time	31.59	−0.63	−0.96	0.01
	Kurtosis	5.04	0.94	0.98	0.01
D	Lag time	24.50	−0.50	−0.86	0.10
	Kurtosis	3.43	1.10	0.97	0.01
λ	Lag time	0.82	0.57	0.98	0.01
	Kurtosis	1.64	−0.19	−0.64	NS
K_i	Lag time	2.12	−1.04	−0.97	0.01
	Kurtosis	0.77	0.18	0.63	NS

Notes: Calculated from data in English units ($ft^3 s^{-1}$ and ft).

The combined impact of increasingly flashy discharges of increasing magnitude with variations in sediment yield is a series of channel changes downstream from the urbanizing area. In the early phases of urbanization, discharges are not radically changed, but more sediment is available. Downstream channels respond by accumulating materials in increased flood-plain deposits and channel filling (Wolman and Schick 1967). After construction is complete, the decline in sediment yield and increase in runoff results in channel enlargement and erosion (Leopold 1973). Hammer (1972) found that in a humid region, channels enlarged 0.7 to 3.8 times their pre-urban sizes with the majority of his 78 examples in the range of 1.0–2.0.

In the semiarid region near Denver, Colorado, Graf (1975) found that in a 5.2 km (3.2 mi) long basin the surface area urbanized increased from 2 per cent to 53 per cent of the total in only 10 years. Channel responses began with a rapid expansion of flood plains in response to initial construction. Before urbanization, flood plains lined 22 per cent of the trunk stream, but afterwards they lined 50 per cent of the trunk stream. Sediment yield amounted to nearly 3,000 tonnes km^{-2} (8,600 tons mi^{-2}). When sediments were deposited during construction the flood plains were nearly at the same level as the channels, but when construction was complete the newly increased runoff and decreased sediment supply resulted in downcutting of the channels, leaving the flood plains elevated as terraces a meter or more above the channel. In other cases, sediment produced by construction may pass through the channel system as a slug or pulse of material (Dawdy 1967).

Applications of generalizations concerning urban impacts on fluvial processes and forms are difficult because almost all the foundation research on the subject used data from humid regions. The relationships between area urbanized and magnitude of hydrologic change are almost exclusively functions of humid region experiences. On the other hand, these conclusions are basic relationships among relatively simple variables that are not likely to change from one environment to another. The movement and storage of sediment in urban environments and geomorphic responses are more problematical.

Management of the problems associated with fluvial impacts of urbanization is a critical issue in protection of lives and property because of flood-related issues. Experience in drylands of the United States and southwest Asia suggests potential coping mechanisms for urban managers and planners including mapping of hazard zones and more precise prediction of runoff potentials (Cooke et al. 1982; Rhoads 1986a). Management of sediment-related problems requires construction regulations that limit the amount and time of exposed land surface during construction (Maddock 1969; Moore and Smith 1968). Water and sediment yield from new urban areas are subject to increased control with the installation of on-site retention structures that can reduce flood peaks and trap sediments.

7.4 Impacts of Agricultural Development

While previous research has dealt with the impact of dams on dryland rivers and with the impact of urbanization on streams in general, little published work exists addressing the impacts of agricultural development on fluvial processes. In dry-

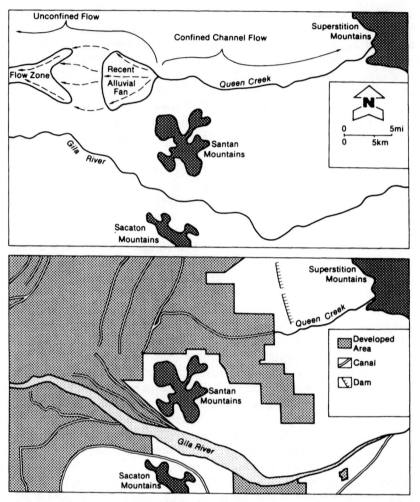

Fig. 7.21. Changes in drainage in the area of lower Queen Creek, southcentral Arizona. Shaded areas are erosional remnants of fault-block mountains, remainder of the mapped area consists of pediments, bajadas, and bolsons. The 1903 configuration represents an interpretation of an irrigation survey by the Salt River Project, an irrigation cooperative. The 1980 configuration is from U. S. Geological Survey maps and field evidence

lands, crop-based irrigation agriculture has transformed many bajadas and bolsons into artificial landscapes with subtle but significantly different configurations from the original natural forms. Successful irrigation schemes in valley locations require canals to provide primary water supply, laterals to deliver the water to individual fields, and drains to accommodate excess water and runoff. Manipulation of field surfaces insures that maintenance of uniform grades to provide optimum distribution of delivered water. The cumulative impacts of these engineered changes are the alteration of preexisting drainage patterns and adjustments in flood-water distribution.

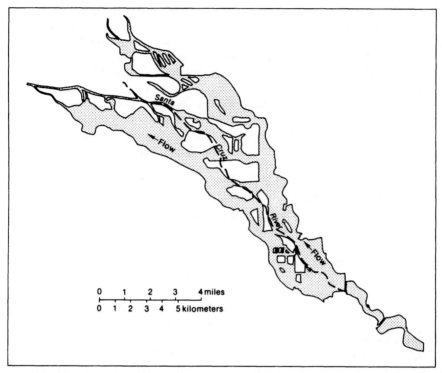

Fig. 7.22. The distribution of floodwaters in 1978 along the lower Santa Cruz River, south-central Arizona, showing the influence of agricultural land management on flood flows. (After Aldridge and Eychaner 1984, plate 3)

Drainage across dryland valley and basin floors under natural circumstances is usually infrequent, locationally unstable, and in some reaches unconfined in the form of flow zones rather than in channels. The imposition on valley surfaces of canals, laterals, drains, and other structures introduces a new stable, confined channel network that frequently conducts water. Changes associated with the development of the Queen Creek area of southcentral Arizona provide an instructive example.

The natural drainage associated with Queen Creek consisted of a complicated set of geomorphic features on the surface of deep alluvial fill in a structural basin (Fig. 7.21). Based on topographic surveys at 0.6 m (2 ft) contour intervals completed by an irrigation cooperative in 1903, Queen Creek flowed intermittently through a confined channel for several kilometers from its mountain source area. The confined channel terminated in a broad alluvial fan on the surface of the valley fill. The recently active fan was no more than a few meters high, but the detail of the survey clearly revealed its outline. In previous millenia, its form was probably different as one location and then another was blanketed by alluvium. Downslope from the fan, flow from the creek that survived transmission losses in travel from the mountains apparently flowed in an unconfined fashion. Eventually the waters recollected in a broad flow zone defined by the topographic surveys as a shallow, elongated swale. In unusually large floods, there was enough water to extend the flow to the Gila River, which was a relatively narrow, slightly meandering channel in the area at that time (in other areas it was a braided channel).

By the 1980s, extensive irrigation development had altered the surface and changed the drainage of lower Queen Creek across the basin fill (Fig. 7.21). Levees lined the lower course of the confined channel to prevent overflows onto adjacent fields. The lower channel required greater than natural capacity because long, low dams on the bajada and pediments above the developed area directed surface runoff to the channel. Under natural conditions this additional water remained dispersed. The new canals also changed the course of Queen Creek so that it flowed around some field areas and made sharp turns rather than gentle sweeping curves of the natural course.

Development of the alluvial fan area converted the natural downslope drainage to a series of irrigation canals that flowed across the gradient almost parallel with the contour lines. Laterals distributed water downslope from the main canals and are not shown on Fig. 7.21. In the broad flow zone the original field construction converted the swale to a planar surface tilted slightly in the original downslope direction, and additional canals crossed it orthogonally to the original drainage. Some drainage systems conducted water through flood channels to the Gila River which in its 1980s configuration was a compound channel with a wide shallow component.

The new drainage configuration for Queen Creek is efficient for low flows, but large floods are powerful enough to adjust the system partially back to its original configuration. In the period 1978–1983, several discharges with return intervals in the 50–100 year range flowed through Queen Creek and followed the original fan-dispersion-recollection sequence rather than following the artificially imposed routes. The result was expensive destruction of irrigation works, fields, and agricultural facilities. Though they were all rebuilt after the floods, they do not represent equilibrium configurations for large flood flows, and they appear to be at risk in such events.

Flood flows across valley and basin surfaces and the erosion and sedimentation processes accompanying the flows take on an altered geographic distribution because of agricultural development. Under natural circumstances flows are confined to channels, to flood plains along the channels, or spread out in dispersed sheets in broad flow zones. Agricultural developments confine the flow in some areas where it would otherwise be dispersed, and in other cases eliminate channel flow. The resulting pattern of flood flow reflects artificial channels and landforms and is difficult to generalize for planners and managers.

The intricate pattern of inundation near the lower Santa Cruz River in southern Arizona during floods in 1977 illustrates the geographic complexity imposed by agricultural development (Fig. 7.22). Under natural conditions the Santa Cruz River in the area of Red Rock had a defined channel, but during flood periods the channel became a broad continuous flow zone. Construction of Greene Canal, other water distribution works, roads, and construction of "leveled" fields radically altered the inundation pattern and resulted in destructive erosion and sedimentation.

7.5 Conclusions

The construction and operation of dams, cities, and agricultural works have a common impact on fluvial processes in drylands. The structures change a preexisting tendency toward equilibrium between forms and processes. The systems operate to reestablish the original tendencies or to develop in a different direction toward a new equilibrium configuration. Because fluvial processes are sporadic in drylands, most adjustment takes place discontinuously, and managers view the adjustments as expensive and destructive, so they are "repaired."

The result is that equilibrium, rare in dryland fluvial systems under natural circumstances, is even more unusual in designed systems. The typical resolution is to design for the low or medium magnitude event; in any case the likely magnitude of rare events is poorly known. The behavior of dryland rivers in rare events where dams, cities, or agricultural developments play a significant role is as much a geomorphic as an engineering problem, but it remains a problem rarely addressed.

8 Generalizations for Dryland Rivers

The objective of all sciences is the creation of generalizations, theoretical statements about the world that have wide applicability. The science of geomorphology has been an inconsistent producer of generalizations, and the progress of the science has therefore been sporadic. Researchers of the 1800s usually sought general lessons in the solution of their specific problems, a method of intellectual operation that has too often been abandoned by twentieth century workers. Investigations of dryland rivers have thus far produced a limited number of generalizations and the construction of a complete integrated theory for the subject is not yet possible.

8.1 Toward an Integrated Theory

An integrated theory for the modern behavior of dryland rivers combines explanations for processes and forms as well as accounting for temporal and spatial changes. Such a theory accounts for process reversals and the influence of human activities. Because a theory can take the form of an assemblage of interrelated statements, this book represents an initial, tentative approximation of an integrated theory. Unfortunately, as a theoretical fabric this statement has more gaps than solid material. Incomplete though it may be, this explanation of modern dryland rivers provides some indications about fundamental precepts, processes and forms, and modifications of processes and forms. This final chapter summarizes the major points of previous chapters. Some speculations are also given below about the relationships between this growing body of generalizations and other endeavors such as engineering, the law, planning, and environmental management.

8.1.1 Fundamental Precepts

In the broadest, most sweeping sense, the most basic explanation for dryland rivers and all their components is that they represent a visible landscape of work produced by the interaction of invisible landscapes of forces and resistances. The changes in forces and resistances across space and through time produce the movement of materials and create the processes and forms of the geomorphic environment.

There are a number of explanatory statements possible on a large scale of generalization about dryland rivers. In some cases the statements presented in Chapter 2 seem so general that they are not helpful, but when geomorphologists communicate with other professionals these generalizations become useful. For example, the comments that rivers are integrated general systems of elements connected by the processes of water and that what happens to one element affects all the elements seems to be such a basic statement as to be self-evident. Yet in every dryland zone in the world there are numerous examples of attempts to alter limited parts of river systems without regard to the potential impacts on other distant elements of the systems.

Among the building blocks of successful theories in the area of fundamental time precepts for dryland rivers, the statements concerning equilibrium and rate laws are most widely applicable. On the scale of applied physics and mechanics, all fluvial processes operate in a manner to adjust the systems into a state of equilibrium between forces and resistances. On the larger scale of analysis of rivers, tendencies toward such a balance are inevitable, but because of the sporadic operation of runoff processes in semiarid, arid, and extremely arid climatic regions, many dryland rivers may not operate as systems in equilibrium for more than a small percentage of the time. Rate laws describe the nature of change from one state to another by specifying rapid change at first, followed by declining rates of change because of the operation of feedback mechanisms in the general system, that result in progressively slower adjustments.

Network laws and the concept of distance decay provide the geomorphologist with useful explanations in dryland streams just as in humid region systems. Stream systems have their particular topologic arrangement, not because they are rivers but because all systems with origins and connecting links must have some common, topologic characteristics. Deviations from the most likely arrangement result from the imposition of restraining controls, usually of geologic materials or structures. The most fundamental geographic law, that things close together influence each other more than things that are far apart, also applies to dryland rivers. The dissipation of energy and materials increases as distance from the source increases because of the exponential increase in area or volume available for dispersion.

For those concepts that explain changes in both time and space, the most significant are those related to thresholds and catastrophe theory. In the more simple application, river systems sometimes adjust rapidly and at other times not at all because of the existence of thresholds of resistance. As long as these thresholds are not exceeded there is no change. If a force exceeds a resistance threshold, far-reaching changes throughout the river system may result. Catastrophe theory adds some understanding and extends the concept of thresholds to a more complex level because it stipulates that all systems controlled by two factors (such as force and resistance) exhibit response variables that sometimes change slowly, at other times rapidly. Under certain conditions more than one state of equilibrium is possible.

8.1.2 Processes and Forms

Fluvial processes in drylands are driven by local precipitation events that are of short duration and spatially limited, so that within any extended system runoff and channel processes are also brief and spatially restricted. In addition, throughflowing rivers in drylands experience strong seasonal changes because they receive their primary input from highland areas where hydrologic rhythms depend on seasonal climatic inputs. At the restricted scale of the individual cross section, streamflow characteristics are products of geometry and flux just as in humid-region streams. At a more general system scale of analysis, dryland rivers experience high rates of downstream transmission losses because of the porous nature of typical channel bed materials.

Precipitation and runoff inputs to dryland channels are sporadic so that the difference between high and low flows is greater than in humid streams. These wide fluctuations prevent the development of an intimate linkage between a particular discharge magnitude and channel geometry related to "bankfull" conditions. Large floods have strong influence on channel geometry, a geometry that then confines subsequent flows. Process therefore controls form in high magnitude events, while form controls process in low magnitude events.

The sparse vegetation and relatively smooth topographic surfaces available in some dryland areas produce conditions conducive to unconfined flows. Sheetflows occur over slopes, and broad flow zones develop instead of clearly defined channels on some basin and valley floors.

At restricted scales of analysis, fundamental balances between force and resistance drive sedimentologic processes in drylands in the same manner as in humid regions. At more general scales, sediment moves through slope and channel systems in pulses over brief periods because of kinematic effects, over intermediate periods because dryland stream profiles are stepped or have bars and pools which promote step-by-step transport, and over long time periods because flows are not continuous and therefore sediment transport is not continuous. Sediment storage is most common in mid-basin locations because of the downstream distribution of stream power whereby power is low in small streams, high in intermediate streams, and low relative to sediment supply in the largest streams. Wide fluctuations in mid-basin channels produce alternative storage and evacuation of materials.

Streams transport heavy metals as bed-load material and concentrate metals in placers because the fluvial processes sort the sediments by weight, with waning flows depositing the heavy particles first, producing placers. These placers therefore are commonly found in locations coincidental with the active channel floor on the waning flows of floods. Streams also transport and sort radionuclides adsorbed onto sedimentary particles. Actinides in particular behave in a fashion similar to heavy metals, so that streams, especially those in drylands with sediment pulses, concentrate radionuclides in particular places.

The interaction of forces and resistances gives rise to a variety of landforms that characterize drylands. Badlands represent terrains controlled by cohesive materials without vegetation cover. Piping frequently associated with badlands develops because of varying permeability in subsurface materials that permit the creation of conduits between upper slopes and lower level outlets. Chemical compounds that reduce cohesion enhance piping and badland development by making regolith particles more mobile.

Pediments are transport-dominated slopes connecting mountain areas to basin zones. The thin covering of sediments is in temporary storage during their journey from mountain sources to basin sinks. Sheet flow may occur on pediments, but did not cause the development of the landform because the generally planar surface is required for the flow. Alluvial fans are also temporary storage areas for sediment in the mountain to basin transition. Fans have their conical shape and radiating distribution of sediments because channels issue from confining mountain slopes and can expand their widths. Depth of flow and transport capacity decline, resulting in the deposition of materials.

River channels assume braided, meandering, or compound configurations because of the interplay between stream power and bank-bed resistance. When stream power is low and bank-bed resistance is high, meandering configurations are common; when power is high and resistance is low, braided channels result. At intermediate levels of power and resistance, dryland river channels assume a compound appearance with components of both meandering and braided channels. During low-flow events, channel morphology controls the flow processes, but in rare large events, flow is so powerful that it alters the channel and near-channel forms. A slow recovery period follows large events with sedimentation and reestablishment of vegetation. In many drylands, geomorphologic flood plains do not exist along braided channels because extreme variability of discharges prevents the repetitive over-bank flows needed for flood-plain construction.

8.1.3 Modifications of Processes and Forms

Vegetation dynamics and human activities modify dryland river processes and forms. Changes in upland vegetation directly affect infiltration and runoff and therefore indirectly influence erosion. Changes in upland vegetation result in changes in sediment yield from small watersheds, but in large drainage basins the operation of basin-wide hydroclimatic factors submerges the effects of localized vegetation changes.

Riparian vegetation, especially phreatophytes, has a direct connection with channel processes through hydrologic and geomorphic processes. The hydrologic connections are through the transpiration of near-surface ground water by plants equipped with effective tap roots, while the geomorphic connections are through hydraulic roughness on channel and near-channel surfaces caused by the growth of the plants. The roughness in turn affects erosion and sedimentation. Exotic vegetation influences dryland rivers because newly introduced plants occupy otherwise empty ecological niches or may replace native species with different growth and physical properties. River processes exert a feed-back influence on vegetation communities by changes in hydrologic regimes that change the availability of water or through erosion and sedimentation that change the physical setting for vegetation communities.

Dams represent the most obvious direct human influence on dryland rivers. Upstream impacts of dams are related to the newly established base level of mean water surface elevation for the reservoir. The stream entering the reservoir establishes a new profile that has a gradient just sufficient to transport material into the reservoir. Deposits include a delta of coarse particles that are too heavy to be transported far from the stream mouth, finer particles carried long distances across the floor of the reservoir by density currents, and finest particles that settle from suspension over the entire reservoir floor. Because the newly established gradient upstream from the reservoir is more shallow than the original stream gradient, effects of the reservoir extend only a short distance upstream from the backwater. Downstream impacts of dams include degradation of channels and development of armored beds produced by relatively sediment-free waters re-leased from reservoirs. Channel patterns, channel dimensions, and vegetation communities change because the release schedules for dam operations are rarely the same as the natural flow regimes that produced the original channels.

The conversion of natural landscapes to urban forms alters the flow regimes of rivers draining the urbanized areas by increasing discharges of all return intervals and reducing lag times between precipitation and runoff peak. The hydrologic changes are caused by increases in impervious area and changes in channel networks such as additions of external links, shortening of individual link lengths, and compacting of drainage areas. Urbanization affects small drainage basins most severely. Large basins are much less affected because urbanization subsumes small areas relative to the total area. Urbanization also impacts smaller, frequent hydrologic events more than larger, infrequent ones because during the large events natural hydrologic conditions approach urbanized conditions.

The conversion of natural landscapes to agricultural uses also alters hydro-logic and fluvial geomorphic processes in drylands. Agricultural development changes unstable, curved channels to hardened straight conduits, while the installation and maintenance of leveled fields, canals, laterals, and drains converts natural flow zones into configurations that concentrate and direct flow. The altered flow patterns change the distributions of erosion and sedimentation.

8.2 Utility of Generalizations

Generalizations or theories are the products of scientific activity and are the most important and lasting items scientists provide to the consumers they serve. Descriptions, even fundamental descriptions, are not enough. A science that offers no useful general theories becomes esoteric to the society that supports it, and the science withers from lack of interest by supporters and practitioners. Therefore, a useful exercise in closing this book is to briefly investigate geomor-phology's theoretical developments as connectors between the science and other end-users of its products: engineers, legal advisors, planner-managers, and the common citizen or nonprofessional.

8.2.1 Geomorphology as a Natural Science

Early in its history, geomorphology emphasized the development of generaliza-tions and theory. In studies of dryland rivers, research addressed questions of large scope and produced explanations with wide applicability. For example, Darwin (1846) designed his comments about the landscapes of arid and semiarid

South America to describe specific places, but he also sought to provide general lessons that applied on a global basis without reference to a particular locale. Gilbert's (1877) work provided the reader with insight into the workings of the landscape of the Henry Mountains, but it also provided an intellectual structure for the interpretation and explanation of dryland river problems in other areas. Davis (1899) constructed a grand geomorphologic theory for landscape evolution that was not limited to one place.

In geomorphology during the early twentieth century the search for generalizations and theories stagnated, especially in studies of dryland rivers. A geomorphologic dark age ensued until the 1950s, when new investigations into geomorphic processes transfused ideas from other fields into the science. In the 1960s and 1970s, mixed success resulted from hundreds of detailed investigations of dryland fluvial processes, though at times description became an end in itself. Nonetheless, some generalizations began to emerge and the trickle of theoretical developments continued into the 1980s. The First International Conference on Geomorphology held in 1985 in Manchester, England, provided evidence that at last geomorphology exists as a definable natural science. Fluvial problems in drylands occupied a prominent position on the conference agenda.

8.2.2 Geomorphology and Engineering

Fluvial geomorphology and hydrologic-hydraulic engineering in dryland river settings have obvious common interests. The geomorphologist deals with system-wide operations and adopts a general system perspective when dealing with fluvial problems. The engineer searches for the specific solution to a specific problem and generally deals with issues more limited in time and space but in greater detail. It would seem that the two specialties are symbiotic, each feeding on the other's strengths in a mutually re-enforcing arrangement.

Unfortunately, geomorphologists have too often disregarded the techniques of the engineer or have used the engineering techniques out of context or in situations for which engineering models were ill-suited. On the other hand, engineers too often have ignored geomorphologic approaches to assessing dryland river problems and have tried to approach long-term problems with short-term solutions. The resolution to these problems is improved communication, whereby each specialist makes the effort to accommodate the other's language and evidence.

An example of the fortuitous marriage of geomorphology and engineering in the solution of practical dryland stream problems is Fookes' (1976) work in southwest Asia. Drainage problems associated with highway construction are susceptible to analysis from both perspectives, and the construction of a stable road bed in the arid and extremely arid parts of the region are more likely to be successful if the interpretation of geomorphic evidence of hazardous locations can be combined with engineering expertise to address the hazard. Combining geomorphology and engineering in this situation insures the wise expenditure of scarce capital in developing nations.

8.2.3 Geomorphology and the Law

The operations of dryland rivers have implications for the legal relationships among citizens, corporate enties, and governments. River courses frequently form jurisdictional and property boundaries, but in drylands rivers are locationally unstable. Because rivers are interrelated cascading systems, actions of upstream users have direct and often detrimental impacts on downstream users. The withdrawal of water from the stream and its consumptive use or degradation of water quality have occupied legal and hydrologic specialists in all the drylands of the world for centuries. More recently, geomorphic questions have appeared with legal ramifications.

The alterations of fluvial processes by urbanization or agricultural development in one location often cause erosion or sedimentation in places distant from the initial disruption. Hardening of the channel boundaries or changes in channel shape, size, or pattern to protect one piece of property causes change in the channel system that impacts property elsewhere. Maintenance of flood channels by vegetation control has far-reaching consequences because if the channels are not maintained overbank flows may be much more extensive than otherwise. In each of these issues lie legal questions of responsibility and the possibility of negligence with harmful consequences.

In the dryland western United States, geomorphologists increasingly deal with legal issues surrounding the management of rivers. In one case the stability of a river-defined boundary in western Wyoming was the issue: whether or not locational changes would reasonably be expected in the course of normal river processes or whether recent changes were unusual. Urbanization of a portion of a southern California watershed and the implications for downstream flooding gave rise to another case where the geomorphologic imput was to provide an understanding of dryland river processes on a basin-wide scale of analysis. In still another case, the precise definition of flood-plain boundaries along the Agua Fria River near Phoenix, Arizona, required explanation from a geomorphologist of river channel change and its implications for the use of engineering models. In each instance the major contribution by the geomorphologist was to outline widely accepted generalizations, to show how the generalization applied to a particular problem, and to resolve observations into a cohesive picture of the behavior of the fluvial system.

8.2.4 Geomorphology and Planning-Management

The identification of potentially hazardous areas and predictions of the likely consequences of projects affecting rivers are contributions that geomorphology makes in the planning and management processes. In hazard assessments, engineering-related predictions using deterministic models are only as effective as the numerical data that support them, and in drylands accurate hydrologic data are usually from short records. In these circumstances interpretation of the geomorphologic evidence produces useful additional insights into the long-term history of potentially hazardous surficial processes. The presence of high terraces with recent flood deposits, vastly over-sized channel geometry relative to recently observed flows, distributions of anomalously large particles, and vegetation-landform associations are all important keys to the geomorphologic definition of hazard areas.

The mapping of geomorphic indicators of hazard zones is an effective method of communicating geomorphologic generalizations to planners and managers. Cooke et al. (1982), for example, used geomorphic indicators to produce a flood hazard map for Suez City, Egypt. The result was a highly detailed representation of those areas subject to high, medium, or low probability of flood hazard. Though the precise return interval of each of these classes was not entirely clear, the physical evidence outlined the exact geographic extent of each area. The planner or manager who includes geomorphic information and maps along with deterministically derived predictions is more likely to be successful than the one who relies on only one type of evidence.

8.2.5 Geomorphology and the Nonprofessional

The true success of geomorphology in the analysis of dryland rivers is probably best measured in the reception of the science by the common citizen whose taxes ultimately support the research. Unlike botanists, biologists, and to a lesser degree general geologists, geomorphologists in the past rarely produced popular or semi-popular monographs explaining local natural systems. These publications, designed for use by the nonprofessional serve to educate and entertain the reader as well as to recruit support through heightened interest. Geomorphologists have yet to effectively reach this most important market of ideas.

Many of the world's most readily recognized landscapes are dryland fluvial systems. Most common citizens of the world appreciate the striking forms and colors of the Grand Canyon, Arizona, stark angles of the Atacama of Peru, desolate wadis of Egypt, intricate slopes and channels of western China, and vast openness of the semiarid steppes of central Australia and the southern Soviet Union. Continued study and construction of viable theories will have practical economic benefits, but the most valuable reward will be general appreciation of the beauty and wonder in the workings of rivers in generally dry country.

References

Abrahams AD (1977) The factor of relief in the evolution of channel networks in mature drainage basins. Am J Sci 277:626–645

Abrahams AD (1984) Channel networks: a geomorphological perspective. Wat Resour Res 20:161–188

Abrahams AD, Flint JJ (1983) Geological controls on the topological properties of some trellis channel networks. Geol Soc Am Bull 94:80–91

Ackers P, Charlton FG (1970a) Meander geometry arising from varying flows. J Hydrol 11:230–252

Ackers P, Charlton FG (1970b) The slope and resistance of small meandering channels. Proc Inst Civ Eng 47:349–370.

Ackoff RL (1964) General systems theory and systems research: contrasting conceptions of systems science. In: Mesarovic MD (ed) Views on general systems theory. John Wiley and Sons, New York, pp. 123–132

Adams J, Zimper GL, McLane CF (1978) Basin dynamics, channel processes, and placer formation: a model study. Econ Geol 73:416–426

Adams R, Adams M, Willens A, Willens A (1979) Dry lands: man and plants. St Martin's, New York

Adriano DC (1986) Trace elements in the terrestrial environment. Springer, Berlin Heidelberg New York Tokoyo

Agarwal VC (1983) Studies on the characteristics of meandering streams. PhD diss, Univ Roorkee, India

Aksoy S (1970) River bed degradation downstream of large dams. Trans 10th Int Congr on Large Dams 3:835–844

Albertson ML, Liu H-K (1957) Discussion of "River-bed degradation below large-capacity reservoirs" by M. G. Mostafa. Trans Am Soc Civil Eng 122:699–701

Aldrige BN (1970) Floods of November 1965 to January 1966 in the Gila River Basin, Arizona and New Mexico, and adjacent basins in Arizona. US Geol Surv Wat-Supply Pap 1850-C

Aldridge BN, Eychaner JH (1984) Floods of October 1977 in southern Arizona and March 1978 in central Arizona. US Geol Surv Wat-Supply Pap 2223

Alekseyev GA (1955) Formulas for the calculation of the confidence of hydrological quantities. Metr Gidr (Leningrad) 6:40–43

Alexander D (1982) Difference between "calanchi" and "biancane" badlands in Italy. In: Bryan R, Yair A (ed) Badland geomorphology and piping. Geobooks, Norwich, pp. 71–88

Allen JRL (1970) Physical processes of sedimentation. Allen & Unwin, London

American Society of Civil Engineers (1978) Environmental effects of large dams. Am Soc Civil Eng, New York

Anderson BW, Higgins AE, Ohmart RD (1977) Avian use of saltcedar communities in the Lower Colorado River Valley. In: Jones RR, Johnson, DA (eds) US For Serv Gen Tech Rep RM-43, Ft Collins, pp. 128–136

Anderson DG (1968) Effects of urban development on floods in northern Virginia. US Geol Surv Unnumbered Open-File Rep

Anderson DG (1970) Effects of urban development on floods in northern Virgina. US Geol Surv Wat-Supply Pap 2001-C

Anderson HW (1949) Flood frequencies and sedimentation from forest watersheds. Trans AmGeophys Un 39:576–584

Anderson HW (1954) Suspended-sediment discharge as related to streamflow, topography, soil and land use. Trans Am Geophys Union 44:268–281

Anderson HW (1975) Sedimentation and turbidity hazards in wildlands. Watershed Management, Proc Symp Irr Dr Div, Am Soc Civil Eng, pp. 347–376

Anderson M, Burt TP (1982) Throughflow and pipe monitoring in the humid temperate environment. In: Bryan R, Yair A (eds) Badland geomorphology and piping. Geobooks, Norwich, pp. 337–354

Andrews ED (1979) Scour and fill in a stream channel, East Fork River, Western Wyoming. US Geol Surv Prof Pap 1117

Andrews ED (1986) Downstream effects of Flaming Gorge Reservoir on the Green River, Colorado and Utah. Geol Soc Am Bull 97:1012-1023

Andropovskiy VI (1972) Criterial relations of types of channel processes. Sov Hydrol 11:371–381

Annandale GW (1987) Reservoir sedimentation. Elsevier, Amsterdam

Anstey RL (1965) Physical characteristics of alluvial fans. US Army Materiel Command. US Army Natick Lab, Tech Rep ES-20

Armstrong CF, Stidd CK (1967) A moisture-balance profile in the Sierra Nevada. J Hydrol 5:258–268

Armstrong PH (1985) Evolution. In: Goudie AS (ed) The encyclopaedic dictionary of physical geography. Blackwell, London, pp. 173–175

Arnold JF, Schroeder WL (1955) Juniper control increases forage production on the Fort Apache Reservation. US For Serv Res Pap RM-18

Atchison GD, Wood CC (1965) Some interactions of compaction, permeability and post-construction deflocculation affecting the probability of piping failure in small earth dams. Proc 6th Int Conf Soil Mech Found Eng 2:442–446

Babcock HM, Cushing EM (1941) Recharge to ground water from floods in a typical desert wash, Pinal County, Arizona. Am Geophys Union Trans 23:49–56

Bagnold RA (1931) Journeys in the Libyan Desert. Geogr J 78: 13–39, 524–535

Bagnold RA (1933) A further journey in the Libyan Desert. Geogr J 82:103–129, 403–404

Bagnold RA (1941) The physics of blown sand and desert dunes. Chapman & Hall, London

Bagnold RA (1954) Experiments on gravity-free dispersion of large solid spheres in a Newtonian fluid under shear. Proc R Soc London Ser A, 225:49–63

Bagnold RA (1960) Some aspects of the shape of river meanders. US Geol Surv Prof Pap 282E:135–144

Bagnold RA (1962) Auto-suspension of transported sediment: turbidity currents. Proc R Soc London Ser A, 265:315–319

Bagnold RA (1963) Mechanics of marine sedimentation. In: Hill MN (ed) The sea. Wiley, New York, pp 507–523

Bagnold RA (1966) An approach to the sediment transport problem from general physics. US Geol Surv Prof Pap 422-J

Bagnold RA (1977) Bed load transport by natural rivers. Wat Resour Res 13:303–312

Bailey RW (1935) Epicycles of erosion in the valleys of the Colorado Plateau Province. J Geol 43:337–355

Baker MB (1984) Changes in streamflow in an herbicide-treated pinyon-juniper watershed in Arizona. Wat Resour Res 20: 1639–1642

Baker VR (1971) Paleohydrology of catastrophic Pleistocene flooding in eastern Washington. Geol Soc Am Abstr with Progr 3:497

Baker VR (1973) Paleohydrology and sedimentology of Lake Missoula flooding in eastern Washington. Geol Soc Am Spec Pap 144

Baker VR (1974) Paleohydraulic interpretation of Quaternary alluvium near Golden, Colorado. Quat Res 4:94–112

Baker VR (1977) Stream channel response to floods with examples from central Texas. Geol Soc Am Bull 88:1057–1071

Baker VR (1982) The channels of Mars. Univ Texas Press, Austin

Baker VR, Pickup G (1987) Flood geomorphology of the Kathrine Gorge, Northern Territory, Australia. Geol Soc Am Bull 98:635–646

Baker VR, Pyne S (1978) G. K. Gilbert and modern geomorphology. Amer J Sci 278:97–123

Baker, VR and Ritter DF (1975) Competence of rivers to transport coarse bedload material. Geol Soc Am Bull 86:975–978

Ballard GA (1976) Evidence to suggest catastrophic flooding of Clarks Fork of the Yellowstone River, northwestern Wyoming. MS Thesis Univ Utah, Salt Lake City

Barclay HV (1916) Report on exploration of a portion of Central Australia by the Barclay-MacPherson expedition, 1904–1905. R Geogr Soc Austral-Asia, S Aust Br, Proc 29:57–102

Barekyan AS (1962) Discharge of channel forming sediments and elements of sand waves. Soviet Hydr (Am Geophys Un) 2:451–466

Barnes HH (1967) Roughness characteristics of natural channels. US Geol Surv Wat-Supply Pap 1849

Barr GW (1956) Recovering rainfall, Pt 1. Arizona Watershed Program. Ariz State Land Dep, Water Div, Salt River Valley Water Users' Assoc, and Univ Ariz

Bartlett BA (1962) Great surveys of the American West. Univ of Oklahoma Press, Norman

Bartley WW (1962) The retreat to commitment. Knopf, New York

Bartley WW (1964) Rationality versus the theory of rationality. In: Bunge M (ed) The critical approach to science and philosophy. Free Press, New York, pp 3–31

Bateman AM (1950) Economic mineral deposits (2nd edn). John Wiley & Sons, New York

Bates CP (1956) Manual of river behavior, control and training. CHIP, India, Publ 60

Bates RL, Jackson JA (1980) Glossary of geology. Am Geol Inst, Falls Church

Baulig H (1950) La notion de profil d'equilibre. Strasbourg, Publ Fac Lett Univ 114:43–86

Baum BR (1967) Introduced and naturalized tamarisks in the United States and Canada. Baileya 15:19–25

Bazin, H. (1897) Etude d'une nouvelle formule pour calculer le debit des canaux decouverts. Ann Ponts Chaussees 14:20–70

Beard LR (1943) Statistical analysis in hydrology. Trans Am Soc Civil Eng 108:1110–1160

Beaty CW (1963) Origin of alluvial fans, White Mountains, California and Nevada. Ann Assoc Am Geogr 53:516–535

Beaty CW (1970) Age and estimated rate of accummulation of an alluvial fan. Am J Sci 268:50–77

Beaty CW (1974) Debris flow, alluvial fans, and a revitalised catastrophism. Z Geomorphol Suppl 21:39–51

Beaumont P (1972) Alluvial fans along the foothills of the Elburz Mountains, Iran. Paleogeogr Paleoclim Paleoecol 12:251–273

Beer CE, Farnham CW, Heineman HG (1966) Evaluation of sedimentation prediction techniques in western Iowa. Trans Am Soc Agric Eng 9:828–831, 833

Begin ZB, Meyer DF, Schumm SA (1981) Development of logitudinal profiles of alluvial channels in response to base-level lowering. Earth Surf Proc Landforms 6:49–68

Bell DA (1956) Information theory. Pitman and Sons, London

Bell GL (1968) Piping in the Badlands of North Dakota. Proc 6th Annu Eng Geol Soils Eng Symp, Boisie, Idaho, Idaho Dept Highways, 242–257

Beltaos, S. (1982) Dispersion in tumbling flow. J Hydrol Div, Am Soc Civil Eng 108(HY4): 591–612

Benedict PC and Matejka DQ (1952) The measurement of total sediment load in alluvial streams. Proc 5th Hydraul Conf, Univ Iowa

Benedict PC and Nelson ME (1944) A study of methods used in measurement and analysis of sediment loads in streams. St Paul US Eng Dist Sub-Office Hydraul Lab, Univ Iowa (unpublished)

Bennett JP (1974) Concepts of mathematical modeling of sediment yield. Wat Resour Res 10:485–493

Bennett RJ, Chorley RJ (1978) Environmental systems: philosophy and control. Methuen, London

Benson MA (1964) Factors affecting the occurrence of floods in the Southwest. US Geol Surv Wat-Supply Pap 1580-D

Benson MA (1971) Uniform flood-frequency estimating methods for federal agencies. Wat Resour Res 4:891–908

Berry L (1970) Some erosional features due to piping and subsurface wash with special reference to the Sudan. Geogr Ann 52A:113–119

Bertalanffy L von (1950) The theory of open systems in physics and biology. Sci 111:23–29

Bertalanffy L von (1960) Principles and theory of growth. In: Nowinsky W (ed) Fundamental aspects of normal and malignant growth. Elsevier, New York, pp 25–51.

Beschta RL (1979) Debris removal and its effects on sedimentation in an Oregon Coast Range stream. Northwest Sci 53:71–77

Bharadwaj OP (1961) The arid zone of India. UNESCO Arid Zone Res 17:143–174

Bigarella JJ, Andrada GO (1965) Contributions to the study of the Brazilian Quaternary. Geol Soc Am Spec Pap 84:433–451

Birkeland PW (1968) Mean velocities and boulder transport druing Tahoe-age floods of the Truckee River, California-Nevada. Geol Soc Am Bull 79:137–142

Birot P (1958) La lecon de Grove Karl Gilbert. Ann de Geogr 362: 289–307

Birot P (1968) The cycle of erosion in different climates. Batsford, London

Birot P, Dresch J (1966) Pediments et glacis dans l'Ouest des Etats-Unis. Ann Geogr 411: 513–552

Bishop WW (1962) Gully erosion in Queen Elizabeth National Park. Uganda J 26:161–165

Biswell HH, Schultz AJ (1957) Surface runoff and erosion as related to prescribed burning. J For 55:372–374

Blaney HF (1961) Consumptive use and water waste by phreatophytes. J Irr Dr Div, Am Soc Civil Eng 87(IR3):37–46

Blanford WT (1872) Note on geological formations seen along the coasts of Baluchistan and Persia from Kerachi to the head of Persian Gulf, and some gulf islands. Geol Surv India Records 5

Blanford WT (1873) On the nature and probable origin of the superficial deposits in the valley sand deserts of central Persia. Geol Soc London Q J 29:495–501

Blanford WT (1876) Geology. In: Persian Boundary Commission (ed) Eastern Persia, an account of the journeys of the Persian Boundary Commission, 1870–72. Oxford, London

Blanford WT (1877) Geological notes on the great desert between Sind and Rajputana. Geol Surv India Rec 10:10–21

Blissenbach E (1954) Geology of alluvial fans in semiarid regions. Geol Soc Am Bull 65: 175–190

Blom G (1958) Statistical estimates and transformed beta-variables. John Wiley & Sons, New York

Bluck BJ (1964) Sedimentation on an alluvial fan in southern Nevada. J Sediment Petrol 34:395–400

Bluck BJ (1971) Sedimentation in the meandering River Endrick. Scott J Geol 7:93–138

Bockemeyer H (1890) Beschreibung der Kuste zwischen Mossamedes und Port Nolloth. Dtsch Kolonialz (Berlin) 10:21–34

Boffey PM (1977) The Teton Dam verdict: a foul-up by the engineers. Sci 195:270–272

Bogardi J (1978) Sediment transport in alluvial streams. Akademiai Kiado, Budapest

Bombicci C (1881) Geologia del 'Appennino bolognese. CAI, Bologna

Boothroyd JC, Nummedal D (1978) Proglacial braided outwash: a model for humid alluvial-fan deposition. In: Miall AD (ed) Fluvial sedimentology. Can Soc Petrol Geol (Calgary), pp 641–668

Bork HR, Rohdenburg H (1979) The behaviour of overland flow and infiltration under simiulated rainfall. Symp Agricultural Soil Erosion in Termperate Climates, Strasbourg, Sept 20–23, 1978

Borland WM (1971) Reservoir sedimentation. In: Shen HW (ed) River mechanics. Shen, Ft Collins, pp 29/1–29/38

Bosworth TO (1922) Geology and paleontology of northwest Peru. Macmillan, London

Boughton WC (1980) A frequency distribution for annual floods. Wat Resour Res 16:347–354

Boughton WC, Stone JJ (1985) Variation of runoff with watershed area in semi-arid location. J Arid Environ 9:13–25

Bouvee KD, Milhous R (1978) Hydraulic simulation in instream flow studies: theory and techniques. US Fish and Wildlife Serv, Instream Flow Information Pap 5

Bowie JE, Kam W (1968) Use of water by riparian vegetation, Cottonwood Wash, Arizona. US Geol Surv Wat-Supply Pap 1858

Bowman D (1977) Stepped-bed morphology in arid gravelly channels. Geol Soc Am Bull 88: 291–298

Boyce RC (1975) Sediment routing with sediment-delivery ratios. Present and prospective technology for predicting seidment yields and sources. US Dep Agric, Agric Res Serv Publ ARS-S-40:61–65

Braithwaite RB (1960) Scientific explanation. Harper, New York

Branson FA (1976) Water use on rangelands. Proc 5th Works US/Australian Rangelands Panel, Utah Water Res Lab, Logan Utah, pp 193–209

Branson FA, Owen JB (1970) Plant cover, runoff, and sediment yield relationships on Mancos Shale in western Colorado. Wat Resour Res 6:783–790

Branson FA, Gifford GF, Renard KG, Hadley, RF (1981) Rangeland hydrology. Kendall/Hunt, Dubuque

Brice JC (1966) Erosion and deposition in the loess-mantled Great Plains, Medicine Creek drainage basin, Nebraska. US Geol Surv Prof Pap 352-H 255–339

Bridge JS (1976) Bed topography and grain size in open channel bends. Sedimentol 23:407–414

Bridge JS (1977) Flow, bed topography, grain size and sedimentary structure in open channel bends: a three-dimensional model. Earth Surf Proc 2:401–416

Bridge JS, Jarvis J (1976) Flow and sedimentary processes in the meandering River South Esk, Glen Clova, Scotland. Earth Sur Proc 1:303–336

Briggs D (1977) Sediments. Butterworths, London

Broecker WS, Orr PC (1958) Radiocarbon chronology of Lake Lahontan and Lake Bonneville. Geol Soc Am Bull 69:1009–1032

Brotherton DI (1979) On the origin and characteristics of river channel patterns. J Hydrol 44:211–230

Brown DE (1971) Evaluating watershed management alternatives. Am Soc Civil Eng, J Irr Dr Div 97 (IR1):93–108

Brown GW (1962) Piping erosion in Colorado. J Soil Wat Cons 17:220–222

Brown GW, Krygier JT (1971) Clearcut logging and sediment production in the Oregon Coast Range. Wat Resources Res 7: 1189–1199

Brown HE, Thompson JR (1965) Summer water use by aspen, spruce, and grassland in western Colorado. J For 63: 756–760

Brune GM (1953) Trap efficiency of reservoirs. Trans Am Geophys Union 34:407–418

Brush LM, Jr, and Wolman MG (1957) Knickpoint behavior in noncohesive materials. Geol Soc Am Bull 71:50–74

Bryan K (1922) Erosion and sedimentation in the Papago Country, Arizona. US Geol Surv Bull 730B:19–90

Bryan K (1926) The San Pedro Valley, Arizona, and the geographical cycle. Geol Soc Am Bull 37:169–170

Bryan K (1927) Channel erosion of the Rio Salado, Socorro County, New Mexico. US Geol Surv Bull 790 17–19

Bryan K (1936) The formation of pediments. Int Geol Congr Rep, 16th Sess 2:765–775

Bryan K (1940) Erosion in the valleys of the southwest. N Mex Q Rev 10:227–232

Bryan K (1941) Pre-Columbian agriculture in the southwest as conditioned by periods of alluviation. Ann Assoc Am Geogr 31:219–242

Bryan RB, Campbell IA (1980) Sediment entrainment and transport during local rainstorms in the Steveville Badlands, Alberta. Catena 7:51–65

Bryan R, Yair A (1982) Badland geomorphology and piping. Geobooks, Norwich

Bryan R, Yair A, Hodges WK (1978) Factors controlling the initiation of runoff and piping in Dinosaur Provincial Park badlands, Alberta, Canada. Z Geomorphol, Suppl 29:151–168

Bryson RA, Wendland WM (1967) Tentative climatic patterns for some late glacial and post glacial episodes in central North America. In: Mayer-Oakes WJ (ed) Life, land, and water. Univ Manitoba Press, Winnipeg, pp 277–278

Bube KP, Trimble SW (1986) Revision of the Churchill Reservoir trap efficiency curves using smoothing splines. Wat Res Bull 22:305–309

Buckham AF, Cockfield WE (1950) Gullies formed by sinking of the ground. Am J Sci 248:137–141

Buckhouse JC, Gifford GF (1976) Sediment production and infiltration rates as affected by grazing and debris burning on chained and seeded pinyon-juniper. J Range Manag 29:83–85

Budel J (1944) Die morphologischen wirkungen des eiszeiklimas im gletscherfreien Gebiet. Geol Rdsch Beitrage 1:34

Budel J (1948) Das system der klimatischen geomorphologie. Dtsch Geogr 27:65–100

Budel J (1963) Climatogentic geomorphology. Geogr Rundsch 15:269–285

Bues SS, Carothers SW, and Avery CC (1985) Topographic changes in fluvial terrace deposits used as campsite beaches along the Colorado River in Grand Canyon. J Ariz Nev Acad Sci 20:111–120

Bull WB (1964) Geomorphology of segmented alluvial fans in western Fresno County, California. US Geol Surv Prof Pap 532-F 89–128

Bull WB (1968) Alluvial fan. In: Fairbridge RW (ed) The encyclopedia of geomorphology. Reinhold, New York, pp 7–10

Bull WB (1975a) Allometric change of landforms. Geol Soc Am Bull 86:1489–1498

Bull WB (1975b) Landforms that do not tend toward a steady state. In: Melhorn WN, Flemal RC (eds) Theories of landform development. SUNY Binghamton, pp 111–128

Bull WB (1977) The alluvial fan environment. Progr Phys Geogr 1: 222–270

Bull WB (1979) The threshold of critical power in streams. Geol Soc Am Bull 90:453–464

Bull WB (1980) Geomorphic thresholds as defined by ratios. In: Coates DR, Vitek JD (eds) Thresholds in geomorphology. Allen & Unwin, London, pp 259–263

Bull WB (1982) Recognition of alluvial-fan deposits in the stratigraphic record. In: Rigby JK, Hamblin WK (eds) Recognition of ancient sedimentary environments. Soc Econ Paleontol Mineral, Tulsa, pp 63–83

Burgy RH, Scott VH (1952) Some effects of fire on the infiltration capacity of soils. Trans Am Geophys Union 33: 405–416

Burkham DE (1970a) A method for relating infiltration rates to streamflow rates in perched streams. US Geol Surv Prof Pap 700-D 226–271

Burkham DE (1970b) Depletion of streamflow by infiltration in the main channels of the Tucson Basin, southeastern Arizona. US Geol Surv Wat-Supply Pap 1939-B

Burkham DE (1972) Channel changes of the Gila River in Safford Valley, Arizona. US Geol SurvProf Pap 655G

Burkham DE (1976) Hydraulic effects of changes in bottom land vegetation on three major floods, Gila River, in southeastern Arizona. US Geol Surv Prof Pap 655-J

Burningham CW (1952) Report to the government of Saudi Arabia on land and water resources. FAO, Rome

Burton I, Kates RW, White GF (1978) The environment as hazard. Oxford Univ Press, New York

Butcher GC, Thornes JB (1978) Spatial variability in runoff processes in an ephemeral channel. Z Geomorphol Suppl 29:83–92

Butzer K (1965) Desert landforms at the Kurkur oasis, Egypt. Ann Assoc Am Geogr 55: 578–591.pn 408

Cailleux A (1945) Distinction des galets marines et fluviatiles. Bull Soc Geol Fr 5:375–404

Cain JM and Beatty MT (1968) The use of soil maps in the delineation of floodplains. Wat Resour Res 4:173–182

California (1923) Flow in California streams. Cal State Dep Pub Works Bull 5: Chap 5

Campbell CJ (1966) Periodic mowings suppress tamarisk growth, increase forage for browsing. US For Serv Res Note RM-76 4

Campbell FE (1977) Stream discharge, suspended sediment and erosion rates in the Red Deer River basin, Alberta, Canada. Int Assoc Hydrol Sci 122:244–259

Campbell FE (1982) Surface morphology and rates of change during a ten-year period in the Alberta badlands. In: Bryan R, Yair A (eds) Badland geomorphology and piping. Geobooks, Norwich, pp 221–238

Campbell FE, Bauder HA (1940) A rating curve method for determination of silt discharge of streams. Trans Am Geophys Union 21:603–606

Campbell IA, Honsaker JL (1982) Variability in badlands erosion: problems of scale and threshold identification. In: Thorn CE (ed) Time and space in geomorphology. Allen & Unwin, London, pp 59–80

Campbell RH (1975) Soil slips, debris flows and rainstorms in the Santa Monica Mountains and vicinity, southern California. US Geol Surv Prof Pap 851

Cant DJ, Walker RG (1978) Fluvial processes and facies sequences in the sandy braided South Saskatchewan River, Canada. Sedimentology 25:625–648

Carson MA (1971) The mechanics of erosion. Pion, London

Carson MA (1986) Characteristics of high-energy "meandering" rivers: the Canterbury Plains, New Zealand. Geol Soc Am Bull 97:886–895

Carson MA, Kirkby MJ (1972) Hillslope Form and Process. Cambridge Univ Press

Carter CE, Greer JD, Braud HJ, Floyd JM (1974) Raindrop characteristics in south central United States. Trans Am Soc Agric Eng 17:1033–1037

Carter HJ (1861) On contributions to the geology of western India including Sind and Baluchistan. Asiat Soc Bombay 6: 161–206

Carter RW (1961) Magnitude and frequency of floods in suburban areas. US Geol Surv Prof Pap 424-B

Central Water Authority (1966) Report on Floods in Southern Jordan on 11 March 1966. Cent Wat Author Hydrol Div, Hashemite Kingdom of Jordan, Amman

Chamberlin TC (1883) Geology of Wisconsin, vol 1. Wisconsin Geol Surv, Madison

Chamberlin TC (1897) The method of multiple working hypotheses. J Geol 5:837–848

Chang HH (1979) Minimum stream power and river channel pattern. J Hydrol 41:303–327

Chang HH, Hill, JC (1977) Minimum stream power for rivers and deltas. J Hydrau Div, Am Soc Civil Eng 103:1375–1389

Chang JH (1972) Atmospheric circulation systems and climates. Oriental Publishing, Honolulu, Hawaii

Chang Y (1939) Laboratory investigations of flume traction and transportation. Trans Am Soc Civil Eng 103:1246–1284

Chapman K (1979) People, pattern and process. John Wiley & Sons, New York

Charney J, Stone PH, Quirk WJ (1975) Drought in the Sahara: a biogeophysical feedback mechanism. Science 187:434–435

Chawner WD (1935) Alluvial-fan flooding, the Montrose, California flood of 1934. Geogr Rev 25:255–263

Cheetham GH (1980) Late Quaternary palaeohydrology: the Kennet Valley case study. In: Jones DKC (ed) The shaping of southern England. Academic Press, London New York, pp 203–223

Cheney ES, Patton TC (1967) Origin of the bedrock values of placer deposits. Econ Geol 62:852–853

Cherkauer DS (1972) Longitudinal profiles of ephemeral streams in southeastern Arizona. Geol Soc Am Bull 83:353–366

Chippen JR, Bue CD (1977) Maximum floodflows in the conterminous United States. US Geol Surv Wat-Supply Pap 1887

Chorley RJ (1962) Geomorphology and general systems theory. US Geol Surv Prof Pap 500-B

Chorley RJ (1978) Bases for theory in geomorphology. In: Embleton C (ed) Geomorphology: present problems and future prospects. Oxford Univ Press, pp 1–13

Chorley RJ, Haggett P (1967) Physical and information models in geography. Methuen, London

Chorley RJ, Kennedy BA (1971) Physical geography: a systems approach. Prentice-Hall, London

Chorley RJ, Dunn AJ, Beckinsale RP (1964) The history of the study of landforms or the development of geomorphology, vol 1: Geomorphology before Davis. Methuen, London, and John Wiley & Sons, New York

Chorley RJ, Beckinsale RP, Dunn AJ (1973) The history of the study of landforms, or the development of geomorphology, vol 2: The life and work of William Morris Davis. Methuen, London

Chow VT (1954) The log-probability law and its engineering applications. Proc Am Soc Civil Eng 80:1–25

Chow VT (1959) Open channel hydraulics. McGraw-Hill, New York

Chow VT (ed) (1964a) Runoff. In: Handbook of applied hydrology, Sec 14. McGraw Hill, New York

Chow VT (ed) (1964b) Statistical and probability analysis of hydrologic data, part I: frequency anlaysis. In: Handbook of applied hydrology, Sec 8. McGraw-Hill, New York

Christensen EM (1962) The rate of naturalization of tamarisk in Utah. Am Mid Nat 68:51–57

Christiansson C (1979) Imagi Dam – a case study of soil erosion, reservoir sedimentation and water supply at Dodoma, central Tanzania. Geogr Ann 61A:113–145

Church M (1978) Palaeohydrological reconstructions from a Holocene valley fill. In: Miall AD (ed) Fluvial sedimentology. Can Soc Petrol Geol, Calgary, pp 743–772

Church M (1981) Reconstruction of the hydrological and climatic conditions of past fluvial environments. In: Starkel L Thrones JB (eds) Paleohydrology of river basins. Br Geomorphol Res Gp, London, pp 50–79

Church M, Gilbert R (1975) Proglacial fluvial and lacustrine environments. Soc Econ Paleontol Mineral Publ 23:22–84

Church M, Kellerhals R (1978) On the statistics of grain size variation along a gravel river. Can J Earth Sci 15: 1151–1160

Church M, Mark DM (1980) On size and scale in geomorphology. Progr Phys Geogr 4:342–390

Churchill MA (1948) Discussion of "Analysis and use of reservoir sedimentation data" by L. C. Gottschalk. Proc Fed Inter-Agency Sediment Conf, Denver, pp 139–140

Churchman CW (1961) Prediction and optimal decision. Prentice-Hall, Englewood Cliffs

Clark AH (1956) The impact of exotic invasion on the remaining New World mid-latitude grasslands. In: Thomas WL Jr (ed) Man's role in changing the face of the earth. Univ Chicago Press, pp 737–762

Cliff AD, Ord JK (1981) Spatial processes: models and applications. Pion, London

Coates DR, Vitek JD (eds) (1980) Thresholds in geomorphology. Allen & Unwin, London

Cohen O, Ben-Zvi A (1979) Regional analysis of peak discharges in the Negev. Symposium on the hydrology of areas of low precipitations. Inter Assoc Hydrol Sci Publ 128:23–32

Coleman JM (1969) Brahmaputra River: channel processes and sedimentation. Sediment Geol 3:129–239

Collings MR, Myrick RM (1966) Effects of juniper and pinyon eradication on streamflow from Corduroy Creek Basin, Arizona. US Geol Surv Prof Pap 491B

Collinson JD (1978) Alluvial sediments. In: Reading HG (ed) Sedimentary environments and facies. Elsevier, New York Amsterdam, pp 15–60

Cooke RU (1964) Planation surfaces or 'matureland' in the southern Atacama Desert? Int Geogr Cong, 20th Meet, London, Abstr Pap, 87

Cooke RU (1984) Geomorphological hazards in Los Angeles. George Allen & Unwin, London

Cooke RU, Reeves RW (1976) Arroyos and environmental change in the American South-West. Clarendon, Oxford

Cooke RU, Warren A (1973) Geomorphology in deserts. Univ Cal Press, Berkeley

Cooke RU, Brunsden D, Doornkamp JC, Jones DKC (1982) Urban geomorphology in drylands. Oxford University Press

Cooley ME, Aldridge BN, Euler RC (1977) Effects of the catastrophic flood of December 1966, North Rim area, eastern Grand Canyon, Arizona. US Geol Surv Prof Pap 980

Cooperider CK, Hendricks BA (1937) Soil erosion and stream flow on range and forest lands ofthe upper Rio Grande watershed in relation to land resources and human welfare. US Dep Agric Tech Bull 567

Copeland OL (1965) Land use and ecological factors in relatin to sediment yield. Proc Fed Inter-Agency Sediment Conf, Agric Res Serv Misc Publ 970:72–84

Cori B, Vittorini S (1974) Ricerche sui fenomeni di erosione accelerata in Val d'Era (Toscana). L'erosione del suolo in Italia e i suoi fattori, I. Inst Geogr Univ Pisa

Costa JE (1983) Paleohydraulic reconstruction of flash-flood peaks from boulder deposits in the Colorado Front Range. Geol Soc Am Bull 94:986–1004

Costa JE, Baker VR (1981) Surficial geology, building with the earth. John Wiley & Sons, New York

Costa JE, Graf WL (1984) The geography of geomorphologists in the United States. Prof Geogr 36:82–89

Cotton CA (1942) Climatic accidents in landscape-making. Hafner New York

Cowan, WL (1956) Estimating hydraulic roughness coefficients. Agric Eng 37:473–475

Crampton FA (1937) Occurrence of gold in stream placers. Min J 20:3–4

Croft AR, Bailey RW (1964) Mountain water. US For Serv Intermountain For Range Exp Stn Unnumbered Publ

Croll J (1976) Is catastrophe theory dangerous? New Sci 70: 630–632

Crowley KD (1981a) Large-scale bedforms in the Platte River downstream from Grand Island, Nebraska: Structure, process, and relationship to channel narrowing. US Geol Surv Open-File Rep 81–1059

Crowley KD (1981b) Hierarchies of bedforms. PhD Diss Princeton Univ

Cubitt JM, Shaw B (1976) The geological implications of steady-state mechanisms in catastrophe theory. Math Geol 8: 657–662

Culler RC, Hanson RC, Myrick RM, Turner RM, Kipple FP (1982) Evapotranspiration before and after clearing phreatophytes, Gila River flood plain, Graham County, Arizona. US Geol Surv Prof Pap 655-P

Dalrymple T (1960) Flood-frequency analysis. US Geol Surv Wat-Supply Pap 1543-A

Daly RA (1945) Biographical memoir of William Morris Davis: 1850–1934. Natl Acad Sci Biogr Mem 23:263–303

Darrah WC (1951) Powell of the Colorado. Princeton Univ Press

Darwin C (1846) Observations on South America. Smith Elder, London

Davidson DA (1978) Science for physical geographers. John Wiley & Sons, New York

Davies BE, Lewin J (1974) Chronosequences in alluvial soils with special reference to historic lead pollution in Cardiganshire, Wales. Environ Pollut 6:49–57

Davies GL (1969) The earth in decay: a history of British geomorphology, 1578 to 1878. Elsevier, New York Amsterdam

Davis WM (1899) The geographical cycle. Geogr J 14:481–504

Davis WM (1902) Base-level, grade, and peneplain. J Geol 10: 77–111

Davis WM (1905) The geographical cycle in an arid climate. J Geol 13:381–407

Davis WM (1918) Grove Karl Gilbert. Am J Sci 46:669–681

Davis WM (1922) Biographical memoir of Grove Karl Gilbert (1843–1918). Natl Acad Sci, Washington, DC

Davis WM (1938) Sheetfloods and streamfloods. Geol Soc Am Bull 49:329–339

Dawdy DR (1967) Knowledge of sedimentation in urban environments. Proc Am Soc Civil Eng Hydrol Div 93(HY6):235–245

Dawdy DR (1979) Flood frequency estimates on alluvial fans. Am Soc Civil Eng, J Hydraul Div 105:1407–1413

DeBano LF (1981) Water-repellent soils: a state-of-the-art. US For Serv Gen Tech Rep PSW-46

DeBano LF, Rice RM (1973) Water-repellent soils: their implications in forestry. J For 71:220–223

DeByle NV, Packer PE (1972) Plant nutrient and soil losses in overland flow from burned forest clearcuts. Proc Am Wat Res Assoc Nat Symp on watersheds in transition, Colo State Univ, Ft Collins, pp 296–307

Decker JP, Gaylor WG, Cole FD (1962) Measuring transpiration of undisturbed tamarisk shrubs. Plant Phys 37:393–397

Delwaulle JC (1973) Resultats de six ans d'observations sur lerosion au Niger. Bois For Trop 150:15–137

Denbigh KG (1951) The thermodynamics of the steady state. Methuen, London

Dendy FE, Bolton GC (1976) Sediment yield-runoff drainage area relationships in the United States. J Soil Wat Cons 31: 264–266

Denevan WM (1967) Livestock numbers in nineteenth-century New Mexico, and the problem of gullying in the Southwest. Ann Assoc Am Geogr 57:691–703

Denny CS (1965) Alluvial fans in the Death Valley region, California and Nevada. US Geol Surv Prof Pap 466

Denny CS (1967) Fans and pediments. Am J Sci 265:81–105

Derbyshire E (1973) Climatic geomorphology. Barnes and Noble, New York

Dietrich RV, Dutro JT Jr, Foose RM (1982) AGI data sheets for geology in the field, laboratory, and office. Am Geol Inst, Falls Church

Dietrich WE (1975) Sediment production in a mountainous basaltic terrain in central coastal Oregon. MS Thes, Univ Washington, Seattle

Dietz RA (1952) The evolution of a gravel bar. Miss Bot Gard Ann 39:250

Diller JS (1911) Major Clarence Edward Dutton. Bull Seismogr Soc Am 1:137–142

Diskin M, Lane LJ (1972) A basinwide stochastic model for ephemeral stream runoff in South-Eastern Arizona. Bull Inter Assoc Hydr Sci 17:62–76

Dixey F (1939) Some observations on the physiographical development of Central and South Africa. Geol Soc S Afr Trans 41:113–171

Dixon JE, Stephenson GR, Lingg AJ, Naylor DV, Hinman DD (1977) Nonpoint pollution control for wintering range cattle. Unpubl Pap Ann Meet Am Soc Agric Eng, Raleigh, NC, June 26–29, 1977

Dodge RE (1902) Arroyo formation. Science 15:746

Doehring DO (1968) The effect of fire on geomorphic processes in the San Gabriel Mountains, California. Contrib Geol, Univ Wyoming 7:43–65

Dolan RA, Howard A, Gallensen A (1974) Man's impact on the Colorado River in the Grand Canyon. Am Sci 62:392–406

Donat J (1929) Uber Sohlangriff und Geschiebetrieb. Wasserwirtschaft 26:27

Doornkamp JC, King CAM (1971) Numerical analysis in geomorphology: an introduction. St Martin's , New York

Dortignac EJ (1956) Watershed resources and problems of the Upper Rio Grande Basin. US For Serv, Rocky Mt Exper Stn, Unnumbered Publ

Downes RG (1946) Tunnelling erosion in north-eastern Victoria. J Council Sci Ind Res Austr 19:283–291

Dresch J (1957) Pediments et glacis d erosion, pediplains et inselbergs. Inf Geogr 22:183–196

Drew DP (1982) Piping in the Big Muddy badlands, southern Saskatchewan, Canada. In: Bryan R, Yair A (eds) Badland geomorphology and piping. Geobooks, Norwich, pp 293–304

Drew F (1873) Alluvial and lacustrine deposits and glacial records of the upper Indus basin. Pt 1: Alluvial deposits. Geol Soc London Q J 29:441–471

DuBoys MP (1879) Le Rhone et les rivieres a lit affouillable. Mem Doc Ann Pont Chauss 5:18

Dunford EG (1949) Relation of grazing to runoff and erosion on bunchgrass ranges. US For Serv Res Note RM-7

Dunne T (1979) Sediment yield and land use in tropical catchments. J Hydrol 42:281–300

Dunne T, Leopold LB (1977) Water in environmental planning. Freeman, San Francisco

Dunne T, Moore TR, Taylor CH (1975) Recognition and prediction of runoff-producing zones in humid regions. Hydrol Sci Bull 20:305–327

Dury GH (1964a) Principles of underfit streams. US Geol Surv Prof Pap 452A

Dury GH (1964b) Subsurface implications of underfit streams. US Geol Surv Prof Pap 452B

Dury GH (1965) Theoretical implications of underfit streams. US Geol Surv Prof Pap 452C

Dury GH (1966a) The concept of grade. In: Dury GH (ed) Essays in geomorphology. Am Elsevier, New York, pp 211–234

Dury GH (1966b) Pediment slope and particle size at Middle Pinnacle, near Broken Hill, New South Wales. Austr Geogr Stud 4:1–17

Dury GH (1970) Morphometry of gibber gravel at Mt. Sturt, New South Wales. J Geol Soc Aust 16:656–666

Dury GH (1973) Magnitude-frequency analysis and channel morphology. In: Morisawa M (ed) Fluvial geomorphology. SUNY Binghamton, pp 91–121

Dury GH (1976) Discharge prediction, present and former, from channel dimensions. J Hydrol 30:219–245

Dury GH, Hails JR, Robbie MB (1963) Bankfull discharge and the magnitude-frequency series. Aust J Sci 26:123–124

Dutton CE (1880) Report on the Geology of the High Plateaus. US Geogr and Geol Surv of the Rocky Mountain Region, Washington, D. C.,

Dutton CE (1882a) The physical geology of the Grand Canyon District. US Geol Surv 2nd Annu Rep, pp 47–166

Dutton CE (1882b) Tertiary history of the Grand Canyon Region. Geol Surv Mongr 2

Dzulynski S, Slaczka A (1958) Directional structures and sedimentation of the Krosno Beds (Carpathian flysch). Ann Soc Geol Pol 28:205–259

Eakin HM, Brown CB (1939) Silting of reservoirs. US Dep Agric Tech Bull 524 168

Eaton EC (1936) Flood and erosion control problems and their solution. Trans Am Soc Civil Eng 101:1302–1330

Eckis R (1928) Alluvial fans of the Cucamonga District, southern California. J Geol 36:224–247

Eggler DH, Larson EE, Bradley WC (1969) Graites, grusses, and the Sherman erosion surface, southern Laramie Range, Colorado-Wyoming. Am J Sci 2:510–522

Egiazaroff IV (1965) Calculation of nonuniform sediment concentrations. Proc Am Soc Civil Eng J Hydraul Div 98(HY4):43–68

Egypt Ministry of Housing and Reconstruction (1978) Suez area subsurface investigation. Sir William Halcrow and Partners, Robert Matthew, Johnson-Marshall and Partners, Economic Consultants Ltd., Hamed Kaddah, Cairo

Ehrenberger R (1931) Direkte Geschiebemessungen an der Donau bei Wien und deren bisherige Ergebnisse. Wasserwirtschaft 34:33-45

Eidemiller DI (1978) The frequency of tropical cyclones in the southwestern United States and northwestern Mexico. Climatol Publ State Climatol Ariz Sci Pap 1

Einstein HA (1948) Determination of rates of bed-load movement. Proc Fed Inter-Agency Sediment Conf, US Dept Agric

Einstein HA (1950) The bed-load function for sediment transportation in open channel flows. US Dept Agric, Soil Cons Serv Tech Bull 1026

Einstein HA (1965) Final report, spawning grounds. Univ Cal Hydraul Eng Lab Unnumbered Rep

Einstein HA, Shen HW (1964) A study of meandering in straight alluvial channels. J Geophys Res 69:5239-5247

Ellison WD (1944) Studies of raindrop erosion. Agric Eng 25: 131-136

Elwell HA (1977) Soil loss estimation system for southern Africa. Rhodesia Dep Cons Extension Res Bull 22

Elwell HA (1978) Compiled report of the multidisciplinary team on soil loss estimation. Rhodesia Dep of Cons Extension, Unnumbered Rep

Elwell HA (1984) Soil loss estimation: a modelling technique. In: Hadley RF, Walling DE (eds) Erosion and sediment yield: some methods of measurement and modelling. Cambridge Univ Press, pp 15-36

Elwell HA, Stocking MA (1976) Vegetal cover to estimate soil erosion hazard in Rhodesia. Geoderma 15:61-70

Emmett WW (1970) The hydraulics of overland flow on hillslopes. US Geol Surv Prof Pap 662-A 68

Emmett WW (1974) Channel aggradation in western United States as indicated by observation at Vigil Network sites. Z Geomorphol Suppl 21:52-62

Emmett WW (1979) A field calibration of the sediment-trapping characteristics of the Helley-Smith bedload sampler. US Geol Surv Open-File Rept 79-411

Emmett WW, Leopold LB, Myrick RM (1983) Some characteristics of fluvial processes in rivers. Proc 2nd Int Symp River Sedimentation, October 11-16, 1983, Nanjing, China, pp 730-754

Ergenzinger PJ (1982) Uber den Einsatz von Magnettracern zur Messung des Grobgeschiebetranportes. Beitr Geol Schweiz-Hydrol 28:483-491

Ergenzinger PJ, Conrady J (1982) A new tracer technique for measuring bedload in natural channels. Catena 9:77-80

Ergenzinger PJ, Custer SG (1983) Determination of bedload transport using magnetic tracers: first experiences at Squaw Creek, Gallatin County, Montana. Wat Resour Res 19: 187-193

Eschner TR (1981) Morphologic and hydrologic changes of the Platte River, south-central Nebraska. Unpubl MS Thes, Col State University, Ft. Collins

Eschner TR, Hadley RF, Crowley KD (1983) Hydrologic and morphologic changes in channels of the Platte River Basin in Colorado, Wyoming, and Nebraska: a historical perspective. US Geol Surv Prof Pap 1277-A

Espenshade ED (1978) Goode's world atlas (15th ed). Rand McNally, Chicago

Espey WH, Morgan CW, Masch FD (1966) Study of some effects of urbanization on storm runoff from a small watershed. Texas Wat Dev Bd Rep

Ethridge FG, Schumm SA (1978) Reconstructing paleochannel morphologic and flow characteristics: methodology, limitations, and assessment. In: Miall AD (ed) Fluvial sedimentology. Can Soc Petrol Geol, Calgary, pp 703-721

Evans R (1980) Mechanics of water erosion and their spatial and temporal controls: an empirical viewpoint. In: Kirkby MJ, Morgan RPC (eds) Soil erosion. Wiley, Chichester, pp 109-128

Everitt BL (1980) Ecology of saltcedar--a plea for research. Environ Geol 3:77-84

Fair IJD (1948) Hillslopes and pediments of the semi-arid Karroo. S Afr Geogr J 30:71-79

Farrand WR (1962) Postgalcial uplift in North America. Am J Sci 260:181-199

Fenneman NM (1916) Physiographic division of the United States. Ann Assoc Am Geogr 6:19-98

Feynman RP, Leighton RB, Sands M (1965) The Feynman lectures on physics. Addison-Wesley, Reading, Mass

Fisher RV (1971) Features of coarse-grained high-concentration fluids and their deposits. J Sediment Petrol 41:916–927

Flammond GM (1899) La traversee de l'erg occidental (grand dunes du Sahara oranais). Ann Geogr 9:231–241

Flaxman EM (1972) Predicting sediment yield in the western United States. Am Soc Civil Eng J Hydraul Div 98(HY12):2073–2085

Flectcher JE, Carroll PH (1948) Some properties of soils associated with piping in southeast Arizona. Soil Sci Soc Am Proc 13:545–547

Fleming G (1969) Suspended solids monitoring: a comparison between three instruments. Wat Wat Eng 73:377–382

Flint RF (1971) Glacial and quaternary geology. John Wiley & Sons, New York

Foggin GT, DeBano LF (1971) Some geographic implications of water-repellent soils. Prof Geogr 23:347–350

Fookes PG (1976) Road geotechnics in hot deserts. J Inst Highway Eng 23:11–23

Fornier F (1960) Climat et erosion. PUF, Paris

Fortescue JAC (1980) Environmental geochemistry: a holistic approach. Springer, Berlin Heidelberg New York

Foster GR, Hakonson TE (1983) Erosional losses of fallout plutonium. Pap Proc Symp Environmental research on the actinide elements, US Dep Energ, December 6, 1983, Mimeo Pap, 30 p.

Foster GR, Hakonson TE (1984) Predicted erosion and sediment delivery of fallout plutonium. J Environ Qual 13:595–602

Foster GR, Wischmeier WH (1974) Evaluating irregular slopes for soil loss prediction. Trans Am Soc Agric Eng 17:305–309

Fraser JC (1972) Regulated dishcarge and the stream environment. In: Oglesby RT (ed) River ecology and man. Academic Press, London New York, pp 263–285

Freeman JR (1922) Flood problems in China. Am Soc Civil Eng Trans 82:1436

Frenguelli J (1928) Acerca del origen de los salares de los dsiertos de la Puna de Atacama. Gaea 3:167–186

Frostick LE, Reid I (1979) Drainage-net controls of sedimentary parameters in sand-bed ephemeral streams. In: Pitty A (ed) Geographical approaches to fluvial processes. Geobooks, Norwich, pp 173–201

Fryberger SG, Ahlbrandt TS, Andrews S (1979) Origin, sedimentary features and significance of low-angle eolian 'sand-sheet' deposits, Great Sand Dunes National Monument and vicinity, Colorado. J Sediment Petrol 49:733–746

Fuller ML (1922) Some unusual erosion features in the loess of China. Geogr Rev 12:570–584

Garde RJ Ranga Raju KG (1977) Mechanics of sediment transportation and alluvial stream problems. Wiley, New York

Gardner JS (1977) Physical geography. Harper, New York

Gardner TW (1983) Paleohydrology and palemorphology of a Carboniferous, meandering fluvial sandstone. J Sediment Petrol 53:991–1005

Garner HF (1959) Stratigraphic-sedimentary significance of climate and relief in four regions of the Andes Mountains. Geol Soc Am Bull 70:1327–1368

Gatewood JS, Robinson TW, Colby BR, Hem JD, Halpenny, LC (1950) Use of water by bottom-land vegetation in lower Safford Valley, Arizona. US Geol Surv Wat-Sup Pap 1103

Gaulton F (1852) Recent expedition into the interior of South West Africa. R Geogr Soc (London) J 22

Gautier EF (1935) Sahara--The great desert. Columbia Univ Press, New York

Gautier EF, Chudeau R (1909) Missions au Sahara. Colin, Paris

Gay FW (1971) Forest climatology studies at Oregon State University. Selected Pap, 29th Annu Meet, Proc Or Acad Sci 7:11–23

Gay LW, Fritschen LJ (1979) An energy budget analysis of water use by saltcedar. Wat Resour Res 15:1589–1592

Gay LW, Hartman RK (1982) ET measurements over riparian saltcedar on the Colorado River. Hydrol and Wat Res Ariz SW 12:9–15

Geddes HJ (1963) Water harvesting. Water resource use and management. Melbourne Univ Press, Melbourne, pp 25–48

Geikie A (1882) Text-book of geology. Readex Microprint (1974 reprint), New York

Gessler J (1970) Self-stabilizing tendencies of alluvial channels. Proc Am Soc Civil Eng, J Waterways, Harb, Coastal Eng Div 96(WW2):235–250

Ghosh A (1952) The Rajputana desert--its archeological aspect. Proc Natl Inst Sci India Bull 1:43–50

Gibbs HS (1945) Tunnel-gully erosion on the Wither Hills, Marlborough. N Z J Sci Tech A-27:135–146

Giessner FW, Price M (1971) Flood of January 1969, near Azusa and Glendora, California. US Geol Surv Hydrol Invest Atlas HA424

Gifford GF (1975) Impacts of pinyon-juniper manipulation on watershed values. The pinyon-juniper ecosystem – A symposium, Utah State Univ, Logan, Utah, May, 1975 127–141

Gifford GF, Williams G, Coltharp GB (1970) Infiltration and erosion studies on pinyon-juniper conversion sites in southern Utah. J Range Manag 23:402–406

Gilbert GK (1875) Report on the geology of portions of Nevada, Utah, California and Arizona, 1871–72. Rep Geogr Geol Explor Surv West of the 100th Mer 3:21–187

Gilbert GK (1876) The Colorado Plateau province as a field for geological study. Am J Sci 12:85–103

Gilbert GK (1877) Report on the Geology of the Henry Mountains. US Geogr Geol Surv of the Rocky Mountain Region, Washington, D C

Gilbert GK (1879) Irrigation lands of the Salt Lake drainage system. In: Powell JW (ed) Report on the lands of the arid region of the United States. US Gov Print Off, Washington, D C, pp 117–132

Gilbert GK (1890) Lake Bonneville. US Geol Surv Monogr 1

Gilbert GK (1914) The transportation of debris by running water. US Geol Surv Prof Pap 86

Gilbert GK (1917) Hydraulic-mining debris in the Sierra Nevada. US Geol Surv Prof Pap 105

Gilead D (1975) A preliminary hydrological appraisal of the Wadi El-Arish flood, 1975. Mimeogr Rep, Israel, Hydrol Serv, Jerusalem

Gilman CS (1965) Rainfall. In: Chow VT (ed) Handbook of applied hydrology, Sec 9. McGraw-Hill, New York

Gilmore R (1981) Catastrophe theory for scientists and engineers. John Wiley & Sons, New York

Girard G, Rodier JA (1979) Application de modeles mathematiques deterministiques a letude des crues et de lecoulement annuel--zones sheliennes. Symp Hydrology of areas of low precipitations. Intl Assoc Hydrol Sci Publ 128: 65–78

Glancy PA, Harmsen L (1975) A hydrologic assessment of the September 14, 1974, flood in Eldorado Canyon, Nevada. US Geol Surv Prof Pap 930

Glennie KW (1970) Desert sedimentary environments. Elsevier, Amsterdam

Glover RE (1964) Dispersion of dissolved or suspended materials in flowing streams. US Geol Surv Prof Pap 433-B

Glymph LM, Holtan HN (1969) Land treatment in agricultural watershed hydrology research. Effects of watershed changes on streamflow. 2nd Symp water resources. Univ Texas Press, Austin, pp 44–68

Goetzmann WH (1959) Army Exploration of the American West, 1803–1863. Yale Univ Press, New Haven, Conn

Gole CV, Chitale SV (1966) Inland delta building activities of Kosi River. J Hydrol Div Am Soc Civil Eng 92:111–126

Gorycki MA (1973) Hydraulic drag; a meander-initiating mechanism. Geol Soc Am Bull 84:175–186

Gottschalk LC (1964) Sedimentation, Pt 1. Reservoir sedimentation. In Chow VT (ed) Handbook of hydrology. McGraw-Hill, New York, pp 17/1–17/34

Goudie A (1985) The encyclopeadic dictionary of physical geography. Blackwell, London

Gould PR (1972) Pedagogic review: entropy in urban and regional modelling. Ann Assoc Am Geogr 62:689–700

Gould SJ (1966) Allometry and size in ontogeny and phylogeny. Biol Rev 41:587–640

Graf WH (1971) Hydraulics of sediment transport. McGraw-Hill, New York

Graf WL (1975) The impact of suburbanization on fluvial geomorphology. Wat Resour Res 11:690–692

Graf WL (1977a) The rate law in fluvial geomorphology. Am J Sci 277:178–191

Graf WL (1977b) Network characteristics of suburbanizing streams. Wat Resour Res 13:459–463

Graf WL (1978) Fluvial adjustments to the spread of tamarisk in the Colorado Plateau region. Geol Soc Am Bull 89:1491–1501

Graf WL (1979a) Development of montane arroyos and gullies. Earth Surf Proc 4:1–14

Graf WL (1979b) Mining and channel response. Ann Assoc Am Geogr 69:262–275

Graf WL (1979c) Catastrophe theory as a model for change in fluvial systems. In: Rhodes DD, Williams GP (eds) Adjustments of the fluvial system. Kendall/Hunt, Dubuque, pp 13–32

Graf WL (1980a) Riparian management: a flood control perspective. J Soil Wat Cons 35:158–161

Graf WL (1980b) The effect of dam closure on downstream rapids. Wat Resour Res 16:129–136

Graf WL (1981) Channel instability in a sand-bed river. Wat Resour Res 17:1087–1094

Graf WL (1982a) Spatial variation of fluvial processes in semiarid lands. In: Thorne CE (ed) Space and time in geomorphology. Allen & Unwin, London, pp 193–217

Graf WL (1982b) Distance decay and arroyo development in the Henry Mountains region, Utah. Am J Sci 282:1541–1554

Graf WL (1983a) Flood-related change in an arid region river. Earth Surf Proc Landforms 8:125–139

Graf WL (1983b) The arroyo problem – paleohydrology and paleohydraulics in the short term. In: Gregory KJ (ed) Background to paleohydrology. John Wiley & Sons, New York London, pp 279–302

Graf WL (1983c) Variability of sediment removal in a semiarid watershed. Wat Resour Res 19:643–652

Graf WL (1984) A probabilistic approach to the spatial assessment of river channel instability. Wat Resources Res 20:953–962

Graf WL (1985a) Magnitude and frequency effects. In: Goudie A (ed) The encyclopaedic dictionary of physical geography. Blackwell, London, pp 272–274

Graf WL (1985b) Mercury transport in stream sediments of the Colorado Plateau. Ann Assoc Am Geogr 75:552–565

Graf WL (1986) Fluvial erosion and federal public policy in the Navajo Nation. Phys Geogr 7:97–115

Graf WL (1987a) Definition of flood plains along arid-region rivers. In: Baker VR, Kochel RC, Patton, P. C. (ed) Flood geomorphology. John Wiley & Sons, (in press)

Graf WL (1987b) Late Holocene sediment storage in canyons of the Colorado Plateau. Geol Soc Am Bull 98:261–271

Gravelius H (1914) Flusskunde, Band 1. Goschenesche Verlagshandlung, Berlin

Gray WA (1968) The packing of solid particles. Chapman & Hall, London

Gregory HE (1917) Geology of the Navajo country. US Geol Surv Prof Pap 93

Gregory HE (1950) Geology and geography of the Zion Park region, Utah and Arizona. US Geol Surv Prof Pap 220

Gregory KJ (1977) River channel change. John Wiley & Sons, New York London

Gregory KJ (1983) Introduction. In: Gregory KJ (ed) Background to paleohydrology. John Wiley & Son, New York Chichester, pp 3–23

Gregory KJ (1985) The nature of physical geography. Arnold, London

Gregory KJ, Park CC (1974) Adjustment of river channel capacity downstream from a reservoir. Wat Resour Res 10:870–873

Griffiths JS (1978) Flood assessment in ungauged semi-arid catchments as a branch of applied geomorphology. King's College, London, Dept Geogr Occasional Pap 8

Gringorten II (1963) A plotting rule for extreme probability paper. J Geophys Res 68:813–814

Grove AT (1978) Desertification. Prog Phys Geogr 2:296–310

Guilke L (1977) Regional geography. Prof Geogr 29:1–7

Gunn CP (1968) Origin of the bedrock values of placer deposits [disc]. Econ Geol 63:86

Gunn RD, Kinzer GD (1949) Terminal velocity of water droplets in stagnant air. J Meteor 6:243–248

Guy HP (1970) Sediment problems in urban areas. US Geol Surv Circ 601-E

Guy HP, Norman VW (1982) Field methods for measurement of fluvial sediment. US Geol SurvTech Wat-Res Invest, Book 3, Chap C2

Hack JT (1960) Interpretation of erosion topography in humid temperate regions. Am J Sci 258-A:89–97

Hadley RF (1961) Influence of riparian vegetation on channel shape, northeastern Arizona. US Geol Surv Prof Pap 424-C: 30–31

Hadley RF (1967) Pediments and pediment-forming processes. J Geol Ed 15:83–89

Hadley RF, Schumm SA (1961) Sediment sources and drainage basin characteristics in upper Cheyenne River basin. US Geol Surv Wat-Sup Pap 1531-B

Haggett P, Cliff AD, Frey A (1977) Locational analysis in human geography. John Wiley & Sons, New York

Haigh MJ (1978) Micro-rills and desiccation cracks: some observations. Z Geomorphol 22:457–461

Haigh MJ, Rydout GB (1985) Catastrophic erosion in southern Arizona: the case of Greene's Canal. Mimeographed manuscript of paper presented at 1st Int Conf Geomorphology, Manchester, England, September 1985

Haight FA (1963) Mathematical theories of traffic flow. Academic Press, London New York

Haines-Young RH, Petch JR (1980) The challenge of critical rationalism for methodology in physical geography. Prog Phys Geogr 4:63–77

Hales ZL, Shindala A, Denson KH (1970) Riverbed degradation prediction. Wat Resour Res 6:549–556

Hall SA (1977) Late Quaternary sedimentation and paleoecologic history of Chaco Canyon, New Mexico. Geol Soc Am Bull 88: 1593–1618

Hammad HY (1972) River bed degradation after closure of dams. Proc Am Soc Civil Eng J Hydraul Div 98(HY9):591–607

Hammer TR (1972) Stream channel enlargement due to urbanization. Wat Resour Res 8:1530–1537

Hanks RJ, Anderson KL (1957) Pasture burning and moisture conservation. J Soil Wat Cons 12:228–229

Hanson CL, Kuhlman AR, Erickson CJ, Lewis, JK (1970) Grazing effects on runoff and vegetation on western South Dakota rangeland. J Range Manag 23:418–420

Happ SC (1975) Genetic classification of valley sediment deposits. In: Vanoni VA (ed) Sedimentation Eng. Am Soc Civil Eng, New York, pp 286–292

Hardie LA (1968) The origin of the Recent non-marine evaporite deposit of Saline Valley, Inyo County, California. Geochim Cosmochim Acta 32:1279–1301

Hardie LA, Smoot JP, Eugster HP (1978) Saline lakes and their deposits: a sedimentological approach. Int Assoc Sediment Spec Publ 2:7–42

Harris DR (1966) Recent plant invasions in the arid and semi-arid southwest of the United States. Ann Assoc Am Geogr 56: 408–422

Harris EE, Rantz SE (1964) Effect of urban growth on streamflow regimen of Permanente Creek, Santa Clara County, California. US Geol Surv Wat-Supply Pap 1591-B

Harrison AS (1950) Report on special investigation of bed sediment segregation in a degrading bed. Univ Calif Berkeley, Inst Eng Res 33:205

Hartshorne R (1939) The nature of geography. Assoc Am Geogr, Lancaster, Penn

Harvey D (1969) Explanation in geography. St Martin's Press, New York

Hassan MA, Schick AP, Laronne JB (1984) The recovery of flood-dispersed coarse sediment particles: a three-dimensional magnetic tracing method. Catena Suppl 5: 153–162

Hastings JR, Turner RM and Warren DK (1972) An atlas of some plant distributions in the Sonoran Desert. Univ Ariz, Instit Atmos Physics, Tech Rep Meteor Clim Arid Reg 21

Hawkes HE (1976) The downstream dilution of stream sediment anomalies. J Geochem Exp 6:345–358

Hayden FV (1876) Notes descriptive of some geological sections of the country about the headwaters of the Missouri and Yellowstone rivers. Bull US Geol Surv Terr 2:197–209

Haynes CV, Jr (1968) Geochronology of late-Quaternary alluvium. In: Morrison RB, Wright HE Jr (eds) Means of correlation of Quaternary successions. Univ Utah Press, Salt Lake City, pp 591–631

Haynes K, Fatheringham AS (1984) Gravity and spatial interaction models. Sage Publ Sci Geogr Ser, London

Hazen A (1930) Flood flows, a study of frequencies and magnitudes. Jon Wiley and Sons, New York

Hedman ER (1970) Mean annual runoff as related to channel geometry of selected streams in California. US Geol Surv Wat-Supply Pap 1999-E

Hedman ER, Osterkamp WR (1982) Streamflow characteristics related to channel geometry of streams in western United States. US Geol Surv Wat-Supply Pap 2193

Heedee BH (1971) Characteristics and processes of soil piping in gullies. US Dept Agri, For Serv, Res Pap RM-68

Heedee BH (1972) Influences of a forest on the hydraulic geometry of two mountain streams. Wat Resour Bull 8:523–530

Heedee BH (1975) Mountain watersheds and dynamic equilibrium. Am Soc Civil Eng, Proc Watershed Manag Symp, Logan, Utah pp 407–420

Heedee BH (1976a) Gully development and control: the status of our knowledge. US For Serv Res Pap RM-109 42

Heedee BH (1976b) Equilibrium condition and sediment transport in an ephemeral mountain stream. Ariz Acad Sci Proc, pp 97–102

Heedee BH (1981) Dynamics of selected mountain streams in the western United States. Z Geomorphol 25:17–32

Heedee BH (1985) Channel adjustments to the removal of log steps: an experiement in a mountain stream. Environ Man 9:427–432

Hefley HM (1935) Ecological studies on the Canadian River flood plain in Cleveland County, Oklahoma. Ecol Monogr 7:345–402

Heggen, RJ, Leonard RE (1980) Arroyo stabilization in an urban environment. J Civil Eng Des 2:321–338

Helley EJ, Smith W (1971) Development and calibration of a pressure-difference bedload sampler. US Geol Surv Open-File Rep

Helvey JD (1973) Watershed behavior after forest fire in Washington. Proc Am Soc Civil Eng, Irr Dr Div Spec Conf, Ft. Collins, pp 403–422

Hem JD (1970) Chemical behavior of mercury in aqueous media. US Geol Surv Prof Pap 713:19–24

Hemple CG (1965) Aspects of scientific explanation. Free Press, New York

Henley S (1976) Catastrophe theory models in geology. Math Geol 8:649–655

Hereford R (1984) Climate and ephemeral-stream processes: twentieth century geomorphology and alluvial stratigraphy of the Little Colorado River, Arizona. Geol Soc Am Bull 95: 654–668

Hereford R (1986) Modern alluvial history of the Paria River drainage basin, southern Utah. Quat Res 25:293–311

Herman P (1908) Beitrage zur Geologie von Deutsch-Sudwestafrika. I: Die geologische Beschaffenheit des mittleren und nordlichen Teils der Deutschen Kalahari. Dtsch Geol Ges 61:259–270

Herschel, C (1897) On the origin of the Chezy formula. J Assoc Eng Sci 18:363–368

Hershfield DM (1967) Rainfall input for hydrological models. Int Assoc Sci Hydrol Pub 78:177–188

Heusch B, Millies-Lacroix A (1971) Une methode pour estimer lecoulement et lerosion dans un bassin: application au Maghreb. Mines Geol (Rabat) 33:24–39

Heward AP (1978) Alluvial fan and lacustrine sediments from the Stephanian A and B (La Magdalena, Cinera-Matallana and Sabero) coalfields, nothern Spain. Sedimentology 25:451–488

Heward AP (1978) Alluvial fan sequence and megasequence models with examples from Westphalian D-Stephanian B coalfields, northern Spain. In: Miall AD (ed) Fluvial sedimentology. Can Soc Petrol Geol, Calgary, pp 669–702

Hewlett JD, Nutter WL (1970) The varying source area of streamflow from upland basins. Proc Symp Interdisciplinary aspects of watershed management. Am Soc Civil Eng, New York, pp 65–83

Hibbert AR (1971) Increases in streamflow after converting chaparral to grass. Wat Resour Res 7:71–80

Hibbert AR (1979) Managing vegetation to increase flow in the Colorado River Basin. US For Serv Gen Tech Rep RM-66

Hibbert AR, Davis EA, Scholl DG (1974) Chaparral conversion potential in Arizona Pt 1: water yield response and effects on other resources. US For Serv Res Pap RM-126

Hibbert AR, Davis, EA, Brown TC (1975) Managing chaparral for water and other resources in Arizona. Watershed management, Proc Am Soc Civil Eng Symp, Logan Utah, Aug 11–13, 1975 pp 445–468

Hickin EJ, Nanson GC (1975) The character of channel migration on the Beatton River, north-east British Columbia, Canada. Geol Soc Am Bull 86:487–494

Higgins CG (1982) Drainage systems developed by sapping on Earth and Mars. Geology 10: 147–152

Higgins RJ (1979) Sediment transport in a river with high induced load. Civil Eng Trans Inst Eng Aust CE21:111–117

Hodges WK, Bryan RB (1982) The influence of material behavior on runoff initiation in the Dinosaur Badlands, Canada. Badland geomorphology and piping. Geobooks, Norwich, pp 13–46

Hogg SE (1982) Sheetfloods, sheetwash, sheetflow, or ...? Earth Sci Rev 18:56–76

Holeman JN (1968) The sediment yield of major rivers of the world. Wat Resour Res 4:737–747

Hollingsworth EG, Quimby PC Jr, Jaramillo DC (1973) Root plow herbicide application. Weed Sci 21:128–130

Holm L, Pancho JV, Herberger JP, Plucknett DL (1979) A geographical atlas of world weeds. John Wiley & Sons, New York

Holt-Jensen A (1981) Geography: its history and concepts. Harper & Row, London

Hong LB, Davies TRH (1979) A study of stream braiding. Geol Soc Am Bull 90, Pt 2:1839–1859

Hooke JM (1977) The distribution and nature of changes in river channel patterns: the example of Devon. In: Gregory KJ (ed) River channel changes. John Wile & Sons, New York Chichester, pp 265–280

Hooke JM (1979) An analysis of the processes of river bank erosion. J Hydrol 42:39–62

Hooke JM (1980) Magnitude and distribution of rates of river bank erosion. Earth Surf Proc 5:143–157

Hooke RLeB (1967) Processes on arid-region alluvial fans. J Geol 75:438–460

Hooke RLeB (1968) Steady-state relationships on arid-region alluvial fans in closed basins. Am J Sci 266:609–629

Hooke RLeB, Rohrer WL (1979) Geometry of alluvial fans: effect of discharge and sediment size. Earth Surf Proc 4:147–166

Horan, JD (1966) Timothy O'Sullivan: America's forgotten photographer. Bonanza, New York

Horton JS (1960) Use of a root plow in clearing tamarisk stands. US For Serv Res Note RM-50

Horton JS (1962) Taxanomic notes on Tamarix Pentandra in Arizona. SW Nat 7:22–28

Horton JS (1964) Notes on the introduction of deciduous tamarisk. US For Serv Res Note RM-16

Horton JS, Campbell CJ (1974) Management of phreatophyte and riparian vegetation for maximum multiple use values. US For Serv Res Pap RM-117

Horton JS, Mounts FC, Kraft JM (1960) Seed germination and seedling establishment of phreatophyte species. US For Serv Res Pap RM-48

Horton RE (1932) Drainage-basin characteristics. Am Geophys Union Trans 13:350–361

Horton RE (1945) Erosional development of streams and their drainage basins; hydrophysical approach to quantitative morphology. Geol Soc Am Bull 56:275–370

Howard AD (1965) Geomorphological systems-equilibrium and dynamics. Am J Sci 263:302–312

Howard AD (1971) Simulation model of stream capture. Geol Soc Am Bull 82:1355–1376

Howard AD (1980) Thresholds in river regimes. In: Coates DR, Vitek JD (eds) Thresholds in geomorphology. Allen & Unwin, London, pp 227–258

Howard AD, Dolan, R (1981) Geomorphology of the Colorado River in the Grand Canyon. J Geol 89:269–298

Howard CS (1947) Suspended sediment in the Colorado River, 1925–1941. US Geol Surv Wat-Supply Pap 998

Howard CS (1953) Density current in Lake Mead. Proc Min Intern Hydrol Convention, Minneapolis, mimeographed manuscript

Hubbell DW (1964) Apparatus and techniques for measuring bedload. US Geol Surv Wat-Supply Pap 1748

Hubbell DW, Sayre WW (1964) Sand transport studies with radioactive tracers. J Hydraul Div, Am Soc Civil Eng Proc 90(HY3):39–68

Hudson HE and Hazen R (1964) Droughts and low streamflow. In: Chow VT (ed) Handbook of applied hydrology. McGraw-Hill, New York, Sec 18

Hudson NW (1961) An introduction to the mechanics of soil erosion under conditions of subtropical rainfall. Rhod Sci Assoc Proc 49:14–25

Hudson NW (1963) Raindrop size distribution in high intensity storms. Rhod J Agric Res 1:6–11

Hugget R (1980) Systems analysis in geography. Clarendon Pr, Oxford

Hugget R (1985) Earth surface systems. Springer, Berlin Heidelberg New York Tokyo

Hughes EE (1966) Research in chemical control of various phreatophytes. Pacific Southwest Interagency Committee unnumbered report

Hunt CB (1969) Geologic history of the Colorado River. US Geol Surv Prof Pap 669-C:59–130

Hunt CB, Averitt P, Miller RL (1953) Geology and geography of the Henry Mountains region, Utah. US Geol Surv Prof Pap 228

Huntington E (1914a) The Pulse of Asia. Houghton Mifflin, Boston

Huntington E (1914b) The climatic factor as illustrated in arid America. Carnegie Inst of Wash Publ 192

Hurst HE (1952) The Nile. Constable, London Hurst HE, Phillips P (1931) The Nile Basin. I: General description of the basin, meteorology, and topography of the White Nile Basin. Government Press, Cairo

Hutter K (1982) Glacier flow – Recent developments. Mitt Versuchsanst Wasserbau Hydrol Glaziol 57:27–33

Hutton J (1788) Theory of the earth; or an investigation of the laws observable in the composition, dissolution, and restoration of land upon the globe. Trans Roy Soc Edinburgh 1(2):209–304

Huxley JS (1924) Constant differential growth-ratios and their significance. Nature (London) 114:895–896

Huxley T (1877) Physiography. Macmillan, London

Hydraulic Engineering Center (1976) HEC-6: Scour and deposition in rivers and reservoirs. US Army Corps Eng User's Manu 723-G2-L2470

Hydraulic Engineering Center (1977) HEC-6, scour and deposition in rivers and reservoirs – user's manual. US Army Corps Eng, Davis, Cal, Gen Comput Progr 723-G2-L2470

Hydraulic Engineering Center (1981) HEC-2: Water surface profiles. US Army Corps Eng User's Manu 723-X6-L202A

Hylckama TEA van (1974) Water use by saltcedar as measured by the water budget method. US Geol Surv Prof Pap 491-E

Iddings A, Olsson AA (1928) Geology of northwest Peru. Geol Soc Am Bull 12:1–40

Ilo GC (1976) Analytical and experimental investigation of river bed degradation. PhD Thes Ind Inst Tech, India

Imeson AC, Kwaad FJPM (1980) Gully types and gully prediction. Geogr Tijdschr 14:430–441

Imeson AC, Kwaad FJPM, Verstraten JM (1982) The relationship of soil physical and chemical properties to the development of badlands in Morocco. In: Bryan R, Yair A (eds) Badland geomorphology and piping. Geobooks, Norwich, pp 48–70

Inbar M (1982) Spatial and temporal aspects of man-induced changes in the hydrological and sedimentological regime of the Upper Jordan River. In: Amiran DH, Schick AP (eds) Research contributions to the physical geography of Israel. Weizmann, Jerusalem, pp 9–22

Inderbitzen AL (1959) Gravels of Alameda Creek, California. J Sediment Petrol 29:212–220

Inglis CC (1949) The behaviour and control of rivers and canals. CWINRS, India, Res Publ 13

International Association of Hydrological Sciences (1979) The hydrology of areas of low precipitation. Intern Assoc Hydr Sci (Canberra)

International Association of Scientific Hydrology (1974) Flash floods. UNESCO/WMO, Paris

Ives RL (1936) Desert floods in Sonoyta Valley. Am J Sci 234:349-360

Jackson RG (1976) Depositional model of point bars in the lower Wabash River. J Sediment Petrol 46:579–594

Jaeger F (1921a) Deutsch Sudwestafrika. Wiss Ges, Breslau

Jaeger F (1921b) Deutsch Sudwestafrika. Hettner Festschr, Breslau

Jaeger F (1923) Die Grundzuge der Oberflachengestalt von Sudwestafrika. Ges Erdk Berlin Z 2:14–24

Jaeger F (1927) Die Diamentenwuste Sudwestafrikas. Geogr Z 33:25–32

Jaeger F and Waibel L (1920) Beitrage zur Landeskunde von Sudwestafrika. Mitt Dtsch Schutzgeb (Berlin)

James LA (1986) Hydraulic mining debris in Bear River, Northern Sierra Nevada, CA: delivery, storage, and transport of episodically introduced sediment [abst]. Assoc Am Geog Abstr, 82nd Annu Meet, Minneapolis, sess 81

James LD (1965) Using a computer to estimate the effects of urban development on flood peaks. Wat Resour Res 1:223–234

Jarvis RS (1976) Stream orientation structures in drainage networks. J Geol 84:563–582

Jarvis RS, Woldenberg MJ (1984) River networks. Hutchinson Ross, Stroudsburg

Jencsok EI (1968) Hydrologic design for highway drainage in Arizona. Ariz Highway Dep, Phoenix

Jennings JN, Sweeting MM (1963) The limestone ranges of the Fitzroy Basin, Western Australia. Bonn Geogr Abh 32:45–59

Jens SW, McPherson MB (1964) Hydrology of urban areas. In: Chow VT (ed) Handbook of applied hydrology, Sec 20. McGraw Hill, New York

Jeppson RW, Ashcroft GL, Huber AL, Skogerboe GB, Mabutt, JA (1968) Hydrologic atlas of Utah. Utah State Univ, Salt Lake City

Jessen O (1936) Reisen und Forschungen in Angola. Dtsch Kolonialz (Berlin)

Johnson DW (1909) Geographic essays by William Morris Davis (reproduction). Dover, New York

Johnson DW (1932a) Rock fans of arid regions. Am J Sci 223: 432–439

Johnson DW (1932b) Rock planes in arid regions. Geogr Rev 22: 656–665

Johnston RJ (1983a) Philosophy and human geography. Arnold, London

Johnston RJ (1983b) Geography and geographers (2nd ed). Edward Arnold, London

Joly F (1965) Hydrology of arid and semiarid regions. In: Chow VT (ed) Handbook of applied hydrology, Sec 25. McGraw-Hill, New York

Jones A (1971) Soil piping and stream channel initiation. Wat Resources Res 3:602–610

Jones NO (1968) The development of piping erosion. PhD Diss, Univ Arizona, Tucson

Jopling AV (1963) Hydraulic studies of on the origin of bedding. Sedimentology 2:115–121

Jordan PR (1977) Streamflow transmission losses in western Kansas. Am Soc Civil Eng J Hydraul Div 103 (HY8):905–919

Judd, HE (1964) A study of bed characteristics in relation to flow in rough high-gradient channels. PhD Thes, Utah State Univ, Logan

Judson S (1952) Arroyos. Sci Am 187:71–76

Judson S, Ritter DF (1964) Rates of regional denudation in the United States. J Geophys Res 69:3395–3401

Justson JT (1934) The physiography (geomorphology) of Western Australia. Geol Surv W Austr Bull 95:101–129

Kadar L (1957) Die Entwicklung der Schwemmkegel. Petermanns Geogr Mitt 101:241–244

Kaiser E (1921) Morphogenetische Ergebnise auf Reisen in Sudwestafrika. Dtsch Geogr Tagesz (Leipzig)

Kaiser E (1923) Abtragung und Auflagerung in der Namib, der sudwestafrikanischen Kusten-wuste. Geol Charakterbilder 10:27–28

Karcz I (1969) Mud pebbles in a flash flood environment. J Sediment Petrol 39:333–337

Karl HH (1976) Depositional history of Dakota formation (Cretaceous) sandstone, southeastern Nebraska. J Sediment Pet 46:124–131

Karl TR, Knight RW (1985) Atlas of Monthly Palmer Hydrological Drought Indicies for the Contiguous United States. US National Climatic Data Center, Asheville, North Carolina

Karlinger MR, Eschner TR, Hadley RF, Kircher, JE (1983) Relation of channel-width mainte-nance to sediment transport and river morphology: Platte River, south-central Nebraska. US Geol Surv Prof Pap 1277-E

Kaselau A (1928) Die naturlichen lanschaften nord- und Zentral-Arabiens. Petermanns Geogr Mitt 74:341–342

Keill J (1708) An account of animal secretion, the quantity of blood in the human body, and muscular motion. Strahan, London

Keller EA (1971) Areal sorting of bed load material; the hypothesis of velocity reversal. Geol Soc Am Bull 82:753–756

Keller EA (1975) Channelization: a search for a better way. Geol 3:246–248

Keller EA (1976) Channelization: environmental, geomorphic, and engineering aspects. In: Coates DR (ed) Geomorphology and engineering. Dowdon, Hutchinson & Ross, Strouds-burg, pp 115–140

Keller EA, Melhorn W (1973) Bedforms and fluvial processes in alluvial stream channels: selected observations. In: Morisawa M (ed) Fluvial geomorphology. SUNY Binghamton, pp 253–283

Keller EA, Swanson FJ (1979) Effects of large organic material on channel form and fluvial processes. Earth Surf Proc 4: 361–380

Keller EA, Tally T (1979) Effects of large organic debris on channel form and fluvial processes in the coastal redwood environment. In: Rhodes DD, Williams GP (eds) Adjustments of the fluvial system. Kendall-Hunt, Dubuque, pp 169–197

Kellerhals R, Neil CR, Bray DI (1976) Classification and analysis of river processes. Am Soc Civil Eng, J Hydraul Div 107:813–329

Kennedy BA (1985a) Entropy. In: Goudie A (ed) The encyclopaedic dictionary of physical geography. Blackwell, London, pp 155–156

Kennedy BA (1985b) General systems theory. In: Goudie AS (ed) The encyclopaedic dictionary of physical geography. Blackwell, London, pp 200–201

Kennedy BA (1985c) Uniformitarianism. In: Goudie AS (ed) The encyclopaedic dictionary of physical geography. Blackwell, London, pp 448–450

Kennedy BA (1985d) Allometric growth. In: Goudie A (ed) The encyclopaedic dictionary of physical geography. Blackwell, London, pp 13–14

Kennedy RG (1895) The prevention of silting in irrigation canals. Inst Civil Eng London Proc 119:281–290

Kesel RH (1985) Alluvial fan systems in a wet-tropical environment, Costa Rica. Nat Geogr Res 1:450–469

Kesseli JE (1941) The concept of the graded river. J Geol 49: 561–588

Kesseli JE, Beaty CB (1959) Desert flood conditions in the White Mountains of California and Nevada. US Army Quartermaster Res Eng Cent, Tech Rep EP-108

Keulegan GH (1957) Thirteenth progress report on model laws for density currents. An experi-mental study of the motion of saline water from locks into freshwater channels. US Natl Bur Stand Rep 5168

Khosla AN (1953) Silting of reservoirs. CPIP Publ 51, New Delhi, India

King LC (1953) Canons of landscape evolution. Geol Soc Am Bull 64:721–752

Kingsbury JW (1952) Pothole erosion in the western part of Molokai Island, Territory of Hawaii. J Soil Wat Cons 7: 197–198

Kircher JE, Karlinger MR (1981) Changes in surface-water hydrology for the South Platte River in Colorado and Nebraska. US Geol Surv Open-File Rep 81–53

Kirkby MJ, Chorley RJ (1967) Throughflow, overland flow and erosion. Int Assoc Sci Hydrol Bull 12:5–21

Kirkby MJ, Morgan RPC (1980) Soil Erosion. John Wiley & Sons, Chichester

Kirpich ZP (1940) Time of concentration of small agricultural watersheds. Civ Eng 10:362

Klingeman PC, Emmett WW (1982) Gravel bedload transport processes. In: Hey RD, Bathurst JC, Thorne CR (eds) Gravel-bed rivers. John Wiley & Sons, Chichester, pp 141–169

Knighton AD (1982) Longitudinal changes in the size of stream bed material: evidence of variable transport conditions. Catena 9:25–34

Knighton AD (1984) Fluvial forms and processes. Arnold, London

Knox JC (1972) Valley alluviation in southwestern Wiconsin. Ann Assoc Am Geogr 62:401–410

Knox JC (1975) Concept of the graded stream. In: Melhorn WN, Flemal RC (ed) Theories of landform development. SUNY Binghamton, pp 169–198

Knox JC (1983) Responses of river systems to Holocene climates. In: Wright HE (ed) Late-Quaternary environments of the United States, vol 2: The Holocene. Univ Minnesota Press, Minneapolis, pp 26–41

Koch GS, Jr, Link RF (1970) Statistical analysis of geological data. John Wiley and Sons, New York

Kochel RC and Baker VR (1981) Paleoflood hydrology. Science 215:353–361

Kolata GB (1977) Catastrophe theory: the emperor has no clothes. Science 196:287, 350–351

Komura S (1971) Prediction of river-bed degradation below dams. Proc 14th Congr Int Assoc Hydrol Res (Paris) 3:257–264

Komura S and Simons DB (1967) River-bed degradation below dams. Proc Am Soc Civil Eng J Hydraul Div 93(HY4):1-14

Kopal Z (1979) The realm of the terrestrial planets. John Wiley & Sons, New York

Kraatz DB (1977) Irrigation and canal lining. UN Food Agric Org (Rome)

Krimgold DB (1946) On the hydrology of culverts. Proc Highway Res Bd 26:214-226

Krumbein WC (1934) Size frequency distributions of sediments. J Sediment Petrol 4:65-77

Krumbein WC (1941) Measurement and geological significance of shape and roundness of sedimentary particles. J Sediment Petrol 11:64-72

Krumbein WC (1955) Experimental design in the earth sciences. Am Geophys Union Trans 36:1-11

Krumbein WC, Graybill FA (1965) An introduction to statistical models in geology. McGraw-Hill, New York

Kuenen PH (1955) Experimental abrasion of pebbles. I: Wet sand blasting. Leidsche Geol Meded 20:131-137

Kuhn F (1922) Fundamentos de fisiografia argentina. Talleres Graficos, Buenos Aires

Kuhn TS (1970) The structure of scientific revolutions, 2nd edn. Univ of Chicago Press, Chicago

Kulesh NP (1971) Experimental study of reservoirs with density currents formed during the flood, calculation principles for silting of such reservoirs. 14th Congr Int Assoc Hydrol Sci 5:13-16

Lacey G (1930) Stable channels in alluvium. Inst Civil Eng Proc 229:259-384

Lacey G (1958) Flow in alluvial channels with sandy mobile beds. Inst Civil Eng Proc 9(ns):185-195

Lagasse PF, Schall JD, Peterson M (1985) Erosion risk analysis for a southwestern arroyo. Am Soc Civil Eng, J Urban Plan Dev Div 111:10-24

Laidler KJ (1965) Chemical kinetics. McGraw-Hill, New York

Lam KC (1977) Patterns and rates of slopewash on the badlands of Hong Kong. Earth Surf Proc 2:319-332

Lamb J, Carleton EA, Free GR (1950) Effect of past management and erosion of soil on fertilizer efficiency. Soil Sci 70: 385-392

Lambert AM, Hsu KJ (1979) Varve-like sediments of the Walensee, Switzerland. In: Schluchter C (ed) Moraines and varves: genesis, classification. Balkema, Rotterdam, pp 287-294

Lane EW (1951) Discussion of "Retrogression on the lower Colorado River after 1935" by J. W. Stanley. Proc Am Soc Civil Eng 76:1-12

Lane EW (1953) River-bed scour during floods. Am Soc Civil Eng Trans 119:1069-1079

Lane EW (1957) A study of the shape of channels formed by natural streams flowing in erodible material. MRD Sed Ser 9, US Army Eng Div, Missouri River

Lane EW, Borland WM (1951) Estimating bed load. Trans Am Geophys Union 32:121-123

Lane EW, Carlson EJ, Hanson OS, Kalinske AA (1941) Engineering calculations of suspended sediment. Trans Am Geophys Un 22:603-607

Lane LJ (1972) A proposed model for flood routing in abstracting ephemeral channels. Proc Hydrol and Wat Res Ariz SW 2:439-453

Lane LJ (1980) Transmission losses. In: US Soil Conservation Service (ed) National Engineering Handbook. US Dep Agric, Washington, Sec 4, Chap 19

Lane LJ (1982) Distributed model for small semi-arid watersheds. Am Soc Civil Eng J, Hydraul Div 108(HY10):1114-1131

Lane LJ, Purtymun WD, Becker NM (1985) New estimating procedures for surface runoff, sediment yield, and contaminant transport in Los Alamos County, New Mexico. Los Alamos Natl Lab Pub LA-10335-MS, UC-11

Langbein WB (1949) Annual runoff in the United States. US Geol Surv Circ 52

Langbein WB, Leopold LB (1968) River channel bars and dunes – theory of kinematic waves. US Geol Surv Prof Pap 422L

Langbein WB, Schumm SA (1958) Yield of sediment in relation to mean annual precipitation. Trans Am Geophys Union 39:1076-1084

Langford-Smith T, Dury GH (1964) A pediment at Middle Pinnacle near Broken Hill, New South Wales. J Geol Soc Austr 11: 79-88

Larmer DC (1974) Woody phreatophytes along the Colorado River from southeast Runnels County to the headwaters in Borden County, Texas. Texas Water Devel Bd Rep 182

Laronne JB (1984) Rhythmic couplets: sedimentology and prediction of reservoir design periods in semiarid areas. Univ Calif Los Angels-BGU Conf Desert Res, Sde Boker Desert Res Inst, mimeographed manuscript

Last Y (1974) Inner-berm discharge and drainage area in the eastern Sinai. MS Thes, Dep Geogr, Hebrew Univ Jerusalem

Laursen EM, Ince S, Pollack J (1976) On sediment transport through the Grand Canyon. Proc 3rd Inter-Agency Sediment Conf, Denver 4/76–4/87

Lawson AC (1915) The epigene profiles of the desert. Univ Cal Bull Dep Geol 9:23–48

Lazenby JF (1976) Lake Powell sedimentation surveys. Proc 3rd Fed Inter-Agency Sediment Conf, Denver 4/52–4/63

Leeder MR (1982) Sedimentology: process and product. Allen & Unwin, Boston

Lefevre M-A (1952) Note sur les pediments du desert Mojave, Californie. Bull Soc Belge Estud Geogr 21:259–268

LeHouerou HN (1968) La desertification du Sahara septentrional et des steppes limitrophes (Libye, Tunisie, Algerie). UNESCO, Paris

Lehre AK (1982) Sediment budget of a small coast range drainage basin in north-central California. In: Swanson FJ, Janda RJ, Dunne, T, Swanston, DN (eds) Sediment budgets and routing in forested drainage basins. US For Serv Gen Tech Rep PNW-141. pp 67–77

Lekach J (1974) Reconstruction of the January 1971 event in Wadi Mikeimin, eastern Sinai. MS Thes Hebrew Univ Jerusalem

Lekach J, Schick AP (1980) Bedload transport and stream power – examples from desert floods. In: Schick AP (ed) Arid zone geosystems. Div Phys Geogr, Hebrew Univ Jerusalem, pp 38–50

Lekach J, Schick AP (1982) Suspended sediment in desert floods in small catchments. Isr J Earth Sci 31:100–156

Lekach J, Schick AP (1983) Evidence for transport of bedload in waves: analysis of fluvial sediment samples in a small upland channel. Catena 10:267–279

Leliavsky S (1955) An introduction to fluvial hydraulics. Constable, London

Leonard RJ (1929) An earth fissure in southern Arizona. J Geol 37:765–774

Leopold LB (1951a) Pleistocene climate in New Mexico. Am J Sci 249:152–167

Leopold LB (1951b) Rainfall frequency: an aspect of climatic variation. Trans Am Geophys Un 32:347–357

Leopold LB (1968) Hydrology for urban land planning: a guidebook on the hydrologic effects of urban land use. US Geol Surv Circ 554

Leopold LB (1969) The rapaids and pools Grand Canyon. US Geol Surv Prof Pap 669-D

Leopold LB (1973) River channel change with time: an example. Geol Soc Am Bull 84:208–216

Leopold LB (1976) Reversal of erosion cycle and climatic change. Quat Res 6:557–562

Leopold LB (1978) El Asunto del Arroyo. In: Embleton C, Brunsden D, Jones DKC (eds), Geomorphology: Present problems and future prospects. Oxford Univ Press, Oxford, pp 25–39

Leopold LB, Bull WB (1979) Base level, aggradation, and grade. Proc Am Phil Soc 123:168–202

Leopold LB, Emmett WW (1976) Bedload measurements, East Fork River, Wyoming. Proc US Natl Acad Sci USA 73:1000–1004

Leopold LB, Langbein WB (1962) The concept of entropy in landscape evolution. US Geol Surv Prof Pap 500-A

Leopold LB, Maddock T (1953) The hydraulic geometry of stream channels and some physiographic implications. US Geol Surv Prof Pap 252

Leopold LB, Miller JP (1954) Postglacial chronology for alluvial valleys in Wyoming. US Geol Surv Prof Wat-Supply Pap 1261:61–85

Leopold LB, Miller JP (1956) Ephemeral streams-hydraulic factors and their relationship to the drainage net. US Geol Surv Prof Pap 282-A

Leopold LB, Wolman MG (1956) Floods in relation to the river channel. Darcy Symp, Assoc Int Hydrol 42:85–98

Leopold LB, Wolman MG (1957) River channel patterns: braided, meandering, and straight. US Geol Surv Prof Pap 282-B

Leopold LB, Wolman MG, Miller JP (1964) Fluvial processes in geomorphology. Freeman, San Francisco

Leopold LB, Emmett WW, Myrick RM (1966) Channel and hillslope processes in a semiarid area, New Mexico. US Geol Surv Prof Pap 352-G 193–253

Levey RA (1978) Bedform distribution and internal stratification of coarse-grained point bars, Upper Congaree River, South Carolina. In: Miall AD (ed) Fluvial sedimentology. Can Soc Petrol Geol (Calgary), pp 105–127

Levinson AA (1980) Introduction to exploration geochemistry. Applied Publ, Wimette

Lewin J (1978) Floodplain geomorphology. Prog Phys Geogr 2: 408–437

Lewin J (1985a) Floodplain. In: Goudie A (ed) The encyclopaedic dictionary of physical geography. Blackwell, London, p 187

Lewin J (1985b) Incised channels: morphology, dynamics and control (book review). Prog Phys Geogr 9:469–471

Lewin J, Davies BE, Wolfenden PT (1977) Interactions between channel change and historic mining sediments. In: Gregory KJ (ed) River channel change. John Wiley & Sons, New York, pp 353–368

Lewis GN, Randal M (1961) Thermodynamics. McGraw-Hill, New York

Lieth H (1975) Primary production of the major vegetation units of the world. In: Lieth H, Whittaker RH (eds) Productivity of the biosphere. Springer-Verlag, New York, pp 203–215

Lieth H (1978) Patterns of primary production in the biosphere, Benchmark Papers in Ecology, 8. Dowden, Hutchenson, and Ross, Stroudsburg

Lighthill MJ, Whitham GB (1955) On kinematic waves. II: A theory of traffic flow on long crowded roads. Proc R Soc London Ser A 229:1101–1117

Limerinos JT (1969) Relation of the Manning coefficient to measured bed roughness in stable natural channels. US Geol Surv Prof Pap 650-D:215–221

Lindgren W (1911) Tertiary gravels of the Sierra Nevada of California. US Geol Surv Prof Pap 73

Lindley ES (1919) Regime channels. Punjab Eng Congr Proc 3:42–49

Lindsay D (1889) An expedition across Australia from south to north, between the telegraph line and the Queensland boundary. R Geogr Soc Proc 11:650–671

Lindsay D (1893) Journal of the Elder scientific exploring expedition 1891–2. Gov Printer, Adelaide

Linton DL (1948) The delimitation of morphological regions. Inst Brit Geogr Publ 14:86–87

Lipe WD, Breed WJ, West J, Batchelder G (1975) Lake Pagahrit, southeastern Utah: a preliminary research report. Four Corners Geol Soc Guidebook, 8th Field Conf, Canyonlands, pp 103–110

Little OH (1925) The geography and geology of Makalla, South Arabia. Geol Surv Egypt, Cairo

Little WC, Mayer PG (1972) The role of sediment gradation on channel armoring. Georgia Ins Tech, Sch Eng, Unnumbered Publ

Livesey RH (1975) Corps of Engineers methods for predicting sediment yields. Present and prospective technology for predicting sediment yields and sources. US Dept Agric, Agric Res Serv Publ ARS-S-40, pp 16–32

Love DW (1983a) Summary of the Late Cenozoic geomorphic and depositional history of Chaco Canyon. In: Wells SG, Love DW, Gardner, TW (eds) Chaco Canyon Country. Am Geomorphol Field Gp, Albuquerque, pp 195–206

Love DW (1983b) Quaternary facies in Chaco Canyon and their implications for geomorphic-sedimentologic models. In: Wells SG, Love DW, Gardner, TW (eds) Chaco Canyon Country. Am Geomorphol Field Gp, Albuquerque, pp 207–219

Love SK, Benedict PC (1948) Discharge and sediment loads in the Boise River drainage basin, Idaho, 1939–40. US Geol Surv Wat-Supply Pap 1048

Lusby GC (1970) Hydrologic and biotic effects of grazing vs. non-grazing near Grand Junction, Colorado. J Range Manag 23:256–260

Lusby GC, Reid VH, Knipe OD (1971) Effects of grazing on the hydrology and biology of the Badger Wash Basin in Western Colorado, 1953–1966. US Geol Surv Wat-Sup Pap 1532-D

Lustig LK (1965) Clastic sedimentation in Deep Springs Valley, California. US Geol Surv Prof Pap 352-F:131–192

Lustig LK (1968) Appraisal of research on geomorphology and surface hydrology of desert environments. In: McGinnies WG Goldman BJ, Paylore, P (eds) Deserts of the World. Univ Ariz Press, Tucson, pp 95–286

Lustig LK (1969) Trend surface analysis of the Basin and Range Province. US Geol Surv Prof Pap 500-D

Lyman T (1878) Mode of forking among Astrophytons. Boston Soc Natl Hist Proc 19:102–108

Lyons HG (1905) On the Nile flood and its variation. Geogr J 26:395–421

Mabbutt JA (1966) Mantle-controlled planation of pediments. Am J Sci 264:78–91

Mabbutt JA (1977) Desert landforms. MIT Press, Cambridge

Macke DL (1977) Stratigraphy and sedimentology of experimental alluvial fans. MS Thes Col State Univ, Ft Collins, Col

Mackin JH (1948) Concept of the graded river. Geol Soc Am Bull 59:463–512

Mackin JH (1953) Rational and empirical methods of investigation in geology. In: Albritton CC, Jr (ed) The fabric of geology. Addison-Wesley, Reading, pp 135–163

Maddock T (1948) Reservoir problems with respect to sedimentation. Proc Fed Inter-Agency Sediment Conf, Denver

Maddock T, Jr (1969) Economic aspects of seidmentation. Proc Am Soc Civil Eng, J Hydraul Div 95(HY9):191–207

Madej MA (1984) Recent changes in channel-stored sediment, Redwood Creek, California. Redwood Natl Park Res and Dev Tech Rep 11

Magura LM, Wood DE (1980) Flood hazard identification and floodplain management on alluvial fans. Wat Resour Bull 16:56–112

Mahmood K, Shen HW (1971) The regime concept of sediment-transporting canals and rivers. In: Shen HW (ed) River mechanics. Wat Resour Publ, Lakewood, pp 30/1–30/39

Mahoney MJ (1976) Scientist as subject: the psychological imperative. Ballinger, Cambridge

Maigels JK (1983) Palaeovelocity and palaeodischarge determination for coarse gravel deposits. In: Gregory KJ (ed) Background to palaeohydrology. John Wiley & Sons, New York, pp 101–139

Makkaveev NI (1970) Effect of major dam and reservoir construction on geomorphologic processes in river valleys. Geomorphol 2:106–110

Malde HE and Scott AG (1977) Observations of contemporary arroyo cutting near Santa Fe, New Mexico, U. S. A.. Earth Surf Proc 2:39–54

Malone JM (1972) Hydrologic design manual for drainage areas under 25 square miles. US Dep Agric, Phoenix

Manning R (1895) On the flow of water in open channels and pipes. Trans Inst Civil Eng Irel 20:161–207

Mammerickx J (1964) Quantitative observations on pediments in the Mojave and Sonoran deserts (southwestern United States). Am J Sci 262:417–435

Mao SW, Rice L (1963) Sediment transport capacity in erodible channels. Am Soc Civil Eng J Hydraul Div 89(HY4):69–96

Marcus A (1980) First-order drainage basin morphology-definition and distribution. Earth Surf Proc 5:389–398

Marcus WA (1983) Copper dispersion in ephemeral stream sediments, Queen Creek, Arizona. MA Thes Ariz State Univ, Tempe

Marston RA (1982) The geomorphic significance of log steps in forest streams. Ann Assoc Am Geogr 72:99–108

Martens LA (1966) Flood inundation and effects of urbanization in Metropolitan Charlotte. US Geol Surv Unnumbered Open-File Rep 54

Martin L (1950) William Morris Davis: investigator, teacher and leader in geomorphology. Ann Assoc Am Geogr 40:172–180

Marvine AR (1874) The stratigraphy of the east slope of the Front Range. US Geogr Geol Surv, Washington, DC

Massey HF, Jackson ML (1952) Selective erosion of soil fertility constituents. Proc Soil Sci Soc Am 16:353–356

Masson C (1843) Narrative of a journey to Kalat, including an account of the insurrection at that place in 1840; and a memoir on eastern Baluchistan. Masson, London

Masson JM (1972) Lerosion des sols par l'eau en climat Mediteranneen, methodes experimentales pour letude des quantites a lechalle du champ. Houille Blanche 8:673–678

Masterman M (1970) The nature of a paradigm. In: Lakatos I Musgrave A (eds) Criticism and the growth of knowledge. MIT Press, Cambridge, pp 431–454

Mayer L, Gerson R, Bull WB (1984) Alluvial gravel production and deposition: a useful indicator of Quaternary climatic changes in deserts. Catena Suppl 5:137–151

McClintock E (1951) Studies in California ornamental plants, 3. the tamarisks. J Cal Hortic Soc 12:76–83

McDonald DC, Hughes GH (1968) Studies of consumptive use of water by phreatophytes and hydrophytes near Yuma, Arizona. US Geol Surv Prof Pap 486-F

McGee WJ (1888a) The geology of the head of Chesapeak Bay. 7th Annu Rep US Geol Surv, pp 545–646

McGee WJ (1888b) The classification of geographic form by genesis. Nat Geogr Mag 1:27–36

McGee WJ (1897) Sheetflood erosion. Geol Soc Am Bull 8:87–112

McGinn RA (1980) Flood frequency estimates on alluvial fans, discussion. Am Soc Civil Eng J Hydraul Div 106:1718–1720

McGowen JH (1979) Alluvial fan systems. In: Galloway WE Kreitler CW, McGowen, JH (eds) Depositional and groundwater flow systems in the exploration for uranium. Univ Texas Press, Austin, pp 43–79

McHenry JR, Coleman NL, Willis JC, Gill AC, Sansom OW, Carroll BR (1970) Effect of concentration gradients on the performance of a nuclear sediment concentration gauge. Wat Resour Res 6:538–548

McIntyre DS (1958) Permeability measurements of soil crusts formed by raindrop impact. Soil Sci 85:185–189

McMahon AH (1897) The southern borderlands of Afghanistan. Geogr J 10:393–415

McNaughton KG, Black TA (1973) A study of evapotranspiration from a Douglas-fir forest using the energy balance approach. Wat Resour Res 9:1579–1590

Meade RH (1985) Wavelike movement of bedload sediment, East Fork River, Wyoming. Environ Geol Wat Sci 7:215–225

Meade RH, Parker RS (1984) Sediment in rivers of the United States. US Geol Surv Wat-Supplu Pap 2275:49–60

Mears B (1963) Karst-like features in badlands of the Arizona Painted Desert. Univ Wyo Contrib Geol 2:7–11

Megahan WF (1975) Sedimentation in relation to logging activities in the mountains of central Idaho. Present and prospective technology for predicting sediment yields and sources. US Dep Agric, Agric Res Serv Rep ARS-S-40, pp 74–82

Megahan WF, Molitor DC (1975) Erosional effects of wildfire and logging in Idaho. Am Soc Civil Eng Irr Dr Div, Watershed Manag Symp, pp 423–444

Megahan WF, Nowlin RA (1976) Sediment storage in channels draining small forested watersheds in the mountains of central Idaho. Proc 3rd Fed Interagency Sediment Conf, Denver, Col, US Soil Cons Serv. pp 1–12

Mehta KM, Sharma VC, Deo PG (1963) Erodibility investigations of soil of Eastern Rajasthan. J Ind Soc Soil Sci 11:23–31

Meigs P (1953) World distribution of arid and semi-arid homoclimates. In: UNESCO (ed) Reviews of research in arid zone hydrology. UNESCO, Paris, pp 203–209

Melton MA (1959) A derivation of Strahler's channel-ordering system. J Geol 67:345–346

Melton MA (1962) Methods for measuring the effect of environmental factors on channel properties. J Geophys Res 67:1485–1490

Melton MA (1965) The geomorphic and palaeoclimatic significance of alluvial deposits in southern Arizona. J Geol 73:1–38

Mendenhall WC (1920) Memorial to Grove Karl Gilbert. Geol Soc Am Bull 23:26–64

Merriam Company (1977) Webster's new collegiate dictionary. Merriam, Springfield

Meyer-Peter E (1949) Quelques problemes concernant le charriage des matieres solides. Soc Hydrotech Fr 2:51–59

Meyer-Peter E (1951) Transport des matieres solides en general et problemes speciaux. Bull Genie Civil Hydrol Fluv Tome

Meyer-Peter E, Muller R (1948) Formulas for bed-load transport. Int Assoc Hydrol Res, Proc 2nd Meet, Stockholm

Michel M (1979) The effect of reservoir implementation on the discharge of the River Duero in central Spain. BS Diss, Dep Geogr, King's College, London

Middleton GV (1966) Experiments on density and turbidity currents, Parts I, II, and III. Can J Earth Sci 3:523–546, 3:627–637, 4:475–505

Miller CR (1972) Runoff volumes from small urban watersheds. Water Resour Res 8:429–434

Miller JP (1958) High mountain streams; effects of geology on channel characteristics and bed material. New Mex State Bur Mines Miner Res Mem 4:1–53

Miraki GD (1983) Sediment yield and deposition profiles in reservoirs. PhD Thes Univ Roorkee, India

Miser HD (1924) The San Juan Canyon: southeastern Utah. US Geol Surv Wat-Supply Pap 538

Mitchell JK, Bubenzer GD (1980) Soil loss estimation. In: Kirkby MJ, Morgan RPC (eds) Soil Erosion. John Wiley & Sons, Chichester, pp 17–62

Mitchell VL (1976) The regionalization of climate in the western United States. J Appl Meteorol 15:920–927

Moore CM, Wood WJ, Renfro GW (1960) Trap efficiency of reservoirs and debris dams. Am Soc Civil Eng J Hydraul Div 86(HY2):69–88

Moore DO (1968) Estimating mean runoff in ungaged semiarid areas. Intern Assoc Sci Hydrol Bull 13:29–39

Moore E, Janes E, Kinsinger F, Pitney K, Sainsbury J (1979) Livestock grazing management and water quality protection (state of the art reference document). US Environ Prot Agency, EPA 910/9–79–67

Moore WR, Smith CE (1968) Erosion control in relation to watershed management. Am Soc Civl Eng J Irr Dr Div 94(IR3):321–331

Morgan CW (1966) Thunderstorms. Texas Wat Dev Bd Rep 33:31–44

Morris RJ, Natalino M (1969) The chemical nature of the organic matrix believed to limit water penetration in granitic soils. Univ Nevada Des Res Inst Proj Pap 13

Mortensen H (1927) Der Formenbschtz de Nordchilenschen Wuste. Weidmannsche Verlagshandlung, Berlin

Mortensen H (1930) Einige Oberflächenformen der Winterregengebiete Düsseldorfer Geogr Vortr Erortg 3:133–157

Mosley MP (1972) Evolution of a discontinuous gully system. Ann Assoc Am Geogr 62:655–663

Mosley MP, Parker RS (1972) Allometric growth: a useful concept in geomorphology? Geol Soc Am Bull 83:3669–3674

Mostaga MG (1957) River-bed degradation below large-capacity reservoirs. Trans Am Soc Civil Eng 122:688–695

Muckel DC (1966) Phreatophytes – water use and potential water savings. Am Soc Civil Eng J Irr Dr Div 88(IR4):27–34

Mueller JE (1968) Introduction to hydraulic and topographic sinuosity indexes. Ann Assoc Am Geogr 58:371–385

Muhlhofer L (1933) Untersuchungen über Schwebstoff- und Geschiebeführung des Inns nach Kirchbichl. Wasserwirtschaft 16:216–233

Murray B, Malin MC, Greeley R (1981) Earthlike planets: Surfaces of Mercury, Venus, Earth, Moon, Mars. Freeman, San Francisco

Murry JA, Bradley H, Craige W A, Onions, CT (1933) The Oxford English dictionary, being a corrected re-issue with an introduction, supplement, and bibliography of a new English dictionary on historical principles. Clarendon, Oxford

Musgrave GW (1947) The quantitative evaluation of factors in water erosion – A first approximation. J Soil Wat Cons 2: 133–138

Nabhan GP (1986) Papago Indian desert agriculture and water control in the Sonoran Desert. Appl Geogr 6:61–76

Nadler CT and Schumm SA (1981) Metamorphosis of South Platte and Arkansas Rivers, eastern Colorado. Phys Geogr 2:95–115

Neev D, Emery KO (1967) The Dead Sea: depositional processes and environments of evaporites. Isr Geol Surv Bull 41:1–147

Neff EL (1967) Discharge frequency compared to long term sediment yields. Publ Int Assoc Sci Hydrol 75:236–242

Nesper F (1937) Die internationale Rheinregulierung: III Schweiz. Bauzeitung 110:43–62

Newberry JS (1861) Colorado River of the West. Am J Sci 33: 387–403

Newberry JS (1876) Geological report of the exploring expedition from Santa Fe, New Mexico, to the junction of the Grand and Green Rivers of the Great Colorado of the West in 1859. US Gov Printing Off, Washington, DC

Newbold D (1924) A desert odyssey of a thousand miles. Sudan Notes Rec 7:104–107

Newell ND (1946) Geological investigations around Lake Titicaca. Am J Sci 244:357–366

Nordin CF, Beverage JP (1965) Sediment transport in the Rio Grande, New Mexico. US Geol Surv Prof Pap 462-F

Oldham RD (1886) On probable changes in the geography of the Punjab and its rivers. Asiat Soc Bengal J 55:322–343

Oldroyd DR (1980) Darwinian impacts: an introduction to the Darwinian revolution. Open Univ Press, Keynes

Oliver JE (1973) Climate and man's environment: an introduction to applied climatology. John Wiley & Sons, New York

Ollier CD (1960) The inselbergs of Uganda. Z Geomorphol 4:43–52

Ollier CD (1968) Open systems and dynamic equilibrium in geomorphology. Austr Geogr Stud 6:167–170

Onstad CA, Larson CL, Hermsmeier LF, Young, RA (1967) A method of computing soil movement throughout a field. Trans Am Soc Agric Eng 10:742–745

Orford JD (1981) Particle form. In: Goudie A (ed) Geomorphological techniques. Allen & Unwin, Boston, pp 86–90

Orth F (1934) Die Verlandung von Staubecken. Bautechnik 30:221-240

Pacific Southwest Interagency Committee (1968) Factors affecting sediment yield and measures for the reduction of erosion and sediment yield. US Dep Agric, Mimeogr rep, Portland, Oregon

Pacific Southwest Interagency Committee (1974) Erosion and sediment yield methods. US Dep Agric, Mimeogr rep, Portland, Oregon

Page AL, Chang AC, Bingham FT (1979) Management and control of heavy metals in the environment. CEP Cons, Edinburgh

Panton HM (1979) Interaction between velocity and effective density in turbidity flow: phase plane analysis, with criteria for autosuspension. Mar Geol 31:59–99

Parker G (1976) On the cause and characteristic scale of meandering and braiding in rivers. J Fluid Mech 76: 459–480

Parker GG (1963) Piping, a geomorphic agent in landform development of drylands. Int Assoc Sci Hydrol 65:103–113

Parker GG, Jenne EA (1967) Structural failure of western US highways caused by piping. US Geol Surv Wat Res Div, 46 Annu Meet, Highway Res Bd, Washington, DC, January 18, 1967

Parker GG, Shown LM, Ratzlaff KW (1964) Officer's Cave, a pseudokarst feature in altered tuff and volcanic ash of the John Day Formation in Eastern Oregon. Geol Soc Am Bull 75: 393–402

Parker RS (1976) Experimental study of drainage system evolution. Unpublished rep. Dept Earth Resour, Colorado State Univ

Pase CP, Ingebo PA (1965) Burned chaparral to grass: early effects on water and sediment yields from two granitic soil watersheds in Arizona. Proc Ann Ariz Watershed Symp 9:8–11

Pase CP, Lindenmuth AW (1971) Effects of prescribed fire on vegetation and sediment in oak-mountain mahogany chaparral. J For 69:800–805

Passarge S (1905a) Die Kalahari. Reimer, Berlin

Passarge S (1905b) Rumpfflache and inselberge. Z Dtsch Geol Ges 56:193–209

Passarge S (1905c) Die Inselberglandschaften im tropischen Afrika. Naturwiss Wochenschr 3:657–665

Passarge S (1930) Ergnebrisse einer Studienreise nach Sudtunisien in jahr 1928. Mitt Geogr Ges, Hamburg 61:96–122

Pattison WD (1964) The four traditions of geography. J Geogr 63:211–216

Patton PC (1977) Geomorphic criteria for estimating the magnitude and frequency of flooding in central texas. PhD Diss, Univ Texas, Austin

Patton PC, Baker VC (1977) Geomorphic response of central Texas stream channels to Catastrophic rainfall and runoff. In: Doehring DO (eds) Geomorphology in arid regions. SUNY Binghamton, pp 189–218

Patton PC, Boison PJ (1986) Process and rates of formation of Holocene alluvial terraces in Harris Wash, Escalante River Basin, south-central Utah. Geol Soc Am Bull 97:369–378

Patton PC, Schumm SA (1981) Ephemeral-stream processes: implications for studies of Quaternary valley fills. Quat Res 15:24–43

Paulhus JL (1965) Indian Ocean and Taiwan rainfall set new records. Mon Weather Rev 93:331–335

Pearthree MS (1983) Channel change in the Rillito Creek system, south-eastern Arizona. MA Thes, Univ Ariz, Tucson

Peebles RW (1975) Flow recession in the ephemeral stream. PhD Thes, Dept of Hydrol Wat Res, Univ of Arizona, Tucson

Peltier LC (1950) The geographical cycle in periglacial regions as it is related to climatic geomorphology. Ann Assoc Am Geogr 50:214–236

Pemberton EL (1976) Channel changes in the Colorado River below Glen Canyon Dam. Proc 3rd Inter-Agency Sediment Conf, Denver, pp 5/61–5/73

Penck A (1910) Versuch einer Klimaklassification auf physiographischer Grundlage. Preuss Akad Wiss Sitz Pys-Math 12:236–246

Penck W (1920) Der Sudrand der Puna de Atacama. Akad Wiss, Leipzig

Penck W (1953) Morphological analysis of land forms. Macmillan, London (transl Czech H, Boswell K)

Pickup G (1975) Downstream variations in morphology, flow conditions and sediment transport in an eroding channel. Z Geomorphol 19:443–459

Pickup G, Reiger WA (1976) A conceptual model of the relationship between channel character-istics and discharge. Earth Surf Proc 4:37–42

Pickup G, Warner RF (1976) Effects of hydrologic regime on magnitude and frequency of dominant dishcarge. J Hydrol 29: 51–75

Pickup G, Higgins RJ, Grant I (1983) Modelling sediment transport as a moving wave – the transfer and deposition of mining waste. J Hydrol 60:281–301

Piest RF, Kramer LA, and Heineman HG (1975) Sediment movement from loessial watersheds. Present and prospective technology for predicting sediment yields and sources. US Dep Agric, Agric Res Serv Publ ARS-S-40, pp 130–141

Pilgrim DH, Cordery I, and Boyd MJ (1982) Australian developments in flood hydrograph modeling. In: Singh VP (ed) Rainfall-runoff relationships. Water Resour Publ, Littleton, pp 37–50

Playfair J (1802) Illustrations of the Huttonian theory of the earth. Creech, Edinburgh

Popper KR (1963) Conjectures and refutation. Harper and Row, New York

Potter KW (1987) Research on flood frequency analysis: 1983–1986. Rev Geophys 25:113–118

Powell JW (1875) Exploration of the Colorado River of the West. US Gov Printing Off, Washington, DC

Powell JW (1876) Biographical notice of Archibald Robertson Marvine. Bull Philos Soc Washing-ton 2:App X

Powell JW (1879) Report on the lands of the arid region of the United States. US Gov Printing Off, Washington, DC

Powers MC (1953) A new roundness scale for sedimentary particles. J Sediment Petrol 23:117–119

Prigogine I (1955) Introduction to thermodynamics of irreversible processes. Thomas, Springfield

Prinz G (1909) Die Vergletscherung des nordlichen Teiles des zentralen Tien-Schan-Gebirges. KK Geogr Ges Mitt 52:10–75

Pumpelly R (1905) Explorations in Turkestan. Carnegie Inst Washington Publ 26

Purtymun WD, Johnson GL, John EC (1966) Distribution of radioactivity in the alluvium of a disposal area at Los Alamos, New Mexico. US Geol Surv Prof Pap 550-D 250–252

Pyne SJ (1980) Grove Karl Gilbert: a great engine of science. University of Texas Press, Austin

Quirk JP, Schofield RK (1955) The effect of electrolyte concentration on soil permeability. J Soil Sci 6:163–178

Rabbitt MC (1979) Minerals, lands, and geology for the common defence and general welfare, vol 1, before 1879. US Geol Surv, Washington, DC

Rabbitt MC (1980) Minerals, lands, and geology for the common defense and general welfare: vol 2, 1879–1904. US Geol Surv, Washington, DC

Rachocki AH (1981) Alluvial fans. John Wiley & Sons, Chichester

Rad U von (1968) Comparison of sedimentation in the Bavarian flysch (Cretaceous) and Recent San Diego Trough (California). J Sediment Petrol 38:1120–1154

Rahn PH (1967) Sheetfloods, streamfloods, and the formation of pediments. Ann Assoc Am Geogr 57:593–604

Rana SA, Simons DB, Mahmood K (1973) Analysis of sediment sorting in alluvial channels. Am Soc Civil Eng J Hydraul Div 99:1967–1980

Range P (1910) Die deutsche Sudkalahari. Ges Erdk Berlin 46:592-619

Range P (1912) The topography and geology of the German South Kalahari. Geol Soc S Afr Trans 15:17–39

Range P (1914) Das artesische Becken in der Sudkalahari. Dtsch Kolonialbl 8:494–509

Rantz SE (1970) Urban sprawl and flooding in southern California. US Geol Surv Circ 601-B

Rantz SE (1971) Suggested criteria for hydrologic design of storm-drainage facilities in the San Francisco Bay Region, California. US Geol Surv Unnumbered Open-File Rep

Rapp A (1972) Conclusions from the DUSER soil erosion project in Tanzania. Geogr Ann 54A:377–379

Rapp A, Murray-Rust DH, Christiansson C, Berry L (1972) Soil erosion and sedimentation in four catchments near Dodoma, Tanzania. Geogr Ann 54A:255–318

Raverty HG (1878) The Mihran of Sind. J Asiat Soc of Bengal 61:14–21

Ree WO, Palmer VJ (1949) Flow of water in channels protected by vegetative linings. US Soil Cons Serv Tech Bull 967

Reitan CH, Green CR (1968) Appraisal of research on weather and climate of desert environments. In: McGinnies WG, Goldman BJ, Palore P (eds) Deserts of the World. Univ Ariz Press, Tucson, pp 21–94

Renard KG (1970) The hydrology of semiarid rangeland watersheds. US Dept Agr Res Serv Pub 41–162

Renard KG (1972) Sediment problems in the arid and semiarid Southwest. Proc 27th Annu Meet Soil Cons Soc Am, pp 225–232

Renard KG (1977) Past, present, and future water resources research in arid and semiarid areas of the southwestern United States. Austr Inst of Eng 1977 Hydrol Symp Proc, pp 1–29

Renard KG, Brakensiek DL (1976) Precipitation on intermountain rangeland in the western United States. 5th US/Austr Range Sci Works, Boise, Idaho, June 1975

Renard KG, Simanton JR (1973) A discussion of: "Predicting sediment yield in western United States," by E. M. Flaxman. Am Soc Civil Eng J Hydraul Div 99(HY13):1647–1649

Renard KG, Simanton JR (1975) Thunderstorm precipitation effects on the rainfall-erosion index of the universal soil loss equation. Hydrol Wat Resour Ariz SW 5:47–55

Renard KG, Stone JJ (1982) Sediment yield from small semiarid rangeland watersheds. Proc Works Estimating erosion and sediment yield on rangelands. US Dep Agric, Agric Res Serv Rev Man ARM-W-26, pp 129–145

Renfro GW (1975) Use of erosion equations and sediment-delivery ratios for predicting sediment yield. Present and prospective technology for predicting sediment yields and sources. US Dep Agric, Agric Res Serv Publ ARS-S-40, pp 33–45

Renner FG (1936) Conditions influencing erosion on the Boise River watershed. US Dep Agric Tech Bull 528

Resnick R, Halliday D (1977) Physics. John Wiley & Sons, New York

Rhoads BL (1986a) Process and response in desert mountain fluvial systems. PhD Diss, Dep Geogr, Ariz State Univ, Tempe

Rhoads BL (1986b) Flood hazard assessment for land-use planning near desert mountains. Environ Manag 10:97–106

Rhoads BL (1987) Stream power terminology. Prof Geogr (in press)

Rich JL (1911) Recent stream trenching in the semi-arid portion of southwestern New Mexico. Am J Sci 32:237–245

Rich JL (1935) Origin and evolution of rock fans and pediments. Geol Soc Am Bull 46:999–1024

Rich LR (1962) Erosion and sediment movement following a wildfire in ponderosa pine forest of central Arizona. US For Serv Res Note RM-76

Rich LR, Reynolds HG (1963) Grazing in relation to runoff and erosion on some chaparral watersheds in central Arizona. J Range Manag 16:322–326

Richards K (1982) Rivers: form and process in alluvial channels. Methuen, London

Richardson HL (1945) Discussion: the significance of terraces due to climatic oscillation. Geol Mag 82:16–18

Riggs HC (1985) Streamflow characteristics. Elsevier, Amsterdam

Riley SJ, Taylor G (1978) The geomorphology of the Upper Darling River system with special reference to the present fluvial system. Proc R Soc Victoria 90:89–102

Ripley PO, Kalbfleisch W, Bourget SJ, Coper DJ (1961) Soil erosion by water: damage, prevention, control. Can Dep Agric, Res Br Publ 1083

Robinson AH (1956) The necessity of weighting values in correlation of areal data. Ann Assoc Am Geogr 46:233–236

Robinson TW (1958) Phreatophytes. US Geol Surv Wat-Supply Pap 1423

Robinson TW (1965) Introduction, spread, and areal extent of saltcedar (Tamarix) in the western states. US Geol Surv Prof Pap 491-A

Robinson TW (1970) Evapotranspiration by woody phreatophytes in the Humbolt River Valley near Winnemucca, Nevada. US Geol Surv Prof Pap 491-D

Rockwell TK, Keller EA, Johnson DL (1984) Tectonic geomorphology of alluvial fans and mountain fronts near Ventura, California. In: Morisawa M, Hack JT (eds), Tectonic geomorphology. Allen & Unwin, Boston, pp 183–208

Rodin LE, Bazilevich NI, Rozor NN (1975) Productivity of the world's main ecosystems. In: Reishle DE, Franklin JF, Goodall DW (eds) Productivity of the world's main ecosystems. Nat Acad Sci, Washington, pp 15–22

Roehl JW (1961) Sediment source area, delivery ratios, and influencing morphological factors. Int Assoc Sci Hydrol Publ 59:202–213

Roglic J (1972) Historical review of morphologic concepts. Springfield M, Herak VT (eds). Elsevier, Amsterdam New York, pp 1–18

Roose E (1977) Use of the universal soil loss equation to predict erosion in West Africa. Soil erosion: prediction and control. Soil Cons Soc Am, Ankeney, pp 60–74

Rouse H, Ince S (1957) History of hydaulics. Iowa Inst Hydraul Res, Iowa City

Rowe PB (1941) Some factors of the hydrology of the Sierra Nevada foothills. Trans Am Geophys Union 22:90–100

Rowe PB (1948) Influence of woodland chaparral on water and soil in central California. US Dep Agric State Cal Dep Nat Res Div For Unnumbered Publ

Rowe PB (1963) Streamflow increases after removing woodland-riparian vegetation from a southern California watershed. J For 61:365–370

Rowe PB, Reiman LF (1961) Water use by brush, grass, and grass-forb vegetation. J For 59:175–181

Rowe PB, Countryman CM, Storey HC (1954) Hydrologic analysis used to determine effects of fire on peak discharges and erosion rates. US For Serv Cal For and Range Exp Stn Unnumbered Publ

Rowell DL, Payne D, Ahmad N (1969) The effect of the concentration and movement of solutions on the swelling, dispersion and movement of clay in saline and alkali soils. J Soil Sci 20:176–188

Rozovskii IL (1961) Flow of water in bends of open channels. Isr Progr Sci Trans, Jerusalem

Rubey WW (1928) Gullies in the Great Plains formed by sinking of the ground. Am J Sci 15:417–422

Rubey WW (1952) Geology and mineral resources of the Hardin and Brussels quadrangles. US Geol Surv Prof Pap 217

Russell IC (1885) Geological history of Lake Lahontan, a Quaternary lake of northwestern Nevada. US Geol Surv Mon 11

Rutter AJ (1968) Water consumption be forests. 2: Water deficits and plant growth. Academic Press, London New York, pp 23–84

Ruxton BP, Berry L (1961) Weather profiles and geomorphic position on granite in two tropical regions. Rev Geomorphol Dyn 12:16–31

Saarinen TF, Baker VR, Durrenburger R, Maddock T, Jr (1984) The Tucson, Arizona Flood of October 1983. Comm on Nat Disasters Comm Eng Tech Syst, US Natl Res Couc

Salisbury NE, Parson CG (1971) Piping and related mass-movement in the Big Badlands of South Dakota. Proc Assoc Am Geogr 3:194 (abstr)

Salomons W, Förstner U (1984) Metals in the hydrocycle. Springer, Berlin Heidelberg New York

Schantz HL (1947) The use of fire as a tool in the management of the brush ranges of California. Cal Div For Rep Unnumbered Publ

Schattner I (1962) The Lower Jordan Valley; a study in the fluvio-morphology of an arid region. Hebrew Univ Scripta Hierosolymitana 11, Jerusalem

Scheidegger AE (1970) Theoretical geomorphology. Springer, Berlin Heidelberg New York

Schick AP (1970) Desert floods. Symposium on the Results of Research on Representative and Experimental Basins, Int Assoc Sci Hydrol, UNESCO, pp 479–493

Schick AP (1971a) A desert flood: physical characteristics; effects on man, geomorphic significance, human adaption. Dep Geogr Univ Jerusalem

Schick AP (1971b) A desert flood. Jerusalem Stud Geogr 2:91–155

Schick AP (1986) Hydrologic aspects of floods in extreme arid environments. In: Baker VR, Kochel RC, and Patton PC (eds) Flood geomorphology (in press)

Schick AP, Sharon D (1974) Geomorphology and climatology of arid watersheds. Hebrew Univ Jerusalem Tech Rep 161

Schick AP, Lekach J, Hassan MA (1985) Bedload transport in desert floods: observations in the Negev. Proc Intern Workshop on Problems of Sediment Transport in Gravel-Bed Rivers. Colo State Univ, Unpubl Mimeogr Manuscript

Schoklitsch A (1914) Über Schleppkraft und Geschiebebewegung. Engelmann, Leipzig

Schoklitsch A (1926) Die Geschiebebewegung an Flussen und an Stauwerken. Springer, Wien

Schraeder FC (1915) Mineral deposits of the Santa Rita and Patogonia Mountains, Arizona. US Geol Surv Bull 582

Schumm SA (1956a) Evolution of drainage systems and slopes in badlands at Perth Amboy, New Jersey. Geol Soc Am Bull 67:597–646

Schumm SA (1956b) The role of creep and rainwash on the retreat of badlands slopes. Am J Sci 254:693–706

Schumm SA (1960) The shape of alluvial channels in relation to sediment type. US Geol Surv Prof Pap 352B:17–30

Schumm SA (1962) Erosion on miniature pediments in Badlands National Monument, South Dakota. Geol Soc Am Bull 73:719–724

Schumm SA (1963) Sinuosity of alluvial rivers on the Great Plains. Geol Soc Am Bull 74:1089–1100

Schumm SA (1965) Quaternary paleohydrology. In: Wright HE, Jr, Frey DG (eds) The Quaternary of the United States. Princeton Univ Press, pp 795–806

Schumm SA (1968) River adjustment to altered hydrologic regime – Murrumbidgee River and paleochannels, Australia. US Geol Surv Prof Pap 598

Schumm SA (1972) Fluvial paleochannels. Soc Econ Paleontol Min Spec Publ 16:98–107

Schumm SA (1973) Geomorphic thresholds and complex response of drainage system. In: Morisawa M (ed), Fluvial geomorphology. SUNY Binghamton, pp 299–310

Schumm SA (1977) The Fluvial System. John Wiley & Sons, New York

Schumm SA, Hadley RF (1957) Arroyos and the semiarid cycle of erosion. Am J Sci 255:161–174

Schumm SA, Khan HR (1972) Experimental study of channel patterns. Geol Soc Am Bull 83:1755–1770

Schumm SA, Lichty RW (1963) Channel widening and flood plain construction along Cimarron River in south-western Kansas. US Geol Surv Prof Pap 352D:71–88

Schumm SA, Lichty RW (1965) Time, space, and causality in geomorphology. Am J Sci 263:110–119

Schumm SA, Harvey MD, Watson CC (1984) Incised channels: morphology, dynamics and control. Wat Resour Publ, Littleton

Scobey FC (1939) Flow of water in irrigation and similar canals. US Dep Agric Tech Bull 652

Scoging H (1982) Spatial variations in infiltration, runoff and erosion on hillslopes in semi-arid Spain. In: Bryan R, Yair A (eds) Badland geomorphology and piping. Geobooks, Norwich, pp 89–112

Scott AG, Kunkler JL (1976) Flood discharges of streams in New Mexico as related to channel geometry. US Geol Surv Open-File Rep 76–414

Scott CH, Culbertson, JK (1971) Resistance to flow in flat-bed sand channels. US Geol Surv Prof Pap 750-B:254–258

Scott KM (1971) Origin and sedimentology of 1969 debris flows near Glendora, California. US Geol Surv Prof Pap 750-C, 242–247

Scott KM (1973) Scour and fill in Tujunga Wash: a fan head valley in urban southern California. Geol Soc Am Bull 64:547–560

Scott KM, Williams RP (1978) Erosion and sediment yields in the Transverse Ranges, southern California. US Geol Surv Prof Pap 1030

Scott VH (1956) Relative infiltration rates of burned and unburned upland soils. Trans Am Geophys Union 37:67–69

Scott VH, Burgy RH (1956) Effects of heat and brush burning on the physical properties of certain upland soils that influence infiltration. Soil Sci 82:63–70

Sebenik PG, Thames JL (1967) Water consumption by phreatophytes. Prog Agric Ariz 19:10–11

Seginer I (1966) Gully development and sediment yield. J Hydrol 4: 236–253

Sellers WD (1965) Physical climatology. University of Chicago Press

Selley RC (1976) An introduction to sedimentology. Academic Press, London

Settergren CD (1967) Reanalysis of past research on effects of fire on wildland hydrology. Missouri Agric Exp Stn Res Bull 954

Shalash S (1974) Facts about degradation. Arab Republ Egypt Minist Irr Dep Hydrol Unnumbered Publ

Shamov GI (1939) Zailenie vodokhranilshch. Lenin, Moscow

Shanan L, Schick AP (1980) A hydrological model for the Negev Desert highlands: effects of infiltration, runoff and ancient agriculture. Hydrol Sci Bull 25:269–282

Shanan L, Evenari M, Tadmor EH (1969) Ancient technology and modern science applied to desert agriculture. Endeavor 28:68–72

Shapere D (1964) The structure of scientific revolutions. Philos Rev 73:383–394

Sharma KD, Chatterji PC (1982) Sedimentation in Nadis in the Indian arid zone. Hydrol Sci J 27:345–352

Sharon D (1972) The spottiness of rainfall in a desert area. J Hydrol 17:161–175

Sharpe RP, Nobels LH (1953) Mudflow of 1941 at Wrightwood, southern California. Geol Soc Am Bull 64:547–560

Shaw EM (1983) Hydrology in practice. Van Nostrand Reinhold, Berkshire

Shea J (1982) Twelve fallacies of uniformitarianism. Geology 10:455–460

Shen HW (1971) River mechanics. Wat Resour Publ, Lakewood

Shen HW (1979) Modeling of rivers. John Wiley & Sons, New York

Shepherd PG, Schumm SA (1974) Experimental study of river incision. Geol Soc Am Bull 85:257–268

Sheridan D (1981) Desertification of the United States. US Counc Environ Qual, Washington, DC

Sherlock RL (1922) Man as a geological agent. Witherby, London

Shields A (1936) Anwendung der Ahnlichkeitsmechanik und der Turbulenzforschung auf die Geschiebebewegung. Mitt Preuss Versuchsanst Wasserbau Schiffbau Berlin (transl Ott WP, Uchelen JC van), Calif Inst Tech Publ 167

Shoneman FW (1914) The design of canal headwork. Punjab Eng Congr Proc 2:712–720

Shoneman FW (1916) The design of head regulators of irrigation distributaries. Punjab Eng Cong Proc 4:428–439

Shreve RL (1966) Statistical law of stream numbers. J Geol 4:17–37

Shreve RL (1967) Infinite topologically random channel networks. J Geol 75:178–186

Shreve RL (1975) The probabilistic-topologic approach to drainage-basin geomorphology. Geology 3:527–529

Simanton JR, Renard KG, Sutter NG (1973) Procedure for identifying parameters affecting storm runoff volumes in a semiarid environment. US Agric Res Serv Publ ARS-W-1

Simons DB, Senturk F (1977) Sediment transport technology. Wat Resour Publ, Littleton

Simons DB, Richardson EV, Nordin CF (1965) Bed-load equation for ripples and dunes. US Geol Surv Prof Pap 462-H

Simons, Li & Associates, Inc. (1981) Sediment transport analysis of Rillito River and tributaries for the Tucson urban study, draft report. Simons, Li & Associates, Ft. Collins

Simpson JE (1972) Effects of the lower boundary on the head of a gravity current. J Fluid Mech 53:759–768

Sittig M (1981) Priority toxic pollutants. Noyes Data Corp, Park Ridge

Slatyer RO, Mabbutt JA (1964) Hydrology of arid and semiarid regions. In: Chow VT (ed) Handbook of applied hydorology. McGraw-Hill, New York, pp 24/1–24/46

Slaymaker O (1982) The occurrence of piping and gullying in the Penticton glacio-lacustrine silts, Okanagan Valley, B.C.. In: Bryan R, Yair A (eds) Badland geomorphology and piping. Geobooks, Norwich, pp 305–316

Smalley IJ, Vita-Finzi C (1969) The concept of 'system' in the earth sciences. Geol Soc Am Bull 80:1591–1594

Smalley JP (1971) Geologic parameters of slope stability in the Upper Santa Ynez drainage system. US Forest Serv, Unpub Rep

Smalley JP, Cappa J (1971) Reconnaissance of geologic hazards related to the Romero Burn. US For Serv, Unnumbered Publ, Los Padres Natl For

Smart JS (1972) Quantitative characterization of channel network structure. Wat Resour Res 8:1487–1496

Smart JS, Werner C (1976) Applications of the random model of drainage basin composition. Earth Surf Proc 1:219–233

Smart JS, Werner C (1978) The analysis of drainage network composition. Earth Surf Proc 3:129–170

Smeins FE (1975) Effect of livestock grazing on runoff and erosion. Watershed Manag. Proc Symp, Logan, Utah, Aug 11–13, pp 267–264

Smith CL (1972) The Salt River Project: a case study in cultural adaptation to an urbanizing community. Univ Ariz Press, Tucson

Smith DE (1981) Riparian vegetation and sedimentation in a braided river. MA Thes, Ariz State Univ, Tempe

Smith DG (1976) Effect of vegetation on lateral migration of anastomosed channels of a glacier meltwater river. Geol Soc Am Bull 87:857–860

Smith GI, Haines DV (1964) Character and distribution of nonclastic minerals in the Searles Lake evaporite deposit, California. US Geol Surv Bull 1181-P

Smith ND (1971) Transverse bars and braiding in the Lower Platte River, Nebraska. Geol Soc Am Bull 82:3407–3420

Smith RE (1972) Border irrigation advance and epehemeral flood waves. Am Soc Civil Eng J Irr Dr Div 98(R2):289–307

Smith WO, Vetter CP, Cummings GB (1960) Comprehensive survey of sedimentation in Lake Mead, 1948–1949. US Geol Surv Prof Pap 295

Snyder CT, Langbein WB (1962) The Pleistocene lake in Spring Valley, Nevada, and its climatic implications. J Geophys Res 67:2385–2394

Soliman MM (1974) Urbanization and the processes of erosion and sedimentation in the River Nile. Int Assoc Sci Hydrol Publ 133:123–129

Spearing DR (1974) Alluvial fan deposits. Summary sheets of sedimentary deposits, sheet 1. Geol Soc Am, Boulder

Spencer B (1896) Report on the work of the Horn Scientific Expedition to Central Australia. Horn Sci Exp, London Melbourne

Stall JB, Huff FA (1971) The structure of the thunderstorm rainfall. Am Soc Civil Eng, Nat Water Resources Eng Meeting, Phoenix, Preprint 1330

Stanley JW (1948) Effect of dams on channel regime. Proc FIASC, US Dep Agric

Stanley JW (1951) Retrogression on the Lower Colorado River after 1935. Trans Am Soc Civil Eng 116:943–957

Stegner W (1954) Beyond the Hundredth Meridian: John Wesley Powell and the second opening of the West. Mifflin, Boston

Stegner W (1977) Introduction. In: Dutton CE and Tertiary History of the Grand Canyon District (Reprint of US Geological Survey Monogr 2). Smith, Santa Barbara, Cal, pp vii-xii

Stelczer K (1981) Bed-load transport: theory and practice. Water Resour Publications, Littleton

Stephenson GR, Street LV (1977) Water quality investigations on the Reynolds Creek Watershed, southwest Idaho: a 3-year progress report. US Dep Agric, Interim Rep on Wat Qual Agreement 12–14–5001–6028

Sternberg H (1875) Untersuchungen uber Langen – und Querprofil Geschiebefuhrende Flusse. Z Bauwes 25:483–506

Stevens JC (1936) The silt problem. Trans Am Soc Civil Eng 101:34–67

Stevens MA, Simons DB, Richardson EV (1975) Non-equilibrium river form. Am Soc Civ Eng J Hydraul Div 101:557–566

Stiffe AW (1873) On the mud craters and geological structure of the Mektran coast. Geol Soc London Q J 30:50–53

Storr DJ, Cork HF, Munn RE (1970) An energy budget study above the forest canopy at Marmot Creek, Alberta, 1967. Wat Resour Res 6:705–716

Strahler AN (1952a) Hypsometric (area-altitude) analysis of erosional topography. Geol Soc Am Bull 63:1117–1142

Strahler AN (1952b) Dynamic basis of geomorphology. Geol Soc Am Bull 63:923–928

Stahler AN (1956) The nature of induced erosion and aggradation. In: Thomas WL, Jr (ed) Man's role in changing the face of the earth. Univ Chicago Press, vol 2, pp 621–638

Strand RI (1975) Bureau of Reclamation procedures for predicting sediment yield. Present and prospective technology for predicting sediment yields and sources. US Dep Agric, Agric Res Serv Publ ARS-S-40, pp 10–15

Straub LG (1935) Missouri River report. House Doc 238, 73rd Congr, 2nd Sess, US Gov Printing Off, Washington, DC

Strickler A (1923) Beitrage zur Frage der Geschwindigkeitsformel und der Rauhigkeitszahlen fur Strome, Kanale und geschlossene Leitungen. Mitt Eidgen Amt Wasserwirtsch, Bern

Stromer von Reichenbach W (1896) Die Geologie der deutschen Schutzgebiete in Afrika. Munchen and Leipzig

Stuckmann G (1969) Les inondations de septembre-octobre 1969 en Tunisie, Pt II: Effets morphologiques. UNESCO Rep, Paris

Sturm M, Matter A (1978) Turbidites and varves in Lake Brienz (Switzerland): deposition of clastic detritus by density currents. Int Assoc Sedidiment Spec Publ 2:145–166

Sturt C (1849) Narrative of an expedition into Central Australia. Boone, London

Sumner GN (1978) Mathematics for physical geographers. Halsted, New York

Swanson FJ, Lienkaemper GW (1978) Physical consequences of large organic debris in Pacific Northwest streams. US For Serv Gen Tech Rep PNW-69

Swanson FJ, Lienkaemper GW, Sedell JR (1976) History, physical effects, and management implications of large organic debris in western Oregon streams. US For Serv Gen Tech Rep PNW-56

Swift TT (1926) Rate of channel trenching in the southwest. Science 63:70–71

Taaffe EJ (1974) The spatial view in context. Ann Assoc Am Geog 64:1–16

Talbot MR, Williams MAJ (1978) Erosion of fixed dunes in the Sahel, central Niger. Earth Surf Proc 3:107–113

Talbot MR, Williams MAJ (1979) Cyclic alluvial fan sedimentation on the flanks of fixed dunes, Janjari, central Niger. Catena 6:43–62

Tanner WF (1961) An alternate approach to morphogenetic climates. SE Geol 2:251–257

Tatum FE (1963) A new method of estimating debris-storage requirements for debris basins. Proc Fed Inter-Agency Sediment Conf US Dep Agric Misc Publ 970:886–897

Taylor PJ (1975) Distance decay models in spatial interactions. GeoAbstr, Norwich

Thakur TR, Mackey DK (1973) Delta processes. Proc Hydrol Symp, Edmunton, Alberta, Natl Res Counc Can, pp 509–530

Thom R (1977) Structural stability and morphogenesis: an outline of a general theory of models. Benjamin, Reading

Thomas DM, Benson MA (1970) Generalization of streamflow characteristics from drainage basin characteristics. US Geol Surv Wat-Supply Pap 1975

Thomas MF (1966) Some geomorphological implications of deep weathering patterns in crystalline rocks in Nigeria. Trans Inst Br Geogr 40:173–193

Thompson JR (1974) Energy budget measurements over three cover types in eastern Arizona. Wat Resour Res 10:145–148

Thornes JB (1976) Semi-arid erosional systems. London Sch Econ Poli Sci Gegr Pap 7

Thornes JB (1978) The character and problems of theory in contemporary geomorphology. In: Embleton C (ed) Geomorphology: present problems and future prospects. Oxford Univ Press, pp 14–24

Thornes JB (1980) Structural instability and ephemeral channel behaviour. Z Geomorphol Suppl 36:233–244

Thornes JB (1983) Discharge: empirical observations and statistical models of change. In: Gregory KJ (ed) Background to palaeohydrology. John Wiley & Sons, Chichester New York, pp 51–67

Thornes JB, Brunsden D (1977) Geomorphology and time. John Wiley & Sons, New York

Thornthwaite CW, Mather JR (1955) The Water Balance. Lab Publ Climatol 9, Drexel Inst Technol Centerton, New Jersey

Thornthwaite CW, Sharpe CFS, Dosch EF (1942) Climate and accelerated erosion in the arid and semi-arid southwest, with special reference to the Polacca Wash drainage basin, Arizona. US Dep Agric Tech Bull 808

Tietkens WH (1891) Journal of the Central Australia Exploration Expedition, 1889. Gov Print, Adelaide

Tilho J (1911) Documents scientifiques de la mission Tilho. Ministere des Colonies, Imprimeure Nationale

Tinkler KJ (1983) Avoiding error when using the Manning Equation. J Geol 90:326–328

Tinkler KJ (1985) A short history of geomorphology. Barnes and Noble, Totowa, New Jersey

Todd N (1844) Report from a journey from Herat to Simla via Candahar and Kabul. Asiat Soc Bengal Calcutta 13:339

Tomanek GW (1958) Annual report on ecological research on salt cedar and other vegetation primarialy at Cedar Bluffs Reservoir, Kansas. Ft Hays State Col Dep Botany Unnumbered Rep

Tomanek GW, Ziegler RL (1961) Ecological studies of saltcedar. Div Biol Sci, Ft Hays Kansas State Coll, Unnumbered Publ

Tricart J (1963) Oscillations et modifications de caractere de la zone aride en Afrique et en Amerique latine lors des periodes glaciaires des hautes latitudes. UNESCO Arid Zone Res 20:415–419

Tricart J, Cailleux A (1972) Introduction to climatic geomorphology. St Martin's Press, New York (transl Kiewiet de Jonge CJ)

Trimble SW (1975) Unsteady state denudation. Science 188:1207

Trimble SW (1977) The fallacy of stream equilibrium in contemporary denudation studies. Am J Sci 277:876–887

Trimble SW (1983) A sediment budget for Coon Creek basin in the Driftless Area, Wisconsin, 1853–1977. Am J Sci 283:454–474

Trimble SW, Lund SW (1982) Soil conservation and the reduction of erosion and sedimentation in the Coon Creek basin, Wisconsin. US Geol Surv Prof Pap 1234

Troeh FR (1965) Landform equations fitted to contour maps. Am J Sci 263:616–627

Tromble JM (1977) Water requirements for mesquite Prosopis juliflora. J Hydrol 34:171–179

Troxell HC, Peterson JQ (1937) Flood in La Canada Valley, California, January 1, 1934. US Geol Surv Wat-Supply Pap 796-C

Truhlar JF (1978) Determining suspended sediment loads from turbidity records. Hydrol Sci Bull 23:409–417

Tuan Y-F (1962) Structure, climate and basin landforms in Arizona and New Mexico. Ann Assoc Am Geogr 52:51–68

Tuck R (1968) Origin of bedrock values of placer deposits [disc]. Econ Geol 63:191–193

Tukey JW (1962) The future of data analysis. Ann Math Statist 33:1–67

Turner RM (1974) Quantitative and historical evidence of vegetation changes along the upper Gila River, Arizona. US Geol Surv Prof Pap 655-H

Turner RM, Karpiscak MM (1980) Recent vegetation changes along the Colorado River between Glen Canyon Dam and Lake Mead, Arizona. US Geol Surv Prof Pap 1132

Turner SF, Halpenny LC (1941) Ground-water inventory in the upper Gila River Valley, New Mexico and Arizona: scope of investigation and methods used. Trans Am Geophys Union 22:738–744

Tweney RD, Doherty ME, Mynatt CR (1981) On Scientific Thinking. Columbia Univ Press, New York

Twidale CR (1967) Origin of the piedmont angle as evidenced in South Australia. J Geol 75:393–411

UNESCO (1950) Flood damage and flood control activities in Asia and the Far East. Flood Control Series 1

UNESCO (1979) Map of the world distribution of arid regions, MAB Technical Notes Number 7. UNESCO, New York

UNESCO (1984) Climate drought and desertification. Nat Res 20:2–8

United Nations (1953) The sediment problem. UN ECAFE, Bankock

Unstead JF (1933) A system of regional geography. Geography 18:175–187

Urvoy Y (1936) Structure et modele du Soudan Francais, Colonie du Niger. Ann Geogr 45:19–49

Urvoy Y (1942) Les bassins du Niger, etude de geographie physique et paleogeographie. Mem Inst Fr Afr Noire 41:444–459

US Bureau of Reclamation (1973a) Evapotranspirometer studies of saltcedar near Bernardo, New Mexico, March, 1973. Pacific Southwest Interagency Subcommittee, Unnumbered Rep

US Bureau of Reclamation (1973b) Progress report: phreatophyte investigations, Bernardo evapotranspirometers. US Bur Recl Middle Rio Grande Off, Unnumbered Rep

US Bureau of Reclamation (1976) 1976 Lake Powell sedimentation surveys, Colorado River Storage Project, Glen Canyon Unit, Montrose Power Operations Office, Unnumbered Rept

US Forest Service (1974) Los Angeles River flood prevention project, mountain and foothill area, review report. US For Serv, Unpubl Rep

US National Park Service (1977a) Draft management plan and environmental impact assessment, Glen Canyon National Recreation Area, Arizona and Utah. US Nat Park Ser, Denver

US National Park Service (1977b) Draft management plan and environmental impact assessment, Dinosaur National Monument, Colorado and Utah. US Nat Park Serv, Denver

US National Park Service (1979) Final environmental statement, proposed Colroado River management plan, Grand Canyon, Arizona. Nat Park Serv, Denver

US Soil Conservation Service (1972) National engineering handbook, Sec 4: Hydrology. US Dep Agric, Washington, DC

US Soil Conservation Service (1975) Soil taxonomy: a basic system of soil classification for making and interpreting soil surveys. US Dep Agric, Washington

US Topographic Corps (1855–1860) Reports of the explorations and surveys to ascertain the most practicable and economical route for a railroad from the Mississippi River to the Pacific Ocean. US Gov Print Off, Washington, DC

US Water Resources Council (1968) The nation's water resources. US Dep Inter, Washington, DC

US Waterways Experiment Station (1935) Studies of river bed materials and their movement, with special reference to the Lower Mississippi River. US Waterways Exp Stn, Vicksburg, Pap 17

Vageler P (1920) Beobachtungen in Sudwest-Angola und im Ambolande. Ges Erdk Berlin 32:179–193

Vajda A, Smallwood TO (1953) Proposal for the Wadi Jizan irrigation development scheme, Saudi Arabia. FAO, Rome

Vanney JR (1960) Pluie et crue dans le Sahara Nord-occidental. Mem Reg Inst Rech Sahara

Veihmeyer FJ, Johnson CN (1944) Soil moisture records from burned and unburned plots in certain grazing areas of California. Trans Am Geophys Union 25:72–88

Visher SS (1913) The history of the bajadas of the Tucson Bolson. Science 37:459

Vittorini S (1977) Observazioni sull-origine e sul ruole di due forme di erosione nelle argille: calanchi and biancane. Boll Soc Geogr Italiana Ser X, VI:25–543

Wagsaff JM (1976) Some thoughts about geography and catastrophe theory. Area 8:319–320

Walker RG (1975) Generalized facies models for resedimented conglomerates of turbidite association. Bull Geol Soc Am 86: 737–748

Walling DE, Kleo AHA (1979) Sediment yhields of rivers in areas of low precipitation. Hydrology of areas of low precipitation, Int Assoc Hydrol Sci Publ 128:479–493

Walling DE, Teed A (1971) A simple pumping sampler for research into suspended sediment transport in small catchments. J Hydr 13:325–337

Walling DE, Webb BW (1983) Patterns of sediment yield. In: Gregory KJ (ed) Background to palaeohydrology. John Wiley & Sons, Chichester New York, pp 69–100

Wallis JR, Anderson HW (1965) An application of multivariate analysis to pediment network design. Int Assoc Sci Hydrol 67:357–378

Warburton PE (1875a) Colonel Warburton's explorations, 1872–3. S Aust Parliam Pap 28

Warburton PE (1875b) Journey across the western interior of Australia. Low & Searle, London

Ward R (1978) Floods: a geographical perspective. John Wiley and Sons, New York

Warnke DA (1969) Pediment evolution in the Halloran Hills, central Mojave Desert, California. Z Geomorphol 13:357–89

Warren DK, Turner RM (1975) Saltcedar (Tamarix Chinensis) seed production, seedling establishment, and response to inundation. J Ariz Acad Sci 10:135–144

Wasson RJ (1974) Intersection point deposition of alluvial fans: an Australian example. Geogr Ann 6:83–92

Wasson RJ (1977) Catchment processes and the evolution of alluvial fans in the lower Derwent Valley, Tasmania. Z Geomorphol 21:147–168

Watters RL, Hakonson TE, Lane LJ (1983) The behavior of actinides in the the environments. Radiochim Acta 32:89–103

Weber JE, Fogel MM, Duckstein L (1976) The use of multiple regression models in predicting sediment yield. Wat Resour Bull 12:1–17

Weibull W (1939) The phenomenon of rupture in solids. Ingenioervetenskapsakad Handl 153:17

Weinberg GM (1975) An introduction to general systems thinking. John Wiley and Sons, New York

Weimer WC, Kinnison RR, Reeves JH (1981) Survey of radionuclide distributions resulting from the Church Rock, New Mexico, uranium mill tailings pond dam failure. US Nuc Regul Comm Rep NUREG/CR-2449, PNL-4122

Weirich FH (1986a) The record of density-induced underflows in a glacial lake. Sedimentology 33:261–277

Weirich FH (1986b) A study of the nature and incidence of density currents in a shallow glacial lake. Ann Assoc Am Geogr 76:396–413

Weise OR (1983) Zur Morphodynamik der Pediplanation. Z Geomorphol Supp 10:64–87

Wells LA (1902) Journal of the Calvert scientific expedition of 1896–7. W Austr Parliam Pap 46

Wells NA, Dorr JA, Jr (1987) Shifting of the Kosi River, northern India. Geol 15:204–207

Wells SG, Gutierriz AA (1982) Quaternary evolution of badlands in the southeast Colorado Plateau, USA. In: Bryan R, Yair A (eds) Badland geomorphology and piping. Geobooks, Norwich, pp 239–258

Wells SG, Bullard TF, Smith LN, Gardner TW (1983) Chronology, rates and magnitudes of Late Quaternary landscape changes in the southeast Colorado Plateau. In: Wells SG, Love DW, Gardner TW (eds) Chaco Canyon Country. American Field Geomorphology Group, Albuquerque, pp 177–186

Wendelaar FE (1978) Applying the universal soil loss equation in Rhodesia. Rhodesia Dep Cons Ext Unnumbered rep

Wentworth C (1922) A scale of grade and class terms for clastic sediments. J Geol 30:377–392

Werrity A (1972) The topology of stream networks. In: Chorley RJ (ed) Spatial analysis in geomorphology. Methuen, London, pp 167–196

Wershaw RL (1970) Sources and behavior of mercury in surface waters. US Geol Surv Prof Pap 713:29–31

Wertz JB (1949) Logarithmic pattern in river placer deposits. Econ Geol 44:193–209

Whicker FW, Schultz V (1982) Radioecology: nuclear energy and the environment. CRC Press, Boca Raton, 2 vols

White CM (1940) The equilibrium of grains on the bed of a stream. Proc Roy Soc London Ser A 174:113–141

Whittaker JG, Jaeggi MNR (1982) Origin of step-pool systems in mountain streams. Am Soc Civil Eng J Hydraul Div 108(HY6) 758–773

Whittaker RH (1975) Communities and ecosystems. Macmillan, New York

Whittlesey D (1956) Southern Rhodesia: an African compage. Ann Assoc Am Geogr 46:1–97

Wiard L (1962) Floods in New Mexico, magnitude and frequency. US Geol Surv Circ 464

Wiitala SW (1961) Some aspects of the effect of urban and suburban development unpon run-off. US Geol Surv Unnumbered Open-File Rep

Wilkens T (1958) Clarence King, a biography. Macmillan, New York

Williams GE (1970) The central Australian stream floods of February-March 1967. J Hydrol 11:185–200

Williams GP (1978) Bankfull discharge of rivers. Wat Resour Res 14:1141–1154

Williams GP (1978) The case of the shrinking channels – the North Platte and Platte Rivers in Nebraska. US Geol Surv Circ 781

Williams GP (1983) Improper use of regression equations in the earth sciences. Geology 11:195–197

Williams GP (1984) Paleohydrologic equations for rivers. In: Costa JE, Fleisher PJ (eds) Developments and applications of geomorphology. Springer, Berlin Heidelberg New York, pp 343–367

Williams GP, Wolman MG (1984) Downstream effects of dams on alluvial rivers. US Geol Surv Prof Pap 1286

Wilm HG (1966) The Arizona watershed program as it enters into its second decade. 10th Annu Ariz Watershed Symp, Proc Ariz Water Comm, pp 9–11

Wilson AG (1970) Entropy in urban and regional modelling. John Wiley & Sons, London

Wilson AG (1981) Catastrophe theory and bifurcation with applications in urban geography. Croom Helm, London

Wilson ED (1961) Gold placers and placering in Arizona. Ariz Bur Mines Bull 168

Wilson KV (1966) Flood frequency of streams in Jackson, Mississippi. US Geol Surv Unnumbered Open-File Rep

Wilson L (1969) Les relations entre les processus geomorphoologiques et le climat modern comme methode de paleoclimatologie. Rev Geogr Phys Geol Dyn 11:303–314

Wilson LG, DeCook KJ, Neuman SP (1980) Final report: regional recharge research for southwest alluvial basins. Wat Resour Res Center Dep Hydrol Wat Resour, Tucson

Winthrop J (1670) An extract of a letter written by John Winthrop, Esq. R Soc London Philos Trans 4:1151–1153

Wischmeier WH (1959) A rainfall erosion index for a universal soil loss equation. Proc Soil Sci Soc Am 23:246–249

Wischmeier WH (1974) New developments in estimating water erosion. Proc 29th Annu Meet Soil Cons Soc Am, pp 179–186

Wischmeier WH (1976) Use and misuse of the universal soil loss equation. J Soil Wat Cons 31:5–9

Wischmeier WH, Mannering JV (1969) Relation of soil properties to its erodibility. Proc Soil Sci Soc Am 33:131–137

Wischmeier WH, Smith DD (1958) Rainfall energy and its relation to soil loss. Trans Am Geophys Union 39:285–291

Wischmeier WH, Smith DD (1965) Predicting rainfall-erosion losses from cropland east of the Rocky Mountains. US Dep Agric Handb 282

Wischmeier WH, Smith DD (1978) Predicting rainfall erosion losses. US Dept Agric Handb 537

Wischmeier WH, Smith DD, Uhland RE (1958) Evaluation of factors in the soil loss equation. Agric Eng 39:458–462

Wischmeier WH, Johnson CB, Cross BV (1971) A soil erodibility nomograph for farmland and construction sites. J Soil Wat Cons 26:189–193

Wise SM, Thornes JB, Gilman A (1982) How old are the badlands? A case study from south-east Spain. In: Bryan R, Yair A (eds) Badland geomorphology and piping. Geobooks, Norwich, pp 259–278

Woldenberg MJ (1966) Horton's laws justified in terms of allometric growth and steady state in open systems. Geol Soc Am Bull 77: 431–437

Woldenberg MJ (1968) Open systems-allometric growth. In: Fairbridge RW (ed) Encyclopedia of geomorphology. Reinhold, New York, pp 776–778

Woldenberg MJ (1969) Spatial order in fluvial systems: Horton's Laws derived from mixed hexagonal hierarchies of drainage basin areas. Geol Soc Am Bull 80:97–112

Woldenberg MJ (1971) The two-dimensional spatial organization of Clear Creek and Old Man Creek, Iowa. In: Morisawa M (ed) Quantitative geomorphology. SUNY Binghamton, pp 83–106

Woldenberg MJ (1972) The average hexagon in spatial hierarchies. In: Chorley RJ (ed) Spatial analysis in geomorphology. Methuen, London, pp 323–354

Wolfenden PJ, Lewin J (1977) Distribution of metal pollutants in floodplain sediments. Catena 4:309–317

Wolfenden PJ, Lewin J (1978) Distribution of metal pollutants in active stream sediments. Catena 5:67–78

Wolman MG (1955) The natural channel of Brandywine Creek, Pennsylvania. US Geol Surv Prof Pap 271

Wolman MG (1967) A cycle of sedimentation and erosion in urban river channels. Geogr Ann 49A:385–395

Wolman MG, Leopold LB (1957) River flood plains: some observations on their formation. US Geol Surv Prof Pap 282-C, pp 87–107

Wolman MG, Miller JP (1960) Magnitude and frequency of forces in geomorphic processes. J Geol 68:54–74

Wolman MG, Schick AP (1967) Effects of construction on fluvial sediment, Urban and suburban areas of Maryland. Wat Resour Res 3:451–464

Womack WR, Schumm SA (1977) Terraces of Douglas Creek, northwestern Colorado: a example of episodic erosion. Geology 5:72–76

Wood A, Smith AJ (1959) The sedimentation and sedimentary history of the Aberystwyth Grits (Upper Llandoverian). Q J Geol Soc London 114:163–195

Woodcock AE, Davis M (1978) Catastrophe theory. Dutton, New York

Woodcock AE, Poston T (1974) A geometrical study of the elementary catastrophes. Springer, Berlin Heidelberg New York

Woodward L (1943) Infiltration capacities of some plant-soil complexes on Utah range watershed lands. Trans Am Geophys Union 24:468–473

Woodyer KD (1968) Bankfull frequency in rivers. J Hydrol 6:114–142

Woolhiser DA and Lenz AT (1965) Channel gradients above gully-control structures. Am Soc Civil Eng J Hydraul Div 93(HY7):165–187

Woolley RR (1946) Cloudburst floods in Utah 1850–1938. US Geol Surv Wat-Supply Pap 994

Wright HA, Churchill FM, Stevens WC (1974) Effect of prescribed burning on sediment, water yield, and water quality from dozed juniper lands in central Texas. Texas Tech Univ US Off Wat Res Tech Compl Rep, Unnumbered Rep

Wu I (1972) Hydrology manual for engineering design and flood plain management within Pima County, Arizona. Pima County Dept of Trans and Flood Control Dis, Tucson

Yair A (1982) Factors affecting the spatial variability of runoff generation over arid hillslopes, southern Israel. Isr J Earth Sci 31:133–143

Yair A (1983) Hillslope hydrology water harvesting and areal distribution of some ancient agricultural systems in the northern Negev desert. J Arid Environ 6:283–302

Yair A (1985) Runoff generation in arid and semi-arid zones. In: Anderson MG, Burt TP (eds) Hydrological forcasting. John Wily & Sons, Chichester New York, pp 183–220

Yair A, Lavee H (1974) Areal contribution to runoff on scree slopes in an extremely arid environment – a simulated rainstorm experiement. Z Geomorphol Suppl 21:106–121

Yair A, Sharon D, Lavee H (1980a) Trends in ruoff and erosion processes over an arid limestone hillside, northern Negev, Israel. Hydrol Sci Bull 25:243–255

Yair A, Goldberg P, Lavee H, Bryan RB, Adar, E (1980) Runoff and erosion processes and rates in the Zin Valley badlands, Northern Negev, Israel. Earth Surf Proc 5:205–225

Yalin MS (1971) On the formation of dunes and meanders. Proc 14th Int Congr Hydrol Res Assoc 3:1–8

Yochelson EL (1980) The scientific ideas of G. K. Gilbert. Geol Soc Amer Spec Pap 183

York JD (1979) Riparian legislation. In: Marcus MG (ed) Proceedings of the 1979 Summer Conference of the Governor's Commission on Arizona Environment. Gov Comm Ariz Environ, Phoenix, pp 48–49

Young AA, Blaney HF (1942) Use of water by native vegetation. Cal Dep Publ Works, Div Wat Res, Bull 50

Young RA (1987) Tertiary history of landscape development in the Colorado Plateau. In: Graf WL (ed) Geomorphic Systems of North America. Geol Soc Am, Boulder

Young T (1809) On the functions of the heart and arteries. R Soc London Philos Trans 99:1–31

Yuretich RF (1979) Modern sediments and sedimentary processes in Lake Rudolf (Lake Turkana) eastern Rift Valley, Kenya. Sedimentology 26:313–332

Zaborski B (1963) Karstlike phenomena in loess areas. Ann Assoc Am Geogr 53:632 (Abstr)

Zahler RS, Sussman HT (1977) Claims and accomplishments of applied catastrophe theory. Nature (London) 269:759–763

Zaslavsky MN (1979) Erozia Pochv. Mysl, Moscow

Zeeman EC (1976) Catastrophe theory. Sci Am 234:65–83

Zeeman EC (1977) Catastrophe theory: selected papers, 1972–1977. Addison-Wesley, Reading

Zimmerman RC, Goodlett JC, Comer GH (1967) The influence of vegetation on channel form of small streams. Int Assoc Hydrol Sci 75:255–275

Zingg AW (1940) Degree and length of land slope as it affects soil loss in runoff. Agric Eng 21:59–64

Zingg T (1935) Beitrag zur Schotteranalyse. Sch Minerol Petrogr Mitt 15:39–140

Zoslavsky D, Kassiff G (1965) Theoretical formulation of piping mechanisms in cohesive soils. Geotechnique 15:305–316

Subject Index

Lightning Source UK Ltd.
Milton Keynes UK

172004UK00001B/11/A